D0983888

Susanne Hummel · Ancient DNA Typing

Methods, Strategies and Applications

Springer

Berlin
Heidelberg
New York
Hong Kong
London
Milan
Paris
Tokyo

Susanne Hummel

Ancient DNA Typing

Methods, Strategies and Applications

With 152 Figures and 34 Tables

Dr. SUSANNE HUMMEL
Historische Anthropologie und Humanökologie
Institut für Zoologie und Anthropologie
Georg August Universität Göttingen
Bürgerstraße 50
37073 Göttingen
Germany

ISBN 3-540-43037-7 Springer-Verlag Berlin Heidelberg New York

Library of Congress Cataloging-in-Publication Data

Hummel, Susanne
Ancient DNA typing : methods, strategies, and applications / Susanne Humel.
p. cm.
Includes bibliographical references and index.
ISBN 3540430377 (alk. paper)
1. DNA, Fossil--Analysis. 2. Mitochondrial DNA--Analysis. 3. Nucleotide sequence. 4. Polymerase chain reaction. I. Title.

Springer-Verlag Berlin Heidelberg New York
a member of BertelsmannSpringer Science+Business Media GmbH
http://www.springer.de

Printed in Germany

Cover design: Design & Production, Heidelberg
Typesetting: Cameraready by the author
SPIN 10645234 31/3150 - 5 4 3 2 1 0 - Printed on acid-free paper

Dedicated to *my parents*

Preface

Who does not remember the genetic identification of the skeletons of the Romanov family or the investigation of the putative clothing of the famous Kaspar Hauser? Those are highlights in the analysis of ancient DNA, a field that has come a long way in the last decade. The topic of ancient DNA analysis and the possible future implications inspire both the public and the scientific community. This book addresses scientific practitioners and graduate students from different disciplines interested in ancient and degraded DNA research. The book is divided into chapters about methods and strategies, protocols and applications, and serves both as a laboratory manual and as a readable textbook in a meanwhile highly specialized field.

It was possible to write this book because of the experience accumulated in ancient DNA typing and the committed efforts of many people who worked in the Göttingen laboratory for ancient DNA research over the last decade. It was mainly these colleagues who carried out experiments for their Diploma and Ph.D. theses; many others worked on their post-doctoral research or were temporary guests in our lab. Therefore, I want to sincerely thank my colleagues, Nancy Banko, Dr. Heike Baron, Andrea Bartels, Ruth Bollongino, Dr. Barbara Bramanti, Dr. Joachim Burger, Dr. Julia Gerstenberger, Birgit Großkopf, Karin Haack, Uta Immel, Oliver Krebs, Dr. Cadja Lassen, Dr. Kristin Launhardt, Dr. Odile Loreille, Annette Müller, Boris Müller, Dr. Gabriele Nordsiek, Dirk Peters, Dr. Jens Rameckers, Dr. Wera Schmerer, Diane Schmidt, Dr. Tobias Schmidt, Dr. Tobias Schultes, Dr. Peter Spencer, Dr. Lingxia Zhao, and Holger Zierdt. For most reliable technical assistance in the laboratory my thanks go to Melanie Kahle, Nicole Weber, and Birgit Zeike as well as to the trainees Annegret Becker, Simone Haffner, and Katrin Lasch. I would also like to thank Sabine Becker for preparing microdissections, Eberhardt George for taxidermic work, Sibylle Hourticolon for object photography, and Gerlinde Tavakolian for the administration of our research grant money.

For cooperation and discussion, I thank Prof. Bernd Brinkmann, Director of the Institute of Legal Medicine, Münster, Dr. Lothar Kaup of the Landeskriminalamt, Hannover, Dr. Martin Oppermann of the Department of Immunolgy, Göttingen, and Dr. Herman Schmitter of the Bundeskriminalamt, Wiesbaden.

Nothing could have been done without sample materials. For supporting our research work through continuous access to the invaluable skeletal materials of the Lichtenstein cave, I am grateful to Dr. Stefan Flindt, archaeologist in Osterode/Harz. Likewise, the support of Prof. Gisela Grupe, Dr. Thomas Kapeghiy, Dr. Lothar Klappauf, Dr. Andje Knaack, Regine Krull, Helmut Mayer, Dr. Helmut Rohlfing, Dr. Reinhold Schoon, Dr. Holger Schutkowski, Dr. Angela Simons, Dr. Susi Ullrich, Christian Velde, Dr. Joachim Wahl, and the Customs Office of the Frankfurt Airport is gratefully appreciated.

The following institutions are acknowledged for funding our research work: the Bundesministerium für Bildung und Forschung ("Neue Technologien in den Geisteswissenschaften," KFA Jülich), the Ministerium für Wissenschaft und Kunst, Niedersachsen, the Deutsche Forschungsgemeinschaft, the Stiftung Volkswagenwerk, and the Niedersächsische Umweltstiftung.

Further, my thanks go to Dr. Peter Spencer, Perth, who spent quite some time "translating" major parts of the manuscript from "German English" to "English English". Also, I am grateful for the advice and help of Dr. Alexander Fabig and Holger Zierdt in handling the "template" for preparing camera-ready manuscripts for Springer-Verlag and finding out more about many special features of additional software that was necessary for this task.

Dr. Dieter Czeschlik of Springer-Verlag, Heidelberg, is acknowledged for incorporating this book into the Springer publishing program, his support, and his patience.

For stimulating scientific discussion throughout the last 2 years and much help in the preparation of the manuscript, I sincerely thank Diane Schmidt.

Last, but not least, I want to express my special thanks to Prof. Bernd Herrmann who has offered me - and still does - constant encouragement and support throughout my scientific career. In the 1980s, it was Prof. Herrmann who initiated the first ancient DNA experiments in Göttingen and who gave me the chance to enter this particular scientific field. His ideas are inspiring and truly belong to the scientific avant-garde.

Contents

1 Introduction

The major work that occurred in scientific research in the field of ancient and degraded DNA in the 1980s laid the groundwork for today's investigations. In the very beginning of that decade, a group of Chinese researchers from the Hunan Medical School proved that DNA is preserved in the tissues of ancient bodies (1980) (Fig. 1.1). Just 1 year later, Anderson and his colleagues published the fully sequenced human mitochondrial genome (1981). The next year, microsatellite DNA, so-called short tandem repeats (STR), was discovered by Hamada and colleagues (1982); it was later recognized that these enable a fingerprinting similar to DNA typing. A crucial event for ancient DNA research was the invention of the polymerase chain reaction, which enables the amplification of DNA into highly sensitive and simple assays better than any other molecular technique (Saiki et al. 1985; Mullis and Faloona 1987). Around the same time, western researchers from very different fields also published their first results on the extraction of DNA from preserved tissues. Higuchi et al. (1984) succeeded in the extraction of DNA from quagga, an extinct member of the horse family (Fig. 1.2), and Johnson et al. (1985) investigated the DNA they had extracted from mammoth remains. Pääbo (1984) demonstrated that DNA was present in a mummified infant from an Egyptian dynasty, while Rogers and Bendich (1985) found DNA to be present in herbarium specimens. At the same time, Jeffreys and his colleagues (1985) recognized that minisatellite DNA patterns are unique to human individuals and coined the term "genetic fingerprinting." The "African Eve" theory by Cann and colleagues (1987), derived from their studies of human mitochondrial DNA, caused new disputes about the origin of humankind. A short time later, the first reports on successful multiplex PCR assays were published (Chamberlain et al. 1988), and PCR-based genetic typing by minisatellites was initiated by Boerwinkle et al. (1989) and Horn et al. (1989). Finally, at the end of the decade, the report on the successful amplification of DNA from archaeological bone by Hagelberg et al. (1989) represented a true breakthrough for anthropology, because human and animal skeletal material is by far the major group of remains throughout the ages.

The first attempts to apply the new methods to the investigation of the ancient DNA of almost any kind of sample material (Herrmann and Hummel 1993) were as varying and inspiring as the successful scientific efforts of the 1980s. Taken together, these efforts have served as the basis for the development of ancient and degraded DNA technology, allowing it to reach the point that has been achieved today: it is now a reliable investigation strategy that serves many scientific fields and interests.

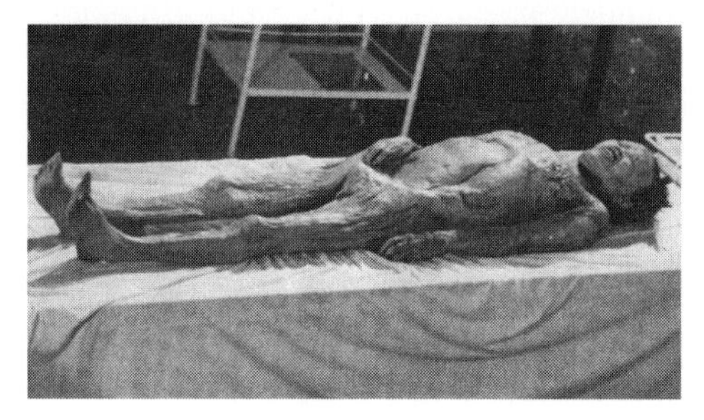

Fig. 1.1. Chinese mummified body, approximately 2,000 years old, from a grave in Mawangtui. Scientists from the Hunan Medical School succeeded in extracting DNA from the corpse

Fig. 1.2. The last living quagga, photographed at the Frankfurt Zoo in the 1880s. Since then, the quagga has been extinct. Higuchi et al. (1984) successfully extracted DNA from conserved skin remains

Many applications of ancient and degraded DNA technology are possible, for example, in the forensic sciences, which demonstrate most impressively the discriminating and validating power of multiplex analysis of autosomal STR typing. Supplements are Y-chromosomal STR and mitochondrial DNA haplotyping, in particular in those cases in which generation gaps have to be overcome. The objectives of the investigations are: the different aspects of typing human and animal DNA at crime scenes (e.g., Hochmeister et al. 1995; Schneider and Neuhuber 1996; Forster and Eberspacher 1999; Scheider et al. 1999; Lederer et al. 2001), the identification of victims in mass disasters such as plane crashes (e.g., Clayton et al. 1995; Corach et al. 1995; Whitaker et al. 1995; Goodwin et al. 1999), the fire catastrophe at the Kaprun cable railway or, most recently, following the terrorist attacks on the World Trade Center. Unknown soldiers were identified long after their deaths by ancient DNA typing techniques (e.g., Ginther et al. 1992), as were victims of war crimes and torture (e.g., Corach et al. 1997; Alonso et al. 2001) and the culprits (Jeffreys et al. 1992; Anslinger et al. 2001). For very different reasons, even the private affairs of a number of American presidents have been the targets of such investigations, as in the cases of Bill Clinton (Woodward 1999), Abraham Lincoln (McKusick 1991; Micozzi 1991) and Thomas Jefferson (Foster et al. 1998), although in the latter case, degraded DNA was not the substance being analyzed.

Indeed, ancient DNA has been analyzed for the identifications of such historic persons as Kaspar Hauser (Fig. 1.3) (Weichhold et al. 1998), Jesse James (Stone et al. 2001), the putative son of Louis XVI (Jehaes et al. 1998) and, most famously, the members of the Romanov family (Fig. 1.4) (Gill et al. 1994; Stoneking et al. 1995). Actually, the identification of the last Russian tsar and tsarina and their daughters was the first time that autosomal STR analysis was employed successfully on historic skeletal remains.

Fig. 1.3. Drawing of Kaspar Hauser (1812–1833), who was believed to be the son of Grand Duchess Stephanie of Baden. Weichhold et al. (1998) investigated mitochondrial DNA extracted from blood stains on clothing that was thought to have once belonged to Kasper Hauser. The sequence polymorphisms that were found did not match with contemporary living descendants of the earls of Baden

Fig. 1.4. Tsarina Alexandra with her four daugthers. Gill et al. (1994) reconstructed the family tree using autosomal STRs from the skeletal remains found at Ekatarinenburg in 1991. By mitochondrial DNA analysis, the relation of the family to Prince Philip of England and members of the Hessian line (Ivanov et al. 1996) was shown

Other objectives of ancient and degraded DNA analysis lie in scientific fields such as conservation biology and conservation genetics; in these fields, the initial works involving genetic investigations of museum animal and plant specimens took place (e.g., Persing et al. 1990; Cooper 1992; Ellegren 1992; Krajewski et al. 1992; Thomas 1994). Contemporary applications focus on mapping the genetic diversity of dolphins, rhinoceroses, marsupials, fish and birds as well as plants (e.g., Spencer et al. 1995; Barriel et al. 1999; Cunningham et al. 1999; Nyakaana et al. 1999; Verma et al. 1999; Spencer and Bryant 2000; McCusker et al. 2000; Landergott et al. 2001; Hamilton et al. 2001; Zawko et al. 2001) or on monitoring the illegal trade of objects and parts from endangered species (e.g., Georgiadis et al. 1990; Yang et al. 1997; Bollongino 2000; Hsieh et al. 2001). Such investigations are particularly interesting and urgent in the cases of highly endangered animal species, such as many primates, large predatory cats, elephants, rhinoceroses, whales, almost all of the Australian marsupials as well as many tropical birds and fish. With the help of the now-established technology of ancient and degraded DNA typing, it is no longer necessary to go directly to the animal for sampling; specimens from feces, hair and feathers are sufficient (e.g., Höss et al. 1992; Leeton et al. 1993; Kohn et al. 1995; Launhardt et al. 1998; Immel et al. 1999; Poinar et al. 2001). Additionally, the often large museum collections and exhibits harbor most valuable reference material, reflecting perfectly the historic status of genetic diversity (Grave and Braun 1992; Roy et al. 1994; Rosenbaum et al. 1997; Fumagalli et al. 1999; Su et al. 1999; Pertoldi et al. 2001). In the cases of investigations on the genetic diversity of historic and contemporary plant species and domestic livestock, the aspects of both conservation biology and the future management of plant and animal breeding play important roles. Also, the insights about historic genetic traits gained in the context of investigations on archaeozoological skeletal material, seeds and crops and historic cultural objects (e.g., Burger et al. 2000; Burger et al. 2001; Brown et al. 1994; Rollo et al. 1994; Goloubinoff et al. 1993; Schlumbaum et al. 1998; Paxinos et al. 2002; Vila et al. 2001; Sorenson et al. 1999; MacHugh et al. 1999; Lalueza-Fox et al. 2000; Troy et al. 2001) might contribute to the genetic stabilization of endangered species.

The scientific and service field of food monitoring also employs ancient and degraded DNA analysis techniques. Here, the technology has been implemented since the early 1990s (Allmann et al. 1995). The applications range from screening for genetically modified organisms, even in minute amounts (e.g., Teuber 1993; Hübner et al. 2001) to monitoring for bacterial contamination in food (e.g., Back et al. 1993; Budu-Amoako et al. 1993; Moura et al. 1993; Siragusa et al. 1999; Brown et al. 2001), to species declarations in highly processed foods (e.g., Meyer et al. 1995; Plath et al. 1997; Wolf et al. 2000). Because of the great differences in national legal regulations, there is also a lot of variation in the international research efforts and goals in this field.

Worldwide, the most intense research focuses on infectious diseases and investigations of their historical records, the information being derived from large collections of formalin-fixed and paraffin-embedded specimens from histopathology. One of the earliest successful reports was the publication on the detection of HIV viruses in respective specimens (Lai-Goldman et al. 1988), followed by detections of such diseases as hepatitis, herpes simplex and HIV in specimens ranging in age from a few years to several decades (e.g., Krafft et al. 1997; Asenbauer et al. 1998; Mizuno et al. 1998; Smith et al. 2000). The specimens that were investigated in the context of paleoepidemiology were of a different kind and even older. For example, there were a number of investigations on the *Mycobacterium tuberculosis* (e.g., Spigelman and Lemma 1993; Salo et al. 1994; Baron et al. 1996; Dixon and Roberts 2001) extracted from mummified soft tissue and bone specimens. Furthermore, there were proofs for *Yersinia pestis* (Herrmann and Hummel 1997; Drancourt et al. 1998; Raoult et al. 2000) and *Mycobacterium leprae* (Rafi et al. 1994). Finally, also endoparasites were the focus of interest (Loreille et al. 2001). Here, it seems very clear that paleoepidemiologic studies may not only serve to gain knowledge about historic situations, but they may also help to better understand the mode and timeframe of the mutational evolution of aggressive pathogens and endoparasites. A comprehensive overview of molecular paleoepidemiological research and its various aspects is given by Greenblatt (1998).

Although the examples derived from paleobotany, archaeozoology and paleoepidemiology are diverse, they all belong to a comprehensive human ecology in historic and prehistoric anthropology and archaeology that addresses questions about how humans shape their environment and vice versa. Still missing in this picture is the human itself. And, of course, human remains were the focus of many ancient DNA-based studies. They range from gender studies (e.g., Lassen et al. 1996; Stone et al. 1996; Faerman et al. 1997; Vernesi et al. 1999; Lassen et al. 2000) to kinship analysis from the closest to the broadest sense. While the latter are represented by phylogenetic investigations of singular famous Neanderthal objects (Krings et al. 1997; Ovchinnikov et al. 2000), questions about population genetics are addressed in various ways. Most recently, the potency of uniparental inherited markers was shown again by Vernesi et al. (2001), who used hypervariable mitochondrial markers in an investigation of the regional origin of the evangelist, Luke. The initial investigation of the genetic characteristics of a historic population was carried out by Zierdt et al. (1996) by the typing of nuclear markers. Later, Bramanti et al. (2000a, 2000b) showed in an elaborated way that STR typing can even be used to solve socio-cultural questions in historic societies. Kinship in its most literal meaning, the genealogical context, was already the focus of Kurosaki et al. (1993) in their promising typing attempts on 2,000-year-old bone remains and by Ramos and colleagues (1995), who most likely only failed because of the wrong choice of STR markers. Later studies, such as the investigation of the genealogy of an early modern noble family by Gerstenberger et al. (1999), were successful because of the close adherence to the knowledge about suitable markers and data validation that had already been made available through investigations carried out by forensic scientists (e.g., Edwards et al. 1992; Kimpton et al. 1994; Lygo et al. 1994; Urquhart et al. 1995). With the works of Schultes et al. (1999 and 2000), who fully typed skeletal remains of Bronze Age human individuals, and Scholz et al. (2001), who carried out autosomal STR typing on an early medieval skeletal series, the true potential of STR-based ancient DNA analysis for prehistoric anthropology became clearly visible: it enables not only the testing but also the reconstruction of genealogies from historic and prehistoric burial sites.

The broad variety of applications in very different scientific fields demonstrates that ancient and degraded DNA analysis has evolved during the 1990s from being the focus of research to being a research tool that is employed to find answers to questions in various disciplines. The technology had to struggle through many very basic discussions on the authenticity of ancient DNA and discussions on data validation. This process was certainly initiated by publications of results that later were discussed as the subjects of contamination events (e.g., Pääbo and Wilson 1991). The process is still not finished, as can be deduced from recent publications that have come up with demanding questions about how an ancient DNA analysis should be carried out (Hofreiter et al. 2001). This is understandable given the background of eventful experiences in the analysis of non-individualizable markers, for which the entire research field suffered. However, it seems that things have started to settle down, which is not unexpected if one has closely followed the developments in forensic science. Those colleagues are more dependent than others on fully reliable and validated techniques. Through the entire last decade, their most urgent need, generating authentic reliable results, has been reflected by a clear development towards multiplex analysis based on autosomal STR markers and supplemented by haplotyping Y-chromosomal and mitochondrial DNA. These steady and very convincing processes and developments in forensic science are further characterized by the almost total absence of publications on spectacular findings in journals with a high impact factor, such as Nature or Science. This is exactly how it used to be in the large number of cases of research work published in anthropology and archaeology. This situation has changed, unfortunately, for quite a number of reasons. On the one hand, the mere summing up by high-impact publications leads to clear misevaluations of research fund applications in scientific fields that come from the humanities rather than natural or medical sciences. On the other hand, the struggle to be published in the high-impact journals will also destroy the culture of publishing the most interesting research work in journals that are dedicated to a certain scientific field, leaving them behind in second and third class positions. Even in a scientific world that so favors interdisciplinary work, intradisciplinary discussion is essential for determining and evaluating the goals of the respective scientific fields. This certainly can only be achieved and supported by dedicated journals.

In order to return to an emphasis on scientific questions, some very basic agreements on technical standards and strategies in ancient DNA analysis are necessary. By focusing on methods and strategies, this book intends and hopes to contribute successfully to these goals. By untangling the very basic technical issues, including the choice of markers, DNA extraction, PCR parameters, primer design and data validation strategies, from the actual application or reviewing tasks as is usually required by journal publications, it may be easier for the readers to consider and evaluate whether the suggested methods apply to their own research goals. The suggested strategies, which are closely based on reliable forensic strategies, are basically transferable to any species and any research background.

One of the most basic questions in the analysis of ancient DNA is about the choice of genetic markers. Chapter 2 gives a short introduction to the structural characteristics of DNA, the different ways of inheriting mitochondrial and chromosomal DNA and the possibilities for its application to scientific issues. A special focus is on multiplex-PCR analysis, which is the best means of data authentication. Therefore, the introduction of DNA markers is accompanied by suggestions about how to run these multiplex analyses. These suggestions were all designed especially for the use in highly degraded DNA analysis. Although most of the multiplex analysis systems were designed to analyze human DNA, the basic analysis strategies are easily transferable to other species and the respective genetic markers.

Chapter 3 is dedicated to DNA extraction, DNA yield and DNA preservation. Knowledge about these topics may be decisive for the success or failure of aDNA analysis. This is shown by the most recent research work, focusing in particular on the comparison of ancient bone DNA-extraction protocols. Other topics are degradation patterns of ancient DNA and the influences of sample and extract storage on DNA degradation.

A fundamental point in aDNA analysis is the PCR process. Chapter 4 touches on some classical questions but mainly focuses on primer design and its consequences, particularly for the success of aDNA analysis. Additionally, the advantages of multiplex PCR are highlighted. Finally, a subchapter deals with the most common reasons for analysis failures and the respective indications to recognize them even if no results are present.

Chapter 5 introduces the different analysis strategies for PCR products. It shows the possibilities of fragment length analysis and discusses some of the most common artifacts in the use of automated sequencing machines and how to overcome them effectively. Also, the pros and cons of the different sequencing strategies are highlighted, as well as the advantages and limits of RFLP analysis.

The authentication of ancient DNA results is the main issue of chapter 6. Therefore, firstly the broad spectrum of possible sources of contamination and the suitable measures to avoid contamination are introduced. In a second part, the different types of control samples and validation strategies are discussed. A primary emphasis is given to multiplex amplified markers that are individual specific and that are most convincing in the authentication process due to their superior power of discrimination.

Chapter 7 presents the range of applications carried out in our department. By focusing on the scientific questions and the results that could be obtained by molecular analysis, the content has an obvious illustrative function. On the other hand, it demonstrates that the suggested strategies and methods introduced throughout the initial chapters are not mere theory, but already serve anthropology, archaeology, archaeozoology, conservation genetics and even food technology as a very reliable tool.

Chapter 8 intends to serve the needs of everyday laboratory use by giving comprehensive and detailed protocols for any analysis step that is necessary in aDNA analysis. This comprises sample collection and preparation, DNA extraction, information on any of the PCR protocols, including primer sequences, and PCR product analyses that were used for all experiments and applications presented throughout the book. Not just the idea of convenience but also the experience that often very minor deviations in the chemical composition of a reagent are fundamental for analysis success or failure determined the decision to present detailed information on the components and the preparation of even the basic chemistry and reagents that are necessary throughout the course of aDNA analysis.

Everyday usefulness is also the purpose of the final section, "PCR troubleshooting," in the appendix. Other than respective sections in PCR manuals, this part of the appendix concentrates on PCR problems that are typical for aDNA analysis. The section, organized in tables, intends to help with the identification of a possible problem and tries to give instant advice whenever possible. Also, it refers to the respective section of the book where the topic is discussed.

The first part in the appendix, "the aDNA laboratory," does not address departments already running a lab and applying ancient and degraded DNA analysis in their research work. It rather intends to give an overview of the organization and the equipment for those who are starting up an ancient DNA laboratory.

The final part of this presentation briefly highlights a most promising and appealing technical perspective for ancient DNA analysis. This is SNP (single nucleotide polymorphism) technology (e.g., Wang et al. 1998; Taillon-Miller et al. 1999; Altshuler et al. 2000; Ohnishi et al. 2001), which needs only very short target sequences. Although even the classical PCR approach can deal with short DNA fragments of approximately 60 base pairs in length, even shorter DNA fragments are sufficient for a SNP analysis. If this technology can be implemented successfully in aDNA analysis, the number of samples that can be investigated may increase remarkably. As the classical PCR assay, the SNP analysis enables multiplex approaches and therefore would perfectly fulfill the authentication requirements for aDNA data (cf. section 6.2). A prerequisite for the successful implementation of SNP technology is the availability of truly DNA-free disposable material (cf. section 6.1.2). If this is not provided by the commercial suppliers of disposable laboratory materials, cleaning strategies as described in section 6.1.3 or the use of alternative materials for the amplification process (section 6.1.4) would meet the demands.

How many of the still unanswered scientific questions can be answered succesfully by aDNA analysis depend mainly on the new discoveries and insights that will be gained from the various genome projects. Certainly, all research fields that focus on the human species will benefit from the ongoing efforts of the Human Genome Project.

References

Allmann M, Höfelein C, Koppel E, Lüthy J, Meyer R, Niederhauser C, Wegmüller B, Candrian U (1995) Polymerase chain reaction (PCR) for detection of pathogenic microorganisms in bacteriological monitoring of dairy products. Res Microbiol 146:85–97

Alonso A, Andelinovic S, Martin P, Sutlovic D, Erceg I, Huffine E, de Simon LF, Albarran C, Definis-Gojanovic M, Fernandez-Rodriguez A, Garcia P, Drmic I, Rezic B, Kuret S, Sancho M, Primorac D (2001) DNA typing from skeletal remains: evaluation of multiplex and megaplex STR systems on DNA isolated from bone and teeth samples. Croat Med J 42:260–266

Altshuler D, Pollara VJ, Cowles CR, Van Etten WJ, Baldwin J, Linton L, Lander ES (2000) An SNP map of the human genome generated by reduced representation shotgun sequencing. Nature 407:513–516

Anderson S, Bankier A, Barrell B, De Bruijn M, Coulson A, Drouin J, Eperon I, Nierlich D, Roe B, Sanger F, Schreier P, Smith A, Staden R, Young I (1981) Sequence and organization of the human mitochondrial genome. Nature 290:457–465

Anslinger K, Weichhold G, Keil W, Bayer B, Eisenmenger W (2001) Identification of the skeletal remains of Martin Bormann by mtDNA analysis. Int J Legal Med 114:194–196

Asenbauer B, McEntagart M, King MD, Gallagher P, Burke M, Farrell MA (1998) Chronic active destructive herpes simplex encephalitis with recovery of viral DNA 12 years after disease onset. Neuropedeatrics 29:120–123

Back JP, Langford SA, Kroll RG (1993) Growth of *Listeria monocytognes* in camembert and other soft cheese at refrigeration temperatures. J Dairy Res 60:421–429

Baron H, Hummel S, Herrmann B (1996) *Mycobacterium tuberculosis* complex DNA in ancient human bones. J Archaeol Sci 23:667–671

Barriel V, Thuet E, Tassy P (1999) Molecular phylogeny of Elephantidae. Extreme divergence of the extant forest African elephant. C R Académie des Sciences 322:447–454

Boerwinkle E, Xiong W, Fourest E, Chan L (1989) Rapid typing of tandemly repeated hypervariable loci by the polymerase chain reaction: application to the apolipoprotein B 3' hypervariable. Proc Natl Acad Sci USA 86:212–216

Bollongino R (2000) Bestimmung humanökologisch relevanter Tierarten aus historischen, musealen und forensischen Materialien durch Sequenzierung mitochondrialer Genorte. Diplomarbeit, Georg August-Universität, Göttingen

Bramanti B, Hummel S, Schultes T, Herrmann B (2000a) STR allelic frequencies in a German skeleton collection. Anthrop Anz 58:45–49

Bramanti B, Hummel S, Schultes T, Herrmann B (2000b) Genetic characterization of an historical human society by means of aDNA analysis of autosomal STRs. In: Susanne C, Bodzar EB (eds) Human populations genetics in Europe. Biennial Books of EAA 1:147–163

Brown CM, Cann JW, Simons G, Frankhauser RL, Thomas W, Parashar OD, Lewis MJ (2001) Outbreak of Norwalk virus in a Caribbean island resort: application of molecular diagnostics to ascertain the vehicle of infection. Epidemiol Infect 126:425–432

Brown TA, Allaby RG, Brown KA, O'Donoghue K, Sallares R (1994) DNA in wheat seeds from European archaeological sites. Experientia 50:571–575

Budu-Amoako E, Toora S, Ablett RF, Smith J (1993) Competitive growth of *Listeria monocytogenes* and *Yersinia enterocolita* in milk. J Food Protection 56:528–532

Burger J, Hummel S, Pfeiffer I, Herrmann B (2000) Palaeogenetic analysis of (pre)historic artifacts and its significance for anthropology. Anthropol Anz 58:69–76

Burger J, Pfeiffer I, Hummel S, Fuchs R, Brenig B, Herrmann B (2001) Mitochondrial and nuclear DNA from (pre)historic hide-derived material. Ancient Biomol 3:227–238

Cann RL, Stoneking M, Wilson AC (1987) Mitochondrial DNA and human evolution. Nature 325:31–36

Chamberlain JS, Gibbs RA, Ranier JE, Nguyen PN, Caskey CT (1988) Deletion screening of the Duchenne muscular dystrophy locus via multiplex DNA amplification. Nucleic Acids Res 16:11141–11156

Clayton TM, Whitaker JP, Maguire CN (1995) Identification of bodies from the scene of a mass disaster using DNA amplification of short tandem repeat (STR) loci. Forensic Sci Int 76:7–15

Cooper A (1992) Seabird 12S sequences using feathers from museum specimens. Ancient DNA Newsletter 1/2:20–21

Corach D, Sala A, Penacino G, Sotelo A (1995) Mass disasters: rapid molecular screening of human remains by means of short tandem repeats typing. Electrophoresis 16:1617–1623

Corach D, Sala A, Penacino G, Iannucci N, Bernardi P, Doretti M, Fondebrider L, Ginarte A, Inchaurregui A, Somigliana C, Turner S, and Hagelberg E (1997) Additional approaches to DNA typing of skeletal remains: The search for "missing" persons killed during the last dictatorship in Argentina. Electrophoresis 18:1608–1612

Cunningham J, Harley EH, O'Ryan CO (1999) Isolation and characterization of microsatellite loci in black rhinoceros (Diceros bicornis). Electrophoresis 20:1778–1780

Dixon RA, Roberts CA (2001) Modern and ancient scourges: the application of ancient DNA to the analysis of tuberculosis and leprosy from archaeologically derived human remains. Ancient Biomol 3:181–193

Drancourt M, Aboudharam G, Signoli M, Dutour O, Raoult D (1998) Detection of 400-year-old *Yersinia pestis* DNA in human dental pulp: An approach to the diagnosis of ancient septicemia. Proc Natl Acad Sci USA 95:12637–12640

Edwards A, Hammond HA, Jin L, Caskey CT, Chakraborty R (1992) Genetic variation at five trimeric and tetrameric tandem repeat loci in four human population groups. Genomics 12:241–253

Ellegren H (1992) Population genetics of museum birds by hypervariable microsatellite typing of genomic DNA prepared from feathers. Ancient DNA Newsletter 1/2:10

Faerman M, Kahila G, Smith P, Greenblatt C, Stager L, Filon D, Oppenheim A (1997) DNA analysis reveals the sex of infanticide victims. Nature 385:212–212

Forster R, Eberspacher B (1999) Evidence of DNA from epithelial cells of the hand of the suspect found on the tool of the crime. Arch Kriminol 203:45–53

Foster EA, Jobling MA, Taylor PG, Donelly P, de Knijff P, Mieremet R, Zerjal T, Tyler-Smith C (1998) Jefferson fathered slave's last child. Nature 396:27–28

Fumagalli L, Moritz C, Taberlet P, Friend JA (1999) Mitochondrial DNA sequence variation within the remnant populations of the endangered numbat (Marsupialia: Myrmecobiidae: *Myrmecobius fasciatus*). Mol Ecol 8:1545-1549

Georgiadis N, Patton J, Western D (1990) DNA and the ivory trade: how genetics can help conserve elephants. Pachyderm 13:45–46

Gerstenberger J, Hummel S, Schultes T, Hack B, Herrmann B (1999) Reconstruction of a historical genealogy by means of STR analysis and Y-haplotyping of ancient DNA. Eur J Hum Genet 7:469–477

Gill P, Ivanov PL, Kimpton C, Piercy R, Benson N, Tully G, Evett I, Hagelberg E, Sullivan K (1994) Identification of the remains of the Romanov family by DNA analysis. Nat Genet 6:130–135

Ginther C, Issel-Tarver L, King MC (1992) Identifying individuals by sequencing mitochondrial DNA from teeth. Nat Genet 2:135–138

Goloubinoff P, Pääbo S, Wilson AC (1993) Evolution of maize inferred from sequence diversity of an Adh2 gene segment from archaeological specimens. Proc Natl Acad Sci USA 90:1997–2001

Goodwin W, Linacre A, Vanzis P (1999) The use of mitochondrial DNA and short tandem repeat typing in the identification of air crash victims. Electrophoresis 20:1707–1711

Grave GR, Braun MJ (1992) Museums: storehouses of DNA? Science 255:1335–1336

Greenblatt CL (ed) (1998) Digging for pathogens. Ancient emerging diseases – their evolutionary, anthropological and archaeological context. Balaban, Rehovot

Hagelberg E, Sykes B, Hedges R (1989) Ancient bone DNA amplified. Nature 342:485–485

Hamada H, Petrino MG, Kakunaga T (1982) A novel repeated element with Z-DNA forming potential is widely found in evolutionary diverse eukaryote genomes. Proc Natl Acad Sci USA 79:6465–6469

Hamilton H, Caballero S, Collins AG, Brownell RL (2001) Evolution of river dolphins. Proc R Soc Lond B Biol Sci 268:549–556

Herrmann B, Hummel, S (1993) Ancient DNA. Recovery and analysis of genetic material from palaeontological, archaeological, museum, medical and forensic specimens. Springer, New York

Herrmann B, Hummel S (1997) Genetic analysis of past populations by aDNA studies. Advances in research on DNA polymorphisms. ISFH-Symposium, Toyoshoten, Hakone. pp 33–47

Higuchi R, Bowman B, Freiberger M, Ryder OA, Wilson AC (1984) DNA sequences from the quagga, an extinct member of the horse family. Nature 312:282–284

Hochmeister M, Haberl J, Borer V, Rudin O, Dirnhofer R (1995) Clarification of a break-in theft crime by multiplex PCR analysis of cigarette butts. Arch Kriminol 195:177–183

Hofreiter M, Serre D, Poinar HN, Kuch M, Paabo S (2001) Ancient DNA. Nat Rev Genet. 2:353–359

Horn GT, Richards B, Klinger KW (1989) Amplification of a highly polymorphic VNTR segment by the polymerase chain reaction. Nucleic Acids Res 17:2140

Höss M, Kohn M, Pääbo S (1992) Excrement analysis by PCR. Nature 359:199

Hsieh HM, Chiang HL, Tsai LC, Lai SY, Huang NE, Linacre A, Lee JC (2001) Cytochrome B gene for species identification of the conservation animals. Forensic Sci Int 122:7–18

Hübner P, Waiblinger HU, Pietsch K, Brodmann P (2001) Validation of PCR methods for quantitation of genetically modified plants in food. J AOAC Int 84:1855–1864

Hunan Medical College (1980) Study of an ancient cadaver in Mawantui tomb no.1 of the Han dynasty in Changsha. Beijing Ancient Memorial Press, Beijing

Immel U-D, Hummel S, Herrmann B (1999) DNA profiling of orangutan (*Pongo pygmaeus*) feces to prove descent and identity in wildlife animals. Electrophoresis 20:1768–1770

Ivanov P, Parsons T, Wadhams M, Holland M, Rhoby R, Weedn V (1996) Mitochondrial DNA variations in the Hessian lineage: heteroplasmy found in Grand Duke of Russia Georgij Romanov ends disputes over authenticity of the remains of Tzar Nicholas II. In: Proceedings of the ISFH, Hakone Symposium on DNA polymorphisms. Toyoshoten, Hakone

Jeffreys AJ, Wilson V, Thein SL (1985) Individual-specific "fingerprints" of human DNA. Nature 316:76–79

Jeffreys AJ, Allen MJ, Hagelberg E, Sonnberg A (1992) Identification of the skeletal remains of Josef Mengele by DNA analysis. Forensic Sci Int 56:65–76

Jehaes E, Decorte R, Peneau A, Petrie JH, Boiry PA, Gilissen A, Moisan JP, Van den Berghe H, Pascal O, Cassiman JJ (1998) Mitochondrial DNA analysis on remains of a putative son of Louis XVII, King of France and Marie-Antoinette. Eur J Hum Genet 6:383–395

Johnson PH, Olson CB, Goodman M (1985) Isolation and characterization of deoxyribonucleic acid from tissue of the woolly mammoth, *Mammuthus primigenius*. Comp Biochem Physiol 81:1045–1051

Kimpton C, Fisher D, Watson S, Adams M, Urquhart A, Lygo J, Gill P (1994) Evaluation of an automated DNA profiling system employing multiplex amplification of four tetrameric STR loci. Int J Legal Med 106:302–311

Kohn M, Knauer F, Stoffela A, Schröder W, Pääbo S (1995) Conservation genetics of the European brown bear – a study using excremental PCR of nuclear and mitochondrial sequences. Mol Ecol 4:95–103

Krafft AE, Duncan BW, Bijward KE, Taubenberger JK, Lichy JH (1997) Optimization of the isolation and amplification of RNA from formalin-fixed, paraffin-embedded tissue: The Armed Forces Institute of Pathology experience and literature review. Mol Diagn 2:217–230

Krajewski C, Driskell AC, Baverstock PR, Braun MJ (1992) Phylogenetic relationships of the thylacine (Mammalia: Thylacinidae) among dasyuroid marsupials: evidence from cytochrome b DNA. Proc R Soc Lond B 250:19–27

Krings M, Stone A, Schmitz RW, Krainitzki H, Stoneking M, Pääbo S (1997) Neanderthal DNA sequences and the origin of modern humans. Cell 90:19–30

Kurosaki K, Matsushita T, Ueda S (1993) Individual DNA identification from ancient human remains. Am J Hum Genet 53:638–643

Lai-Goldman M, Lai E, Grody WW (1988) Detection of human immunodeficiency virus (HIV) infection in formalin-fixed, paraffin-embedded tissues by DNA amplification. Nucleic Acids Res 16:8191

Lalueza-Fox C, Bertranpetit J, Alcover JA, Shailer N, Hagelberg E (2000) Mitochondrial DNA from *Myotragus b alearicus*, an extinct bovid from the Balearic Islands. J Exp Zool 288:56–62

Landergott U, Holderegger R, Kozlowski G, Schneller JJ (2001) Historical bottlenecks decrease genetic diversity in natural populations of *Dryopteris cristata*. Heredity 87:344–355

Lassen C, Hummel S, Herrmann B (1996) PCR-based sex determination in ancient human bones by amplification of X- and Y-chromosomal sequences. A comparison. Ancient Biomol 1:25–33

Lassen C, Hummel S, Herrmann B (2000) Molecular sex identification of stillborn and neonate individuals ("Traufkinder") from the burial site Aegerten. Anthrop Anz 58:1–8

Launhardt K, Epplen C, Epplen JT, Winkler P (1998) Amplification of microsatellites adapted from human systems in faecal DNA of wild Hanuman langurs (*Presbytis entellus*). Electrophoresis 19:1356–1361

Lederer T, Betz P, Seidl S (2001) DNA analysis of fingernail debris using different multiplex systems: a case report. Int J Legal Med 114:263–266

Leeton P, Christidis L, Westerman M (1993) Feathers from museum birds skins – a good source of DNA for phylogenetic studies. Condor 95:465–466

Loreille O, Roumat E, Verneau O, Bouchet F, Hanni C (2001) Ancient DNA from Ascaris: extraction amplification and sequences from eggs collected in coprolites. Int J Parasitol 31:1101–1106

Lygo JE, Johnson PE, Holdaway DJ, Woodroffe S, Whitaker JP, Clayton TM, Kimpton CP, Gill P (1994) The validation of short tandem repeat (STR) loci for use in forensic casework. Int J Legal Med 107:77–89

MacHugh DE, Troy CS, McCormick F, Olsaker I, Eythorsdottir E, Bradley DG (1999) Early medieval cattle remains from a Scandinavian settlement in

Dublin: genetic analysis and comparison with extant breeds. Phil Trans R Soc Lond B 354:99–108

McCusker MR, Parkinson E, Taylor EB (2000) Mitochondrial DNA variation in rainbow trout (*Oncorhynchus mykiss*) across its native range: testing biogeographical hypotheses and their relevance to conservation. Mol Ecol 9:2089–2108

McKusick VA (1991) Advisory statement by the panel on DNA testing of Abraham Lincoln's tissue. Caduceus 7:43–47

Meyer R, Höfelein C, Lüthy J, Candrian U (1995) Polymerase chain-reaction restriction fragment length polymorphism analysis: a simple method for species identification in food. J AOAC Int 78:1542–1551

Micozzi MS (1991) When the patient is Abraham Lincoln. Caduceus 7:34–42

Mizuno T, Nagamura H, Iwamoto KS, Ito T, Fukuhara T, Tokunaga M, Tokuoka S, Mabuchi K, Seyama T (1998) RNA from decades-old archival tissue blocks for retrospective studies. Diagn Mol Pathol 7:202–208

Moura SM, Destro MT, Franco BDGM (1993) Incidence of Listeria species in raw and pasteurized milk produced in Sao Paulo, Brazil. Int J Food Microbiol 19:229–237

Mullis KB, Faloona FA (1987) Specific synthesis of DNA in vitro via a polymerase-catalysed chain reaction. Meth Enzymol 155:335–350

Nyakaana S, Arctander P (1999) Population genetic structure of the African elephant in Uganda based on variation at mitochondrial and nuclear loci: evidence for male-biased gene flow. Mol Ecol 8:1105–1115

Ohnishi Y, Tanaka T, Ozaki K, Yamada R, Suzuki H, Nakamura Y (2001) A high-throughput SNP typing system for genome-wide association studies. J Hum Genet 46:471–477

Ovchinnikov IV, Gotherstrom A, Romanova GP, Kharitonov VM, Lidén K, Goodwin W (2000) Molecular analysis of Neanderthal DNA from the northern Caucasus. Nature 404:490–493

Pääbo S (1984) Über den Nachweis von DNA in altägyptischen Mumien. Das Altertum 30:213–218

Pääbo S, Wilson AC (1991) Miocene DNA sequences – a dream come true? Curr Biol 1:45–46

Paxinos EE, James HF, Olson SL, Sorensen MD, Jackson J, Fleischer RC (2002) mtDNA from fossils reveals a radiation of Hawaiian geese recently derived from the Canada goose (*Branta canadensis*). Proc Natl Acad Sci USA 99:1399–1404

Persing DH, Telford SR, Rys PN, Dodge DE, White TJ, Malawista SE, Spielman A (1990) Detection of *Borrelia burgdorferi* DNA in museum specimens of Ixodes dammini ticks. Science 249:1420–1423

Pertoldi C, Hansen MM, Loeschke V, Madsen AB, Jacobsen L, Baagoe H (2001) Genetic consequences of population decline in the European otter (*Lutra lutra*): an assessment of microsatellite DNA variation in Danish otters from 1883–1993. Proc R Soc Bio Sci 268:1775–1785

Plath A, Krause I, Einspanier R (1997) Species identification in dairy products by three different DNA-based techniques. Z Lebensm Unters Forsch 205:437–441

Poinar HN, Kuch M, Sobolik KD, Barnes I, Stankiewicz AB, Kuder T, Spaulding WG, Bryant VM, Cooper A, Paabo S (2001) A molecular analysis of dietary diversity for three archaic Native Americans. Proc Natl Acad Sci USA 98:4317–4322

Rafi A, Spigelman M, Stanford J, Lemma E, Donoghue H, Zias J (1994) *Mycobacterium leprae* DNA from ancient bone detected by PCR. Lancet 343:1360–1361

Ramos MD, Lalueza C, Girbau E, Pérez-Pérez A, Quevedo S, Turbón D, Estivill X (1995) Amplifying dinucleotide microsatellite loci from bone and tooth samples of up to 5,000 years of age: more inconsistency than usefullness. Hum Genet 96:205–212

Raoult D, Aboudharam G, Crubezy E, Larrouy G, Ludes B, Drancourt M (2000) Molecular identification by "suicide PCR" of *Yersinia pestis* as the agent of medieval black death. Proc Natl Acad Sci USA 97:12800–12803

Rogers S O, B endich AJ (1985) E xtraction o f DNA from milligram a mounts o f fresh, herbarium and mummified plant tissues. Plant Mol Biol 5:69–76

Rollo F, Asci W, Antonini S, Marota I, Ubaldi M (1994) Molecular ecology of a Neolithic meadow: the DNA of the grass remains from the archaeological site of the Tyrolean iceman. Experientia 50:576–584

Rosenbaum HC, Egan MG, Clapham PJ, Brownell RLJ, DeSalle R (1997) An effective method for isolating DNA from historical specimens of baleen. Mol Ecol 6:677–681

Roy MS, Girman DJ, Taylor AC, Wayne RK (1994) The use of museum specimens to reconstruct the genetic variability and relationships of extinct populations. Experientia 50:551–557

Saiki RK, Scharf S, Faloona F, Mullis KB, Horn GT, Erlich HA, Arnheim N (1985) Enzymatic amplification of P-globulin genomic sequences and restriction site analysis for diagnosis of sickle cell anemia. Science 230:1350–1354

Salo WL, Aufderheide AC, Buikstra J, Holcomb TA (1994) Identification of *Mycobacterium tuberculosis* in a pre-Columbian Peruvian mummy. Proc Natl Acad Sci USA 91:2091–2094

Schlumbaum A, Neuhaus JM, Jacomet S (1998) Coexistence of tetraploid and hexaploid naked wheat in a Neolithic lake dwelling of central Europe. Evidence from morphology and ancient DNA. J Archaeol Sci 25:1111–1118

Schneider H, Neuhuber F (1996) Detection of saliva traces on perpetrator masks and their attribution to a particular criminal. Arch Kriminol 198:31–37

Schneider PM, Seo Y, Rittner C (1999) Forensic mtDNA hair analysis excludes a dog from having caused a traffic accident. Int J Legal Med 112:315–316

Scholz M, Hengst S, Broghammer M, Pusch CM (2001) Intrapopulational relationships in ancient societies: a multidisciplinary study. Z Morph Anthrop 83:5–21

Schultes T, Hummel S, Herrmann B (1999) Amplification of Y-chromosomal STRs from ancient skeletal material. Hum Genet 104:164–166

Schultes T (2000) Typisierung alter DNA zur Rekonstruktion von Verwandtschaft in einem bronzezeitlichen Skelettkollektiv. Dissertation, Georg August-Universität, Göttingen. Cuvellier, Göttingen

Siragusa GR, Nawotka K, Spilman SD, Contag PR, Contag CH (1999) Real-time monitoring of *Escherichia coli* O157:H7 adherence to beef carcass surface tissues with a bioluminiscent reporter. Appl Environ Microbiol 65:1738–1745

Smith KM, Crandall KA, Kneissl ML, Navia BA (2000) PCR detection of host and HIV-1 sequences from archival brain tissue. J Neurovirol 6:164–171

Sorenson MD, Cooper A, Paxinos EE, Quinn TW, James HF, Olson SL, Fleischer RC (1999) Relationships of the extinct moa-nalos, flightless Hawaiian waterfowl, based on ancient DNA. Proc R Soc Lond B Biol Sci 266:2187–2193

Spencer PB, Odorico DM, Jones SJ, Marsh HD, Miller DJ (1995) Highly variable microsatellites in isolated colonies of the rock-wallaby (*Petrogale assimilis*). Mol Ecol 4:523–525

Spencer PB, Bryant KA (2000) Characterization of highly polymorphic microsatellite markers in the marsupial honey possum (*Tarsipes rostratus*). Mol Ecol 9:492–494

Spigelman M, Lemma E (1993) The use of the polymerase chain reaction to detect *Mycobacterium tuberculosis* in ancient skeletons. Int J Osteoarchaeol 3:137–143

Stone AC, Milner GR, Pääbo S, Stoneking M (1996) Sex determination of ancient human skeletons using DNA. Am J Phys Anthropol 99:231–238

Stone AC, Starrs JE, Stoneking M (2001) Mitochondrial DNA analysis of the presumptive remains of Jesse James. J Forensic Sci 46:173–176

Stoneking M, Melton T, Nott J, Baritt S, Roby R, Holland M, Weedn V, Gill P, Kimpton C, Aliston G (1995) Establishing the identity of Anna Anderson Manahan. Nat Genet 9:9–10

Su B, Wang Y, Lan H, Wang W, Zhang Y (1999) Phylogenetic study of complete cytochrome B genes in musk deer (Genus Moschus) using museum samples. Mol Phylogen Evol 12:241–249

Taillon-Miller P, Piernot EE, Kwok PY (1999) Efficient approach to unique single-nucleotide polymorphism discovery. Genome Res 9:499–505

Teuber M (1993) Genetic engineering techniques in food microbiology and enzymology. Food Rev Int 9:389–409

Thomas RH (1994) Analysis of DNA from natural history museum collections. In: Schierwater B, Streit B, Wagner GP, DeSalle R (eds) Molecular ecology and evolution: Approaches and applications. Birkhäuser, Basel. pp 311–321

Troy CS, MacHugh DE, Bailey JF, Magee DA, Loftus RT, Cunningham P, Chamberlain AT, Sykes BC, Bradley DG (2001) Genetic evidence for Near-Eastern origins of European cattle. Nature 410:1088–1091

Urquhart A, Oldroyd NJ, Kimpton CP, Gill P (1995) Highly discriminating heptaplex short tandem repeat PCR system for forensic identification. BioTechniques 18:116–121

Verma SK, Khanna V, Singh N (1999) Random amplified polymorphic DNA analysis of Indian scented basmati rice (*Oryza sativa L.*) germplasm for identification of variability and duplicate accession, if any. Electrophoresis 20:1786–1789

Vernesi C, Caramelli D, Carbonell i Sala S, Ubaldi M, Rollo F, Chiarelli B (1999) Application of DNA sex tests to bone specimens from three Etruscan archaeological sites. Ancient Biomol 2:295–305

Vernesi C, Di Benedetto G, Caramelli D, Secchieri E, Simoni L, Katti E, Malaspina P, Novelletto A, Marin VT, Barbujani G (2001) Genetic characterization of the body attributed to the evangelist Luke. Proc Natl Acad Sci USA 98:13460–13463

Vila C, Leonard JA, Gotherstrom A, Marklund S, Sandberg K, Liden K, Wayne RK, Ellegren H (2001) Widespread origins of domestic horse lineages. Science 291:474–477

Wang DG, Fan JB, Siao CJ, Berno A, Young P, Sapolsky R, Ghandour G, Perkins N, Winchester E, Spencer J, Kruglyak L, Stein L, Hsie L, Topaloglou T, Hubbell E, Robinson E, Mittmann M, Morris MS, Shen N, Kilburn D, Rioux J, Nusbaum C, Rozen S, Hudson TJ, Lander ES (1998) Large-scale identification, mapping, and genotyping of single-nucleotide polymorphisms in the human genome. Science 280:1077–1082

Weichhold GM, Bark JE, Korte W, Eisenmenger W, Sullivan KM (1998) DNA analysis in the case of Kasper Hauser. Int J Legal Med 111:287–291

Whitaker JP, Clayton TM, Urquhart AJ, Millican ES, Downes TJ, Kimpton CP, Gill P (1995) Short tandem repeat typing of bodies from a mass disaster: high success rate and characteristic amplification patterns in highly degraded samples. BioTechniques 18:670–677

Wolf C, Burgener M, Hübner P, Lüthy J (2000) PCR-RFLP analysis of mitochondrial DNA: differentiation of fish species. Lebensm Wiss Technol 33:144–150

Woodward B (1999) Shadow: Five presidents and the legacy of Watergate. Simon and Schuster, New York

Yang H, Golenberg EM, Shoshani J (1997) A blind testing design for authenticating ancient DNA sequences. Mol Phylogenet Evol 7:261–265

Zawko G, Krauss SL, Dixon KW, Sivasithamparam K (2001) Conservation genetics of the rare and endangered *Leucopogon o btectus* (Ericaceae). Mol Ecol 10:2389–2396

Zierdt H, Hummel S, Herrmann B (1996) Amplification of human short tandem repeats from medieval teeth and bone samples. Hum Biol 68:185–199

2 DNA markers

Cells are the basic units of any living organism. They consist of various cell organelles, including the nucleus and thousands of mitochondria. Both of these organelles harbor deoxyribonucleic acid (DNA), which is a macromolecule with a helical structure. This codes for all genetically determined traits in a living organism, from metabolism to phenotypic and immunogenetic characteristics. DNA is built up of nucleotides that are formed by three biochemically distinct molecules: nucleobases, sugars and phosphates. While sugar molecules and phosphates form the backbone of the molecule, it is the succession of the four nucleobases, adenine (A), thymine (T), cytosine (C) and guanine (G), that represents the genetic code.

The helical DNA molecule is organized as a double strand, maintained by hydrogen bonds between complementary bases. Two hydrogen bonds form between adenine and thymine, while three hydrogen bonds develop between cytosine and guanine. As a result, the forces keeping cytosine and guanine held together are somewhat stronger (Fig. 2.1).

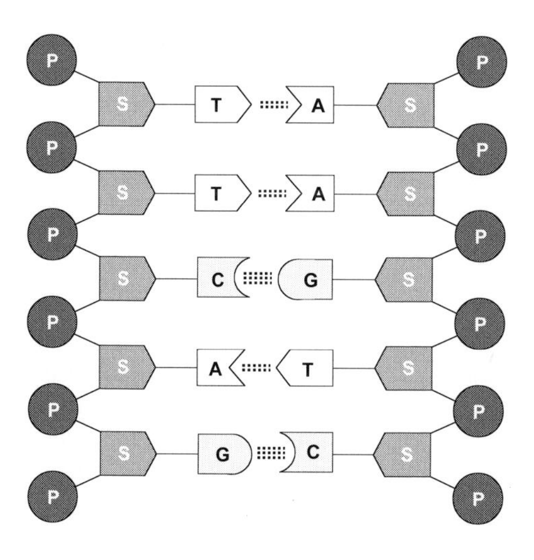

Fig. 2.1. Hydrogen bonds form between the nucleobases, keeping the DNA molecule in a hybridized double-strand formation. Three hydrogen bonds are built up between cytosin (C) and guanine (G), two between adenine (A) and thymine (T). Sugar molecules (S) and phosphates (P) form the backbone of the molecule

The structure of a double-strand DNA formation, which is referred to as a hybridization, is a characteristic of DNA. The two strands may be broken up either by heat or chemical treatment, which is called denaturation. If denaturation is induced by heat, it is fully reversible as soon as the temperature decreases, causing renaturation or reannealing. Both strands of the DNA reveal an anti-parallel directionality, which is determined by the positions of the carbon atoms in the sugar molecules (5' and 3'). The nomenclatorial agreement is to list nucleotide sequences from the 5'-end to 3'-end. This is exactly the same direction in which polymerases synthesize RNA, carry out DNA replication and repair in the cell. All of these properties and characteristics of DNA are used by the polymerase chain-reaction technique (see chapter 4).

2.1 Mitochondrial DNA

A single somatic human cell harbors some 3×10^9 nucleotide base pairs; however, only a minor part of the total genome is located in the mitochondria, where the DNA molecules have a circular configuration (Fig. 2.2). In humans, each mitochondrial genome consists of approximately 16,570 base pairs, coding for 37 genes. Additionally, there are several non-coding regions, so-called control regions, showing a comparatively high rate of polymorphism as a result of earlier mutation events. Depending on the particular region, the variability is characteristic at the population level or, for certain mitochondrial markers with high mutation rates, even down to the family level.

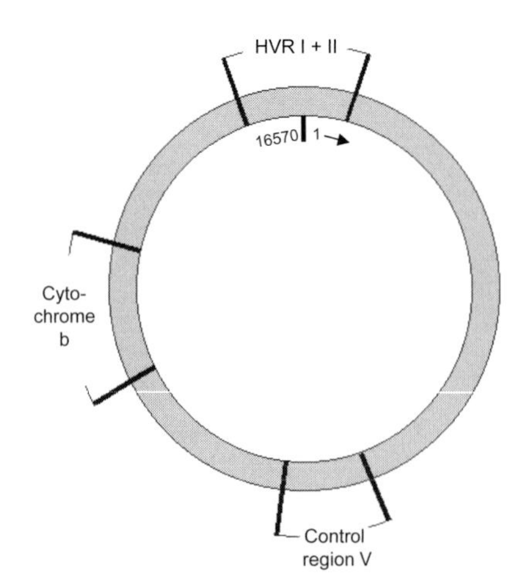

Fig. 2.2. Circular organization of the mitochondrial DNA revealing the locations of the hypervariable regions I and II, control region V and the cytochrome b gene

In contrast to chromosomal DNA, which is present in each somatic cell as a double set, mitochondrial DNA (mtDNA) is present as a single copy only, since it is inherited only through the females of the previous generation. Therefore, mtDNA represents a haplotype. This is due to the fact that oocytes contain a full complement of mitochondria, while there are few mitochondria present in sperm; these are located close to the tail. They are used to provide the energy for movement of the sperm cells but usually do not penetrate the oocyte at the moment of fertilization. If, however, mitochondrial DNA of the sperm enters the oocyte, the mitochondrial DNA of the developing individual may reveal a mosaic in its mitochondria, and the respective sequence polymorphism is called heteroplasmy (Gill et al. 1994; Ivanov et al. 1996; Hühne et al. 1999; Pfeiffer et al. 1999).

Through the regular mechanism, in which only the female mitochondria contribute to the newly developing individual's genetic makeup, all members of a maternal lineage show the same mtDNA. Depending on the particular mtDNA sequence, all members will reveal the same traits and form a haplogroup (Fig. 2.3).

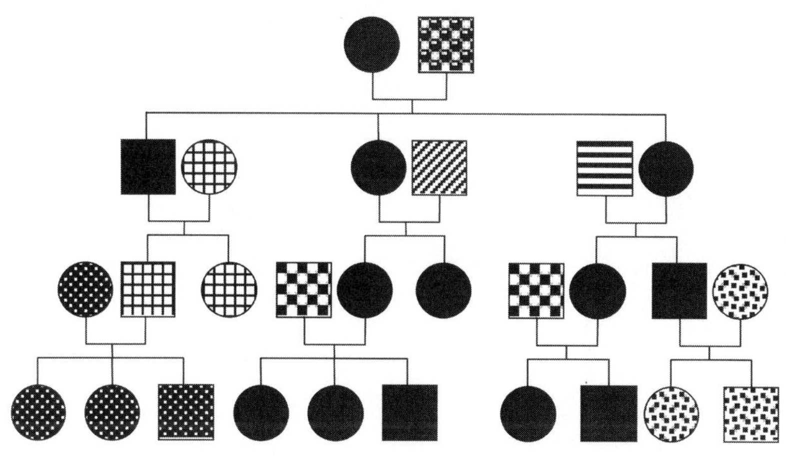

Fig. 2.3. Diagram of a family tree spanning four generations. Individuals showing the same mitochondrial sequence polymorphisms and thus forming haplogroups are marked with the same pattern

From the analysis of isolated chromosomal sperm DNA, a particular type of mutational event is known that may complicate mtDNA analysis. This is the transfer of mitochondrial DNA to the chromosomal DNA, the so-called nuclear insertions. Since chromosomal DNA has significantly lower mutation rates than mitochondrial DNA, nuclear insertions that happened early in evolution represent a kind of ancestral mitochondrial DNA haplotype (Fig. 2.4) (Zischler 1999; Greenwood and Pääbo 1999). In living organisms, this can be studied by investigating the mitochondrial DNA of somatic cells and nuclear DNA extracted from sperm separately. Obviously, primers that are designed to target mitochondrial DNA and that are known to amplify also sequences from sperm DNA should not be used in aDNA analysis. In modern DNA analysis the effective co-amplification of nuclear insertions can be avoided by keeping to low cycle numbers; this is supported by the fact that mitochondrial genomes far outnumber the nuclear genome. Because of the degradation process in ancient DNA (see chapter 3), the ratio of mitochondrial DNA to nuclear DNA may be shifted towards the nuclear genomes, and the strategy of discrimination by cycle numbers no longer applies. Amplifications by primers targeting mitochondrial DNA and nuclear insertions are therefore particularly likely to show ambiguous results in degraded DNA extracts.

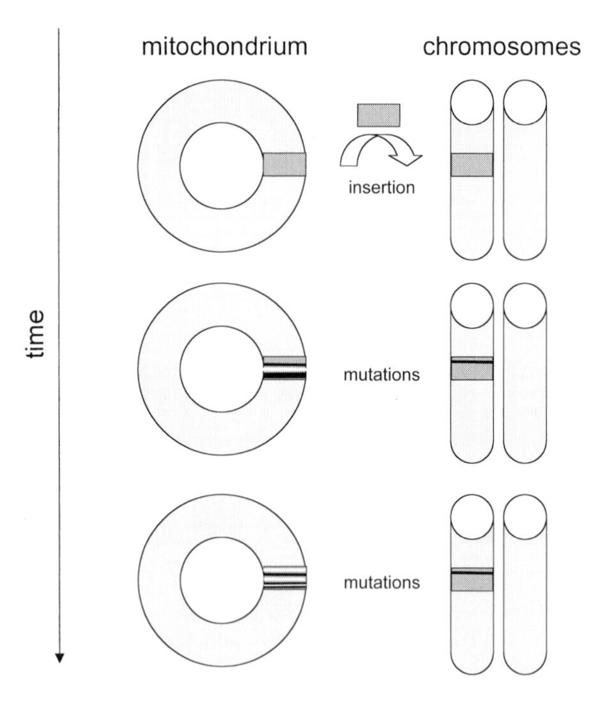

Fig. 2.4. Nuclear insertions represent an ancestral mitochondrial haplotype because of the lower mutation rates of the chromosomal compared to the mitochondrial genome. If the primers that were intended to amplify a mitochondrial sequence co-amplify the nuclear insertion, this may lead to ambiguous results at particular nucleotide positions

2.1.1 The hypervariable regions and control region V

By far, most sequence polymorphism shows the d-loop control region located near the nomeclatorial origin of the mitochondrial genome. This region mainly consists of about 600 base pairs making up the hypervariable regions I and II (HVR I and HVR II). In total, this region spans about 900 base pairs. In the human mitochondrial genome, HVR I is commonly found between the nucleotide position (np) 16,024 to np 16,365 and HVR II from np 73 to np 340 (Fig. 2.5).

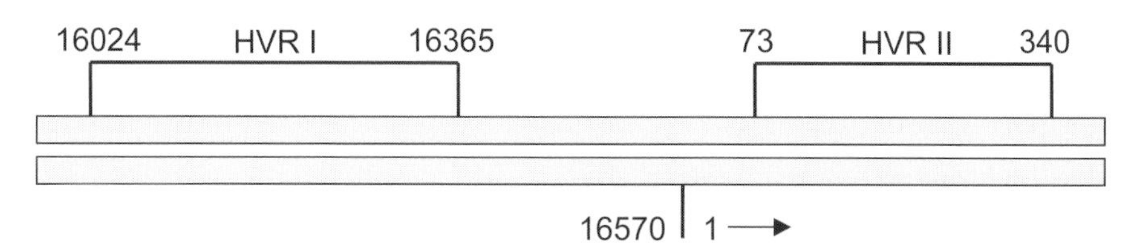

Fig. 2.5. Locations of the human hypervariable regions I and II (HVR I and HVR II). The nucleotide position nomenclature follows the human mitochondrial reference sequence (Anderson et al. 1981)

The HVRs reveal about 3% variability between individuals; i.e., two randomly selected individuals who are not related show differences in 3 out of 100 nucleotide postions (e.g., Stoneking 2000). Within the HVRs, the polymorphic sites are not distributed uniformly but cluster in so-called hotspots (Fig. 2.3). Because of the high number of polymorphic sites, the analysis of HRV I and II enables investigators to identify individuals at the family lineage level (e.g., Ginther et al. 1992; Holland et al. 1993; Butler and Lewin 1998; Jehaes et al. 1998; Schneider et al. 1999). Of course, the HVRs are also highly valuable markers in terms of population genetics or phylogenetic studies (Cann et al. 1984; Boles et al. 1995; Hardy et al. 1995; Harihara et al. 1995; Zischler et al. 1995; Bailey et al. 1996; Parr et al. 1996; Richards et al. 1996; Stoneking and Soodyall 1996; Cavalli-Sforza 1998; Eizirik et al. 1998; Encalada et al. 1998; Stone and Stoneking 1998). Publicly accessible databases are available providing HVR I and II data from worldwide human populations (cf. appendix). Due to the comparatively high levels of polymorphism, the HVRs are in general suitable for aDNA analysis without the additional use of individualizable chromosomal markers (Bär et al. 2000). Since 900 bp usually cannot be targeted in aDNA extracts by a single amplification process, there are two different strategies to amplify the HVRs: either by generating overlapping fragments (e.g., Krings et al. 1997) or by a multiplex assay (Schultes 2000) (cf. section 5.2.2). The method using the multiplex approach followed by direct sequencing reactions requires less aDNA extract and allows for an improved data validation process (see chapter 6). The regions targeted by the three primer pairs using the multiplex approach are shown in Fig. 2.6. The specific applications with this approach are given in sections 7.3.3 and 7.4.1.

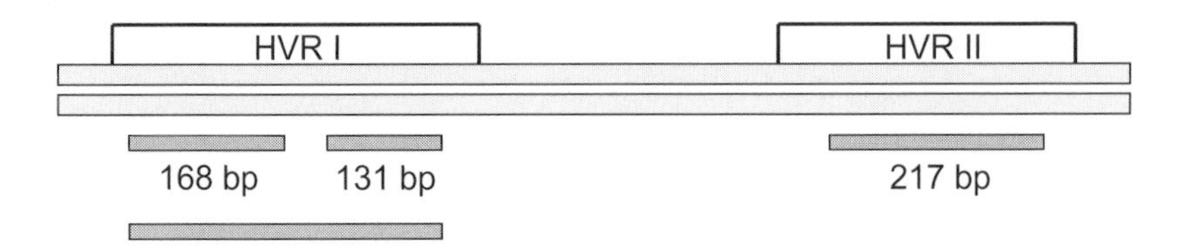

Fig. 2.6. Multiplex amplification of the human HVR I and II (Schultes 2000). In aDNA extracts having average DNA preservation, two or three products (131 bp, 168 bp and 217 bp) should be observed. In the case of exceptionally good aDNA preservation, a fourth product of approximately 312 bp in length in humans results from the multiplex approach

Other control regions on the mitochondrial DNA reveal less variability. One of those is the control region of the cytochrome c oxidase subunit II (CoII/tRNA^{Lys}) (cf. Fig. 2.2). This region is usually used in studies at the population level. Here, a 9-bp deletion spanning from np 8,272 to np 8,280 was found in native American individuals, Pacific islanders and many Southeast Asians. Together with other mitochondrial markers, the frequency of the 9-bp deletion was used to investigate the peopling of the Pacific islands and the colonization of the American continents (Wrischnik et al. 1987; Harihara et al. 1992; Hagelberg et al. 1993; Turbon et al. 1994; Merriwether 1999; Yao et al. 2000). While the 9-bp deletion is absent in Caucasian and many African populations (e.g., Soodyall et al. 1996), a frequency of 10–60% exists in Southeast Asia, the Pacific islands and native North American populations. However, in South American native populations, frequencies of between 90–100% for the deletion are found (Wallace and Torroni 1992; Torroni et al. 1994). In aDNA analysis this may be used, for example, to identify mummies of uncertain geographical origin. However, the marker lacks any discrimination power at the individual level, and since amplifiable fragments are present and widespread in the undeleted form in most of the plastic of the amplification reaction tubes (section 6.1.2), it can only be used for aDNA analysis purposes in combination with individualizable markers.

Additionally, certain cleaning treatments for the amplification tubes or the use of glass tubes are recommended (section 6.1.3 and 6.1.4).

2.1.2 Cytochrome b

The cytochrome-b region is one of the 37 genes that are found in the mitochondrial genome. In human mitochondria, it is located between np 14,747 and np 15,887 (cf. Fig. 2.2). Cytochrome b is a protein that is involved in cellular oxidation, photosynthesis and other biochemical processes where redoxcatalysis is necessary. The region is particularly interesting for aDNA analysis, because animal species show considerable polymorphism throughout the approximately 1,000 base pairs (Fig. 2.7), making it ideal for interspecies investigations.

```
                           14380                     14390                     14400
14371   C G A G A C G T A A A T T A T G G C T G A A T C A T C C G C   human
211     . . . . . . . . G . . C . . C . . . . . . . . . . . . . . A   cattle
211     . . . . . T . . . . . . . . . . . . . . . . . . . . . . . A   goat
211     . . . . . . . . G . . C . . . . . . . . . . . T . . . . . A   sheep
212     . . . . . . . . . . . . . . C . . A . . . G . T . . T . . .   pig
214     . . G A . . . . . C . A . . C . . . . . . C . . . . . . . G   chicken
214     . . . A . . . . . C . A . . C . . T . . . C . . C . . . A T   turkey
211     . . . . . . . . T . . C . . C . . . . . . . . T . . . . . .   dog
211     . . . . . . . . T . . C . . . . . . . . . . . T . . . . . .   fox
330     - - - . . T . . G . . G . . C . . . . . . . . T . . T . . A   elephant
211     . . . . . . . . . . . C . . C . . . . . . T . A . . . . . .   kangoroo

                           14410                     14420                     14430
14401   T A C C T T C A C G C C A A T G G C G C C T C A A T A T T C   human
241     . . . A . A . . . . . A . . C . . A . . T . . . . . G . . T   cattle
241     . . . A . A . . . . . A . . C . . A . . A . . . . . . . . .   goat
241     . . T A . A . . . . . A . . C . . G . . A . . . . . . . . T   sheep
242     . . T . . A . . T . . A . . C . . A . . A . . C . . . . . .   pig
244     A . T . . C . . . . . A . . C . . . . . . . . . T . C . . .   chicken
244     A . . . . C . . T . . G . . . . . G . . . . . . T . C . . .   turkey
241     . . T A . G . . . . . A . . . . . . . . T . . C . . . . . .   dog
241     . . . A . A . . T . . A . . C . . A . . A . . T . . . . . T   fox
357     C . A . . A . . . T . A . . C . . A . . A . . C . . T . . .   elephant
241     A . T . . A . . . . . . . . C . . A . . A . . C . . . . . .   kangoroo
```

Fig. 2.7. Alignments of parts of the cytochrome-b sequence of some selected animal species show a high interspecies variability. Dots in the animal sequences indicate that the same nucleotide bases are present as in the human sequence, which was taken as a reference sequence

The polymorphisms that are characteristic of species and subspecies (e.g., Irwin et al. 1991) can be detected either by direct sequencing (section 5.2) or RFLP analysis (section 5.3). Examples for using analysis of the cytochrome-b region for modern degraded DNA analysis are common in criminal investigations (Parson et al. 2000), conservation biology and food biotechnology (see also sections 7.2.7-7.2.9). Specific aDNA applications have included archaeozoological species, such as discriminations between sheep and goats (Loreille et al. 1997), and investigations on the pocket gopher (Hadley et al. 1998) or the Tasmanian wolf (Krajewski et al. 1992) (section 7.2.3). Other investigations have identified components in historic objects and materials such as glues or stains (Burger et al. 2000). Because of the non-individualizable nature of cytchrome-b amplification products, aDNA investigations on domestic animals (cf. section 6.1.2) should be accompanied, if possible, by the analysis of a HVR or STR markers for data validation purposes.

2.2 Chromosomal DNA

In contrast to the small mitochondrial genome, the vast majority of DNA is found in the cell nucleus. This DNA, which is referred to as the nuclear genome, is packed densely together with very stable proteins called histones and divided up into chromosomes. The number of chromosomes may vary considerably even between comparatively closely related species, e.g., between 40 and 80 chromosomes are found in mammals. The normal nuclear genome of humans consists of 46 chromosomes in the diploid status. This means that a double set of 23 homologous chromosomes are present, each set inherited through the haploid gametes from either parent; this is known as Mendelian inheritance. This mode of passing the genetic information to the next generation leads to unique recombinations in each individual, with the exception of homozygotic twins, who develop from a single fertilized egg. Each gamete (oocyte or sperm cell) receives a random haploid set combination derived from the 23 pairs of chromosomes; in other words, 2^{23} different combinations result already from meiosis. Through union with the opposite gamete, $2^{23\times}2^{23}\approx 70$ trillion possible combinations are possible, neglecting the possibility of crossing-over events, i.e., chromosomes exchanging sequence regions in the course of mitosis.

Estimates from the Human Genome Project suggest that as much as 95% of the human nuclear genome are non-coding sequences, and only 5% are made up of the protein-coding genes. The non-coding sequences are particularly interesting for identification purposes, since they reveal considerable individual variability.

In the 1980s, Alec Jeffreys and his co-workers first discovered the potential of these polymorphisms in terms of human identification. They recognized that total human DNA digested by enzymes reveals random patterns for different individuals (Gill et al. 1985; Jeffreys et al. 1985). The patterns proved empirically to be unique to each individual. Through these discoveries, the term genetic fingerprint was coined. However, the method of typing individuals by analyzing sequence polymorphisms with the RFLP-based technique presented a number of disadvantages in the application to degraded forensic evidence material. Most important, the technique required large amounts of intact DNA with a high molecular weight.

The discovery that polymorphic tandem repeat structures proved to be successfully amplified by PCR made them perfectly suitable for the purpose of typing minute amounts of degraded DNA. In the early stages, only the larger, so-called minisatellites or VNTRs (variable number of tandem repeats), consisting of about 20–70-bp long, tandemly repeated sequences, were applied to forensic evidence material (Jeffreys et al. 1988; Boerwinkle et al. 1989; Horn et al. 1989; Weiß and Turner 1992; Elliot et al. 1993; Jeffreys et al. 1995). The degree of polymorphism was defined by the number of alleles that were found in a population.

Soon afterwards, microsatellites or STRs (short tandem repeats) that had been discovered back in the early 1980s (Hamada et al. 1982) were recognized to be better suited to the needs of degraded DNA. With these findings, the systematic application of genetic typing for forensic evidence material was initiated (Kimpton et al. 1994; Edwards et al. 1992; Urquhart et al. 1995; Hammond et al. 1994; Kimpton et al. 1993). Similar to VNTRs, the basis of the STRs' power of discimination is length polymorphism that is found in human populations. STRs are more amenable to degraded DNA analysis than VNTRs due to the short length of their tandemly repeated core units, which are only 2–6 bp. This enabled amplification products that range between 100–300 bp (Fig. 2.8).

Meanwhile, STRs are also known for many mammalian genomes, where they are used in studies of conservation biology and livestock breeding. The properties of STRs make them highly suitable for ancient DNA analysis. Examples of anthropological and archaeological applications are the identification and assignment of skeletal elements in disturbed burial sites, reconstructions of historic genealogies and population genetics (Chakraborty and Jin 1993; Queller and Strassmann 1993) (also cf. chapter 7). Because of their capacity to genetically profile a sample at the individual level, they have a very special value in validating aDNA amplification results (section 6.2). In particular, this holds true for multiplex analysis of STRs (Chakraborty et al. 1999; Butler 2001; Moretti et al. 2001).

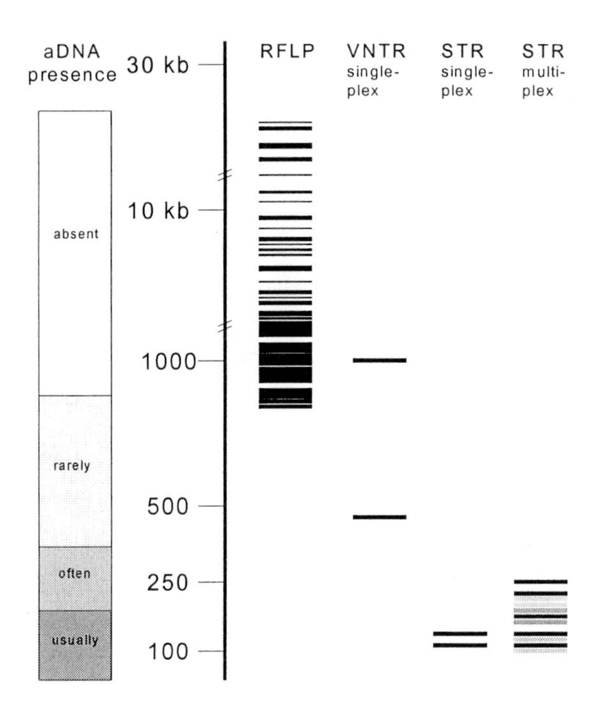

Fig. 2.8. The scheme shows the lengths of the different average analysis products of the original RFLP-based fingerprinting technique, the VNTR singleplex amplifications, singleplex STRs and multiplex STR amplifications in relation to the presence of aDNA fragment lengths measured in base pairs or kilobases (kb), respectively

Beside STRs, there are other chromosomal markers that are used to characterize historic individuals at the DNA level. One of the earliest applications, and still the most widespread, is the analysis of a particular locus of the X- and Y-chromosomal amelogenin gene, which is used for sex determination. Other loci that remarkably improve the determination of sex are X- and Y-chromosomal STRs. Additionally, the Y-chromosomal STRs have been used to identify family lineages and in population genetics.

Questions about selective advantages because of a particular genetic makeup include the analysis of the ABO blood group genes, the CCR5 locus, which is a chemokine receptor and part of the immunogenetic response to viral infection, and the ΔF 508 deletion, which codes for cystic fibrosis. Recently, all of these loci have been shown to successfully amplify in ancient human DNA (section 7.4).

2.2.1 Amelogenin

The amelogenin gene is a single-copy gene located on Yp11.2 of the Y-chromosome and the homologous region Xp22.31-p.22.1 of the X-chromosome (cf. Figs. 2.10 and 2.11). The gene codes for a protein of the tooth enamel. Early in the 1990s, it was discovered that a region on intron 1 of the gene reveals a 6-bp deletion on the X-chromosome (Sullivan et al. 1993; Manucci et al. 1994).

The amplification with one of the originally published primer sets leads to a single amplification product for females (106 bp) and two amplification products for males (106 bp and 112 bp) (Fig. 2.9). In rare cases the Y-chromosomal region may also pass the deletion (Santos et al. 1998), leading to an incorrect sex determination.

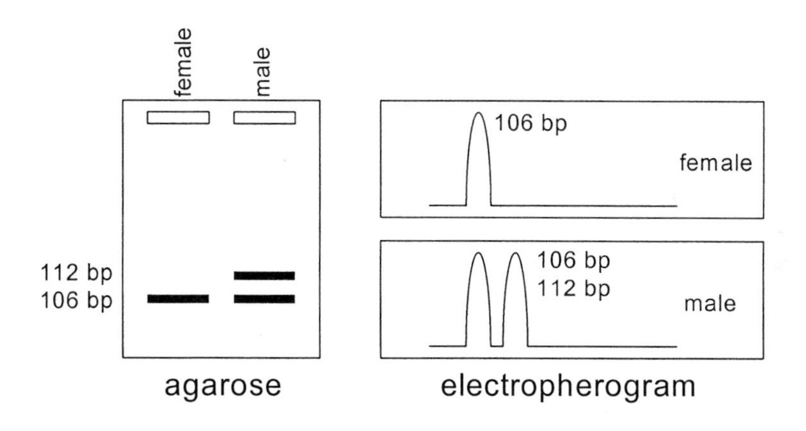

Fig. 2.9. Scheme of the expected results for male and female individuals by the amplification of the amelogenin gene locus

In aDNA analysis, this rare event is not as much of a problem as allelic dropout, in which one of two alleles is not amplified by chance (cf. section 2.2.2). If it is the X-chromosomal allele, a correct sex determination is still possible. If it is a Y-dropout, the result will not distinguish from a female result. Therefore, singleplex amplifications of the amelogenin locus must be carried out at least three times in order to ensure the determination, and this still does not reach the desired level of data validation.

The risk of receiving incorrect female results is reduced if the amelogenin result is derived from an amplification in conjunction with STRs, such as presented by many typing kits. In such cases the number of heterozygotic results gives a hint of the general allelic drop out rate. Applications of this are found throughout chapter 7.

However, in the end, amelogenin amplifications cannot reveal positive proofs for female individuals. Therefore, a multiplex approach employing the amelogenin locus together with X-chromosomal markers and Y-chromosomal STR markers is introduced (section 2.2.4); this most probably enables the gain of a positive proof for females through the high heterozygosity rates of the X markers. Furthermore, this approach avoids misinterpretations in case of a deleted Y-chromosomal amelogenin sequence.

2.2.2 Autosomal STRs

Short tandem repeats (STRs), commonly referred to as microsatellites, are present in the non-coding regions of the human, animal and plant genomes. In humans, they are estimated to make up about 20% of the genome. STRs are found in all of the 22 different autosomes and on both sex chromosomes. The autosomal STRs enable the genetic typing of an individual. Many terms are used synonymously, including STR typing, STR profiling, genetic fingerprinting, DNA profiling or genetic profiling. Different from the original genetic fingerprint technique discovered by Jeffreys and co-workers (1985), STR typing can be carried out successfully on highly fragmented DNA because of the short length of the tandemly repeated structure (cf. Fig. 2.8). This also enables the use of the PCR technique, allowing the target DNA not only to be fragmented but also to be analyzed using minute amounts of target DNA, which are sufficient for an analysis (chapter 4).

STRs are characterized by the tandemly repeated arrays of 2–6 bp core units. Most common are dinucleotide $[CA]_n$ repeats; the second largest group is tetranucleotide repeats. These are preferred in aDNA analysis if possible, because they reveal fewer amplification artifacts than the dinucleotide repeats. Tetranucleotide repeats are not a uniform group. While some of them always reveal the same core unit throughout the entire repeat structure (e.g., $[AGAT]_n$), others reveal combinations (e.g., $[AGAT]_n[ACAT]_m$). Furthermore, many tetranucleotide repeats reveal so-called interalleles, which do not even consist of four bases (e.g., $[AGAT]_n[AT]_m[AGAT]_l[AGA]_k$).

In the early STR nomenclature, the shortest allele that was found within a population was called "allele 1," the next longer one "allele 2" and so on. This was soon abandoned, because it made renaming necessary and thus led to confusion about the nomenclature whenever an individual was found to have an even shorter allele than the one that previously had been recognized. Today, the nomeclature of STR alleles complies with the number of repeat units that are present. Interalleles are named as fractions, e.g., ".2" or ".3" (Fig. 2.10).

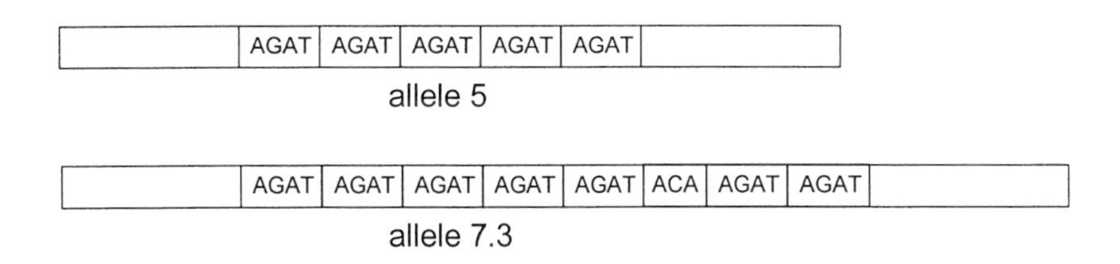

Fig. 2.10. Structure and nomenclature of short tandem repeats (STRs)

In order to serve fingerprinting needs, the autosomal STRs of choice should reveal a high level of polymorphism and must not be linked to other STRs under investigation. Also, a suitable STR must not be linked to genes that may underlie different selective pressures in different environmental conditions. Furthermore, their own allele expression must not itself be the cause of a genetic disorder, as is the case for such neurodegenerative disorders as Huntington's disease that are linked to extended trinucleotide repeat structures (e.g., Nasir et al. 1996; Li and el-Mallakh 1997).

As a rule of thumb, the suitability of a particular STR to be used in human or animal identification can be deduced from the heterozygosity rate and its chromosomal location. The heterozygosity rate itself is determined by the number of alleles and their frequency. Thus, the allele frequency also gives a rough impression if the no-gene-linkage criterion is fulfilled. This is most likely the case in all Gauss distributions, which are interpreted as an indication that there is no selective pressure against one or more of the alleles (Fig. 2.11) (Edwards et al. 1992; Deka et al. 1995; Chakraborty et al. 1999). This ensures a stable intra- and interpopulation power of discrimination (P_d) of the respective STR locus. The calculation of the power of discrimination of combined STR analysis is the basis for the demand that STRs must not be linked to each other. This is fulfilled as soon as they are located on different chromosomes. For most STRs, the chromosomal location is indicated by a systematic nomenclature (Fig.

2.12). All these criteria that are essential for autosomal STR typing have been accounted for in modern, commercially available kits.

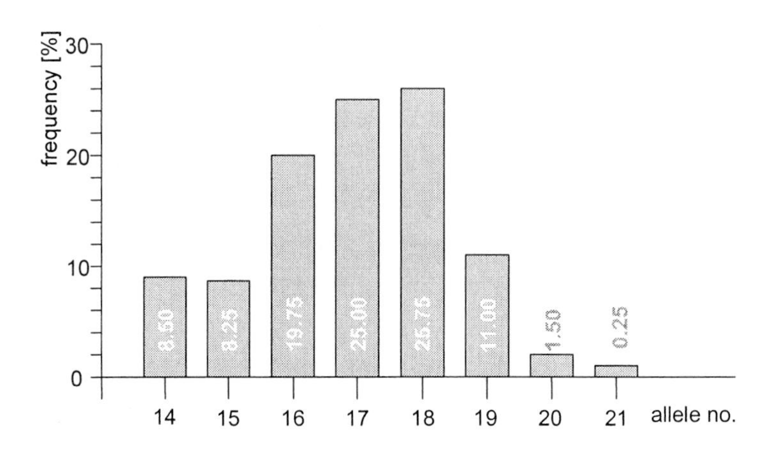

Fig. 2.11. The allele frequencies of STR locus VWA show a distribution close to the Gauss distribution, as do most STRs employed in forensic typing kits. This ensures that there is no impact of selective pressure against particular alleles, which is a prerequisite for a stable intra- and interpopulation power of discrimination

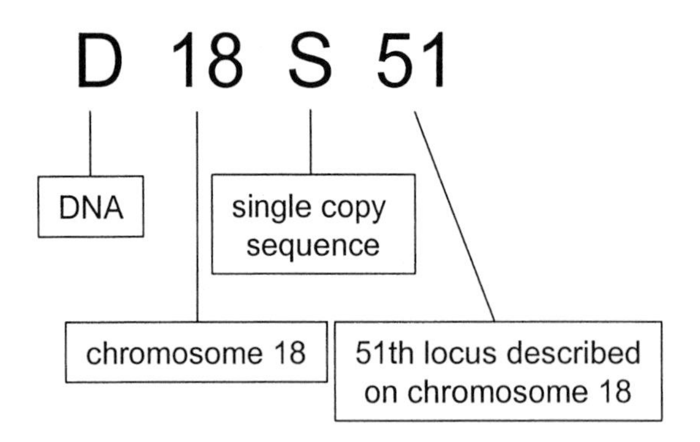

Fig. 2.12. Systematic nomenclature of most STRs

The P_d values of autosomal STRs strongly depend on the heterozygosity and typically range from 0.6 to 0.9. The P_d value is a measure of the average probability that two individuals will show different genotypes.

Population-related and population-specific data for many common autosomal and gonosomal STRs in forensic use are given, for example, in Huckenbeck et al. (1997) and Häntschel et al. (1999) and are found in STR databases (e.g., http://www.cstl.nist.gov/div831/strbase/index.htm).

The matching probability (P_m) (e.g., Brenner and Morris 1990; Brenner 1993) is another value often mentioned in STR contexts. This value is also referred to as the probablilty of identity (P_i) or discrimination index (D_i) (e.g., Jones et al. 1972; Han et al. 2000) (for population genetic and forensic statistics also cf. http://www.dna-view.com/). Basically, the P_m value is just the inverse information compared to the P_d, since it indicates the statistical probability that two unrelated individuals will reveal the same genotype in a STR analysis (P_m=1-P_d). If the average overall matching probability of multiple STRs ($P_{m\ total}$) is indicated, the average P_m values of all STRs must simply be multiplied ($P_{m\ total}=P_{m\ STR1}\times P_{m\ STR2}\times\ldots\times P_{m\ STRn}$).

Of course, it is also possible to calculate $P_{m\ total}$ for a particular genotype. Thus, the calculation for the P_m values for each STR must consider the heterozygous ($P_{m\ het}$=2pq) or homozygous ($P_{m\ homo}=p^2$) status. The results for a particular $P_{m\ total}$ may be remarkably different from the average $P_{m\ total}$ values depending on the random prevalence of either rare or common alleles as demonstrated by two fictitious examples (Table. 2.1).

Table 2.1. Matching probabilities (P_m) for individual STR typing results

STRs	**Genotype individual A "common"**	**Allele frequencies of individual A [p/q]**	$P_{m\ (A)}$	**Genotype indvdiual B "rare"**	**Allele frequencies of individual B [p/q]**	$P_{m\ (B)}$
D3S1358	15/16	0.28/0.22	0.1232	14/18	0.11/0.15	0.0163
VWA	16/17	0.20/0.25	0.100	14/15	0.09/0.08	0.0070
FGA	20/20	0.16/0.16	0.0256	18/19	0.02/0.06	0.0009
D8S1179	13/14	0.35/0.19	0.1330	8/15	0.02/0.13	0.0023
D21S11	29/30	0.21/0.26	0.1092	27/27	0.04/0.04	0.0015
D18S51	14/16	0.17/0.14	0.0476	11/19	0. 02/0.04	0.0001
D5S818	12/12	0.33/0.33	0.1089	10/10	0.07/0.07	0.0015
D13S317	11/12	0.31/0.28	0.1736	13/13	0.10/0.10	0.0066
D7S820	10/11	0.24/0.23	0.1104	12/13	0.16/0.03	0.0044
	$P_{m\ total(A)} = 4.551\times10^{-10}$			$P_{m\ total(B)} = 1.543\times10^{-24}$		

Typically, commercially available kits for human DNA typing consist of five to 16 STR loci, with average P_m values ranging from 10^{-6} to 10^{-18}. For example, the AmpFlSTR ProfilerPlus kit (Applied Biosystems), consisting of nine autosomal STRs, reveals an average overall P_m-value of 10^{-11}, and the Powerplex kit (Promega) reveals 10^{-18} with its 15 autosomal STRs. Both values mean that no two humans share their STR-based DNA profile, with the prerequisite that they are not identical twins or members of a strongly inbreeding group. But even genetically isolated populations proved to be remarkably similar in the allele distributions and heterozygosity compared to other European populations (Perez-Lezaun et al. 1997a). Those parameters can be tested by the use of respective allele frequency data on the population or the group level (e.g., Nei et al. 1983; Chakraborty et al. 1999) (cf. section 7.4.2).

Other values characterizing the discrimination properties of a STR are the mean exclusion chance (MEC) (Krüger et al. 1968) and the polymorphism information content (PIC) (Botstein et al. 1980). Those values are necessary to evaluate the suitability of a particular STR for typing purposes prior to use, for example in a validated typing kit.

In certain cases, the interpretation of typing results derived from archaeological sample material on the basis of modern population data may be questionable. If, for example, only a very a small number of skeletons were tested for whether they represent a two generation family, it would not be possible to ensure positive data at the same statistical power, because in the respective case in a modern randomly mating (panmictic) population, it cannot be excluded that the individuals were members of an inbreeding group. Therefore, the information value of respective paternity and maternity indices calculated on the basis of modern P_m values can only be interpreted as an estimate (section 7.3). However, as soon as the number of historical individuals that are analyzed by STR typing exceeds 20 individuals, it is possible to perform statistical testing for allelic distribution (section 7.4.2). This will give insight into possible inbreeding and therefore enable an evaluation of the validity of P_m values. Indicators for an inbreeding population would show strong shifts in allele distribution and uncommonly high frequencies for a few particular alleles, combined with considerably lower values in the observed heterozygosity rate compared to those expected (Fig. 2.13).

In case of low heterozygosity rates but normal allele distribution, artifacts are usually responsible for the deviation because of DNA degredation (section 7.4.1). In aDNA analysis, one of the two autosomal STR alleles may fail to amplify, which is called allelic dropout. Stochastic effects on the allele being present as an intact target or not are the reason for this phenomenon. Expected allelic dropout occurs comparatively often in aDNA extracts, which generally consist of few intact targets. A rough measure for the amount of intact target may be the number of amplification cycles that are necessary to amplify STRs at all. But stochastic effects may even play a role in aDNA extracts of better-than-average quality. Here, the reason for a seeming allelic dropout could result from one of the two products not reaching the detection limit of the machine (Fig. 2.14).

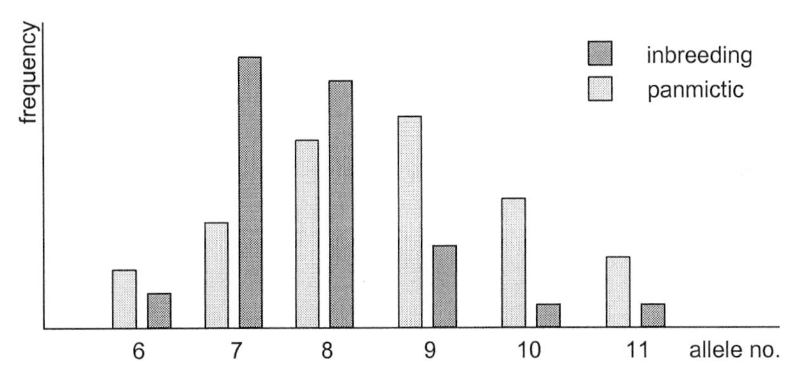

Fig. 2.13. Schematic representation of characteristic patterns observed in an inbred group compared to a panmictic group

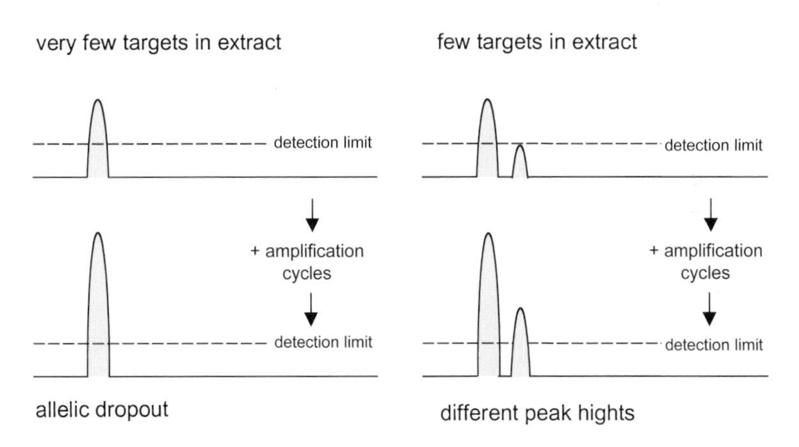

Fig. 2.14. In electrophoresis, allelic drop out because of the lack of intact target sequences or intact targets only being underrepresented look alike. The difference would only be detected by running more amplification cycles

Theoretically, in the second case, i.e., in which one allele did not reach the detection limit, the process could be improved by running a few more amplification cycles or analyzing a larger amount of the sample. In practice this conclusion would give the investigator the negative consequences of overamplification of earlier perfectly detectable alleles (section 4.6) and strong pull-up-artifacts as a result of the detection mode (section 5.1.2). Therefore, whenever allelic dropout is thought to have occurred, the solution will lie in repeated analysis, a process that would have to be carried out in any case for the purpose of data validation.

If the allele distribution within a larger number of samples gives no indications of an inbred population, another possibility that may cause a lower heterozygosity rate is the presence of null alleles. An allele is called a null allele if the hybridization site for one or both primers reveals so many mismatches as a result of mutation events that the STR, although indeed present, cannot be amplified because of primer mismatch. In modern populations the occasional presence of null alleles has been detected by employing different amplification kits that use different primers for the same STR (Butler 2001).

Another factor that also represents only a minor problem in STR typing of ancient DNA is the occurrence of mutational events concerning the repeat structure itself. Studies of forensic populations on more than 10,000 parent-offspring pedigrees revealed the frequency of one unit mutations to be less than 0.1% (Brinkmann et al. 1998).

In contrast, an artifact that particularly concerns aDNA analysis is the generation of stutter bands, which are also known as shadow bands (Hauge and Litt 1993; Litt et al. 1993; Walsh et al. 1996). They are prevalent in the amplification of dinucleotide repeats and are much less intense in tetranucleotide repeats (Fig. 2.15). In the few penta- and hexanucleotide repeats that are known for the human genome, they are effectively absent (Bacher and Schumm 1998). The mechanism behind the generation of stutter bands is slippage events. Slippage may happen as a forward or a backward slippage, although forward slippage leading to a shorter product is more common (Fig. 2.16).

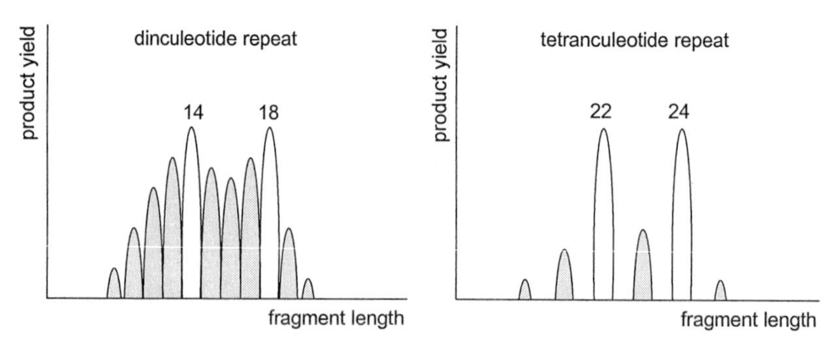

Fig. 2.15. Generation of stutter bands in dinucleotide and tetranucleotide STR amplifications

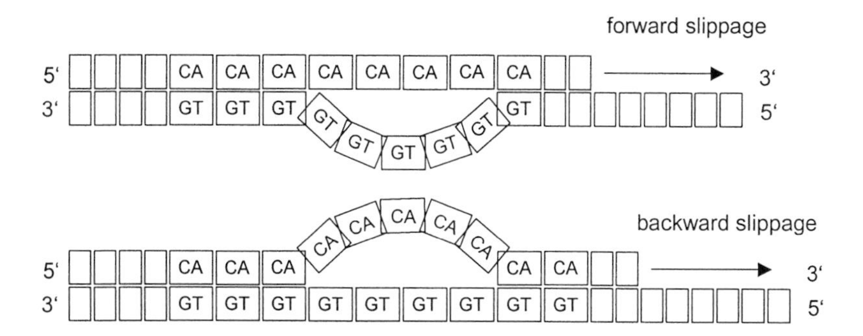

Fig. 2.16. Slippage events during the elongation phase of the amplification, which result in stutter bands. Forward slippage generates an amplification product that is one repeat unit shorter than the matrix sequence; backward slippage generates a product that is one repeat unit longer

Basically, slippage is a type of artifact that is virtually characteristic for any amplification of di-, tri- and tetranucleotide STRs, which is also true in modern DNA analysis. This is best explained by an effect that occurs during the elongation phase called "waving." It indicates the permanent change between denaturation and renaturation of the matrix sequence and the elongated sequence. Since all repeat units reveal the same core sequence, a shift (slippage) may happen in the final renaturation as soon as the repeat structure is fully elongated. Additionally, waving may be favored by the fact that the energy profiles of the repeat structures are uniform compared to those of the neighboring sequences (Fig. 2.17). This theory is supported by empirical findings where repeat sequences that are fully uniform show more intense generation of stutter bands than those STRs that reveal many polymorphisms within the repeat structure. Also, longer repeat structures reveal more intense stutter bands than short repeat structures (Fig. 2.18).

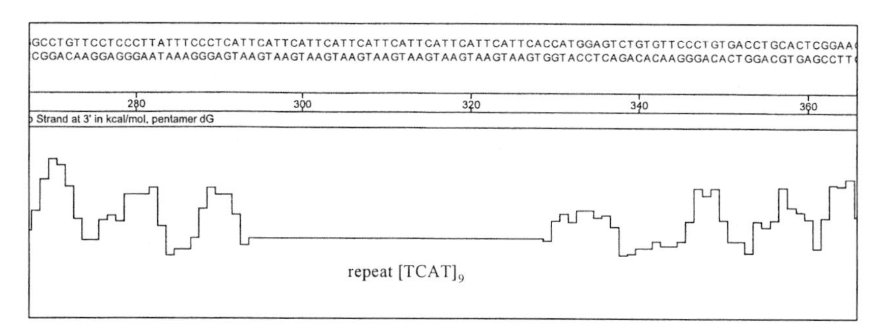

Fig. 2.17. The energy profile of a repeat structure is considerably different from the neighboring sequences. The uniform energy profile favors slippage events

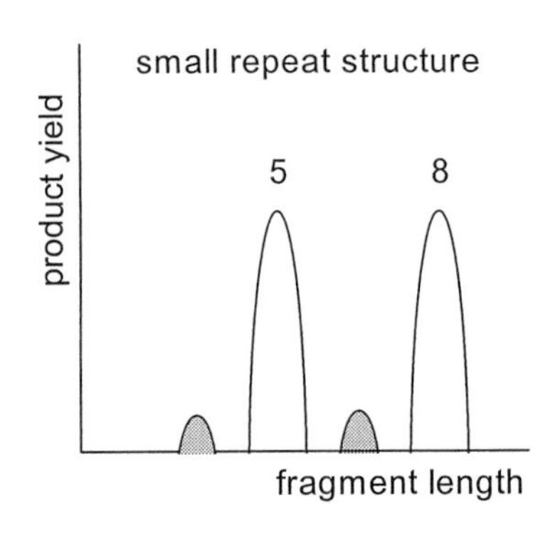

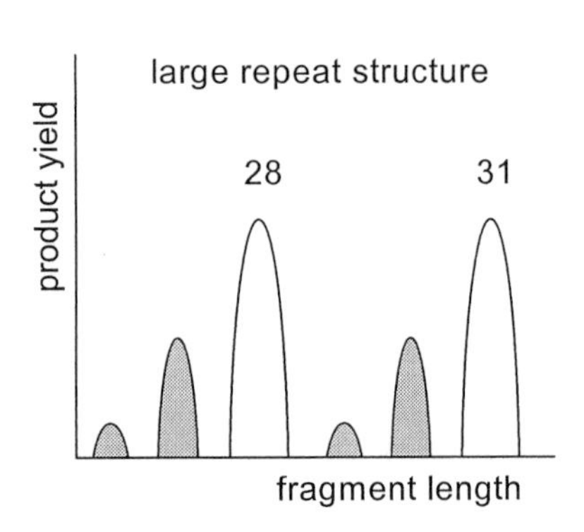

Fig. 2.18. Large STR repeat structures generate more intense stutter bands than small repeat structures because of an increased likelihood of slippage events

Although stutter bands are typical for STR amplifications, in aDNA analysis they may become a more significant problem. Whenever an amplification is started on very few targets, an early slippage event may lead to a majority of the amplification products being of incorrect size. Potentially suspect for carrying the artifact are all of those amplifications that also draw the investigator's attention because of numerous allelic dropout events. False allele determinations because of this type of artifact can again only be excluded through repeated analysis. If possible, it is advisable to employ tetranucleotide repeats in ancient and degraded DNA analysis. Particularly in investigations of animal genomes in which there is little genomic information and limited markers are known, this recommendation may be difficult to follow. Many animals apparently have only few tetranucleotide repeat structures, or, what seems to be much more likely, the tetranucleotides have not yet been identified. However, there are suggestions from experimental data that the intensity of stutter bands may be minimized by reducing the elongation temperature (section 4.3.3).

The applications of autosomal STR typing to ancient human skeletal material presented in this book (chapter 7) have mainly been carried out using the AmpFlSTRProfilerPlus kit (Applied Biosystems). The kit enables a multiplex analysis of nine autosomal STRs and the amelogenin locus and has been validated through several forensic population genetics studies (e.g., LaFountain et al. 2001). The arrangement, length ranges and dye labels of the STR loci and the amelogenin locus are given in Fig. 2.19.

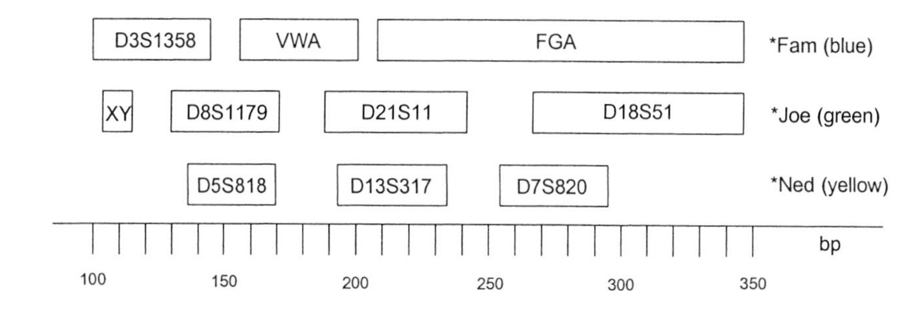

Fig. 2.19. Fragment lengths and dye labels of the STRs and the amelogenin locus (*XY*) amplified by the AmpFlSTRProfilerPlus kit (Applied Biosystems)

2.2.3 Y-chromosomal STRs

The structure of Y-chromosomal STRs is the same as for autosomal STRs (cf. Fig. 2.10). In forensic applications, the use of tetranucleotide repeats is predominant, although di- and trinucleotides are also known (e.g., Roewer et al. 1992; Kayser et al. 1997; de Knijf et al. 1997; Jobling et al. 1997; Gill et al. 2001). In contrast to autosomal STRs, the Y-chromosomal STRs represent a haplotype, because there are no homologues on the X chromosome. This haplotype is passed from father to son as a set. Therefore, Y-chromosomal STR typing enables the follow-up of paternal lineages (Fig. 2.20) analogous to mtDNA HVR-haplotyping that supports the recognition of maternal lineages.

The Y-chromosomal STRs may be used in human identification, but their most important application is in population genetics (Deka et al. 1996; Perez-Lezaun et al. 1997b; Lander and Ellis 1998; Nebel et al. 2000). Because of regional variation of the Y-chromosomal haplotype frequencies, it is a valuable tool for migration research (e.g., Skorecki et al. 1997; Carvalho-Silva et al. 1999; Chakraborty et al. 1999; Seielstadt et al. 1999; Wilson et al. 2001). Most interestingly, it also serves the non-scientific research field of genealogists. This is due to the fact that surnames have been passed along in the same fashion as Y-chromosomal STRs, i.e., through the paternal lineage. Consequently, Y-haplotyping may be an interesting additional tool for certain cases in the reconstruction of family histories (Foster et al. 1998; Sykes and Irven 2000). However, our own studies have shown that even in cases where paternal familial relatedness with sound historical records is known, completely different Y-haplotypes occur, which cannot be explained by mutation events. This indicates that female infidelity is neither a modern phenomenon nor restricted to noble families (Gerstenberger et al. 1999) (also cf. section 7.3.2) but is characteristic to any primate society (Lauenhardt et al. 1998).

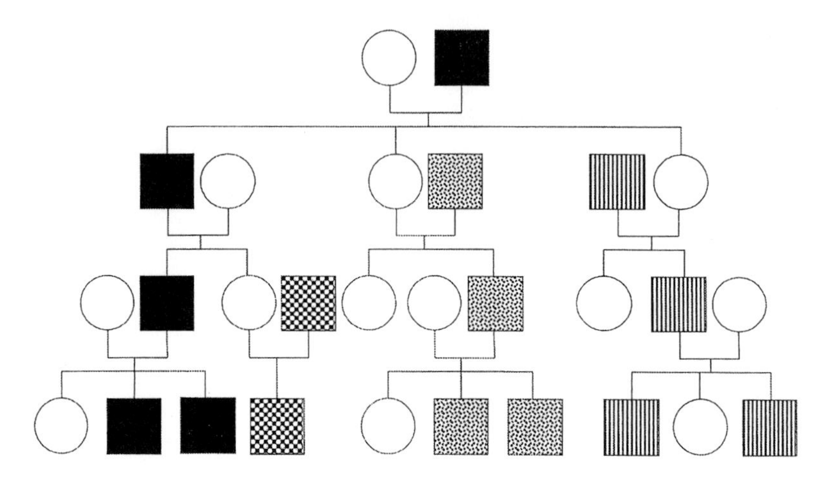

Fig. 2.20. Y-chromosomal haplotypes in a family tree spanning four generations. The individuals forming haplogroups are marked by the same pattern

However, Y-haplotyping truly supports the recognition of biological relatedness; this was demonstrated, for example, by the reconstruction of genealogical features in a group of Bronze Age individuals (section 7.3.3). In this study, as well as in the reconstruction of paternal relatedness for noble families, a quadruplex amplification of Y-chromosomal STRs proved invaluable (Fig. 2.21).

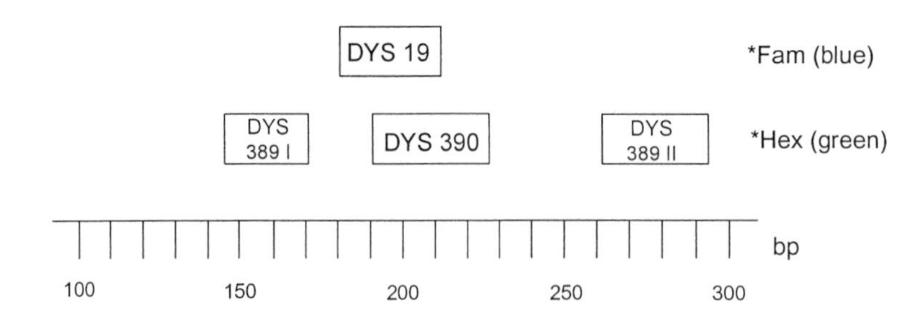

Fig. 2.21. Fragment lengths and dye labels of the quadruplex amplification of Y-chromosomal STRs (Schultes et al. 1999) employing three primer sets. DYS 389 I and II are duplicated sequences that are targeted by a single primer set

Most recently, we redesigned the primers and enlarged the multiplex approach to seven primer sets amplifying nine Y-chromosomal STRs.[1] The Y-STRs employed match the ones in the only Y-STR database, which contains worldwide Y-haplotypes (http://ystr.charite.de).[2] The difference between primer sets and amplified loci was due to the fact that two of the primers amplify two loci each. In those cases, the repeat structure, including the primer hybridization site and further neighboring sequences, has been duplicated in human evolution and evolved separately as far as the number of repeat units is concerned. The first results for the application of this enlarged multiplex approach to ancient skeletal material are given in section 7.3.5. The arrangement, length ranges and dye labels of the multiplexed Y-STR loci are given in Fig. 2.22.[3]

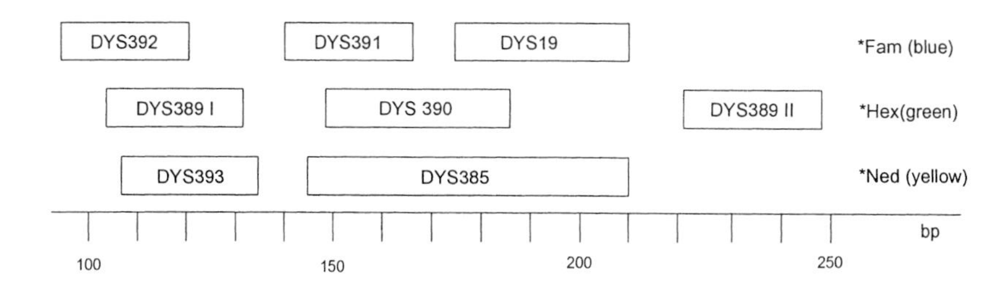

Fig. 2.22. Fragment lengths and dye labels of the multiplex amplification of nine Y-chromosomal STRs employing seven primer sets. DYS 389 I and II and DYS 385 are duplicated sequences that are targeted by a single primer set each

[1] Our sincerest thanks go to Prof. Dr. Bernd Brinkmann, Institute of Legal Medicine at the University of Münster, and his staff for typing a number of modern DNA samples by different Y-chromosomal STRs. This was necessary for establishing the Y-chromosomal multiplex due to the unclear nomenclature of reference alleles at the cstl.nist.gov/biotech/strbase-database.

[2] Other web sites interesting for Y-STR analysis are:
http://www..medfac.leidenuniv.nl/~fldo/hptekst.html
http://www.cstl.nist.gov/biotech/strbase

[3] The primer design and arrangement of Y-chromosomal STRs to a multiplex approach was part of the diploma thesis of Boris Müller.

2.2.4 X-chromosomal STRs

As in Y-chromosomal STRs, the structure of X-chromosomal STRs is the same as for autosomal STRs (cf. Fig. 2.10). Males reveal single amplification products from one allele only. However, this does not represent a true haplotype because of the possible crossing-over of their mothers' X-chromosomal sequence fragments during mitosis. Female individuals reveal two alleles with comparatively high heterozygosity rates of 0.6–0.8 for the X-chromosomal STR markers in modern forensic studies (e.g., Edelmann et al. 2001). The high heterozygosity rates for females suggest the use of X-chromosomal markers to improve sex determination in ancient DNA analysis (cf. also Cali et al. 2002). Differing from the amelogenin amplification (section 2.2.1), X-chromosomal STRs enable a positive proof for female individuals, as has been shown by Pascal et al. (1991) and Tun et al. (1999).

In order to increase the likelihood of receiving at least one heterozygous result for a female, primers for two X-chromosomal STR markers were designed to use in a new multiplex approach. For forensic identification purposes, X-chromosomal STRs should not to be linked (e.g., Edelmann and Szibor 2001), as is demanded for forensic human identification on the basis of autosomal STR analysis. However, we choose markers that are linked in order to generate the pattern of a kind of a X-chromosomal haplotype. This may simplify the identification of kinship of isolated father-daughter relations, which would otherwise be difficult to identify because of the lack of true haplotype inheritance.

Further, two Y-chromosomal markers were built in purposely as a feedback for female amelogenin and X-chromosomal STR results and, alternatively, to discover possible Y-chromosomal deletions on the amelogenin gene, although these are reported to be rare (section 2.2.1). Through this approach the multiplex amplification enables a strongly improved, reliable sex determination even on ancient DNA, as is shown by the first applications (section 7.1). The arrangement, length ranges and dye labels of the multiplexed loci are given in Fig. 2.23. The chromosomal map (Fig. 2.24) shows the locations of the different X-chromosomal markers that are used in forensic typing.

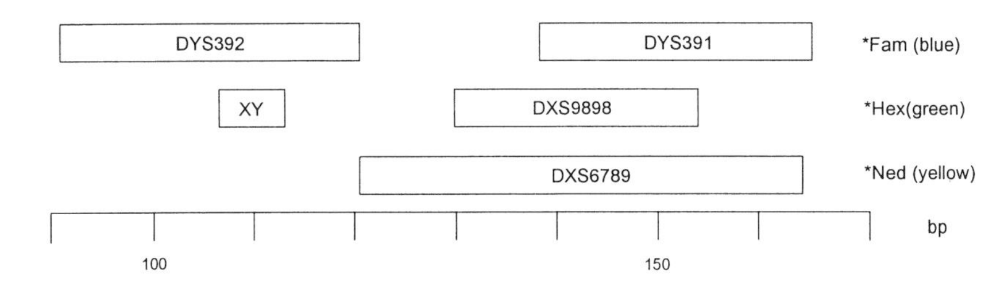

Fig. 2.23. Fragment lengths and dye labels of the multiplex PCR for sexing purposes using the amelogenin locus (*XY*), two X- and two Y-chromosomal STRs

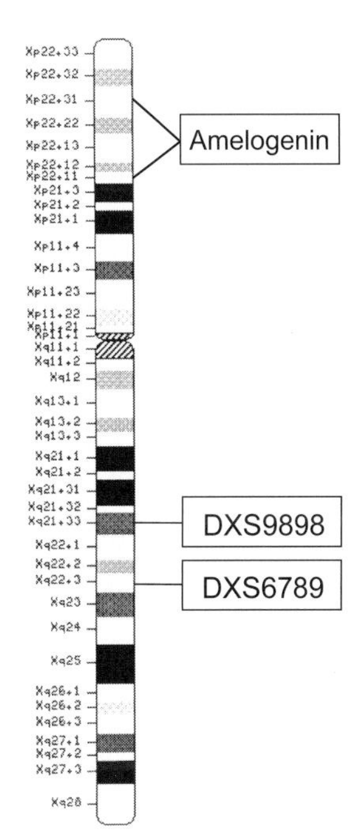

Fig. 2.24. X-chromosomal map showing the two STRs and the amelogenin locus of the multiplex assay

2.2.5 ABO blood group genes

The genes coding for the different amino acids that determine the AB0 serological traits are located on chromosome 9 at exon 6 and exon 7 (Fig. 2.25). They are characterized by a number of sequence polymorphisms, of which at least 13 are well investigated (e.g., Yamamoto et al. 1990; Lee et al. 1992; Grunnet et al. 1994; Nishimukai et al. 1996; Tsai et al. 2000; Yip

2000). Four sequence polymorphisms that enable the discrimination among alleles coding for A, B, 0_1, 0_{1v} and 0_2 were chosen for aDNA analysis (Table 2.2). The primer design was carried out in such a way that only two short amplification products and in total four RFLP analyses of the amplification products are necessary for allele determination.

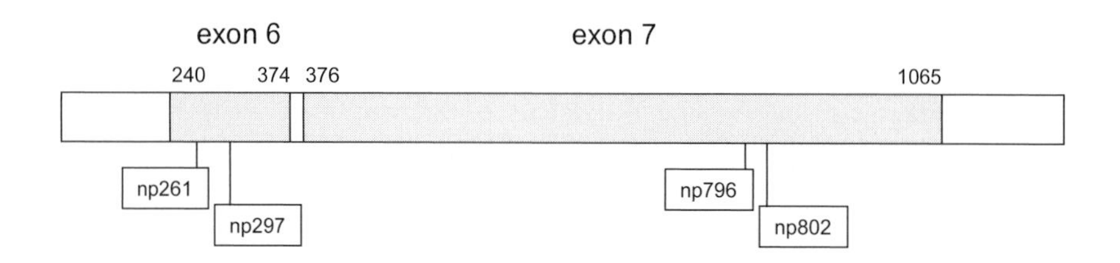

Fig. 2.25. Exon 6 and exon 7 on chromosome 9 revealing polymorphisms at four nucleotide positions determining the ABO blood group characteristics

Table 2.2. Analysis strategy for ABO sequence polymorphisms (*del* deletion)

Chromosome 9	Exon 6		Exon 7	
PCR product	103/104 bp		64 bp	
Endonuclease	Rsa I	HpyCH4IV	Nla III	Mnl I
Product length	66+37 bp	73/72+31 bp	37+27 bp	42/41+23/22 bp
Nucleotide pos.	261	297	796	802
A	G	A	C	G
B	G	G	A	G
0_1	del	A	C	G
0_{1v}	del	G	C	G
0_2	G	G	C	A

The most common genotypes (AA, $A0_1$, 0_10_1, 0_10_{1v} and $0_{1v}0_{1v}$) can already be identified by two RFLP analyses of the 103/104-bp amplification product derived from exon 6. It is necessary to perform the second PCR and the respective RFLP analysis only if genotype $A0_{1v}$ or any genotype involving a B- or an 0_2-allele are present. This is due to the fact that alleles 0_2 and B are not distinguishable by the analysis of exon 6.

In order to simplify the interpretation of RFLP results, the restriction patterns are given in Table 2.3. Their application to ancient skeletal material is given in section 7.4.3.

Table 2.3. Restriction patterns of the ABO blood group genotypes (-/- PCR product remains undigested; +/- one allele of the PCR product is digested, the other remains undigested; +/+ PCR product is completely digested; --- not necessary for analysis, since the Rsa I/HpyCH4IV patterns are already unique. If carried out, the pattern for both of the enzymes is -/-)

ABO blood groups		Exon 6		Exon 7	
Phenotype	Genotype	Rsa I	HpyCH4IV	Nla III	Mnl I
A	AA	-/-	-/-	---	---
A	$A0_1$	+/-	-/-	---	---
0	0_10_1	+/+	-/-	---	---
0	0_10_{1v}	+/+	+/-	---	---
0	$0_{1v}0_{1v}$	+/+	+/+	---	---
A	$A0_2$	-/-	+/-	-/-	+/-
AB	AB	-/-	+/-	+/-	-/-
0	0_20_2	-/-	+/+	-/-	+/+
B	$B0_2$	-/-	+/+	+/-	+/-
B	BB	-/-	+/+	+/+	-/-
A	$A0_{1v}$	+/-	+/-	-/-	-/-
0	0_10_2	+/-	+/-	-/-	+/-
B	$B0_1$	+/-	+/-	+/-	-/-
0	$0_{1v}0_2$	+/-	+/+	-/-	+/-
B	$B0_{1v}$	+/-	+/+	+/-	-/-

The worldwide differences in ABO blood group frequencies are considered to result from the differential immunogenetic response of the carriers of the respective alleles to infectious diseases (Vogel and Motulski 1997). Actually, modern epidemiological studies show that individuals belonging to blood group O indeed have a considerably higher mortality risk in cholera epidemics, for example, than do A-, AB- or, in particular, B-individuals (e.g., Van Loon et al. 1991; Faruque et al. 1994). On the other hand, it seems that O-individuals may have a little more overall fitness (longevity, lower frequencies for many cancers) in the absence of infectious epidemics (Jörgensen 1980). This is supported by the fact that the blood group O is predominant throughout native South American populations, where most of the severe infectious diseases have only been imported during the last few hundred years by European colonizers. On the other hand, B- and AB-frequencies are the highest in central and east Asia, where, for example, the great medieval plague pandemics are thought to have originated geographically.

The study of ABO alleles from historic skeletal material will therefore not only be interesting for identification purposes but also, above all, in the context of paleoepidemiology.

2.2.6 CCR5

The chemokine receptors are cell-surface molecules that play a major role in the control of inflammatory infections. Recently, the chemokine receptor gene CCR5 that is located on chromosome 3 at p21.3 has been identified as an important co-receptor in HIV-1 infections (Samson et al. 1996a). At the same time, it was discovered that individuals may be carriers of a mutant allele Δccr5, which is characterized by a 32-bp deletion. Individuals having a homozygous status for Δccr5 are obviously resistant to infection by HIV-1, while heterozygous individuals show a retarded progression in disease (e.g., Samson et al. 1996b; Liu et al. 1996). There is general agreement that the most probable other cause of heterozygote advantage against developing rheumatoid arthritis or multiple sclerosis led to the spread of the mutant allele throughout Europe (e.g., Blanpain et al. 2000). Most recent findings about climatic factors correlating with the allele distribution (Limborska et al. 2002) do not conflict but rather support the heterozygote advantage if climate is interpreted as a proxidata of inflammatory diseases.

In order to enable investigations on ancient skeletal material concerning the spread of Δ32ccr5,[4] primers amplifying considerably shorter products were designed, resulting in 130 bp for CCR5 and 98 bp for Δ32ccr5. For data validation, the amplification was built into a multiplex assay consisting of four STRs and the amelogenin gene. STR typing data and sex determination may therefore be acquired in addition to the chemokine receptor information.

The arrangement of the markers, their length ranges and the dye labels are provided in Fig. 2.26. Applications to two skeletal series are introduced in section 7.4.4.

---

[4] We are most grateful to Dr. Martin Oppermann of the Department of Immunology at the University of Göttingen, who approached us with the idea to investigate the CCR5/Δccr5 directly from ancient skeletons.

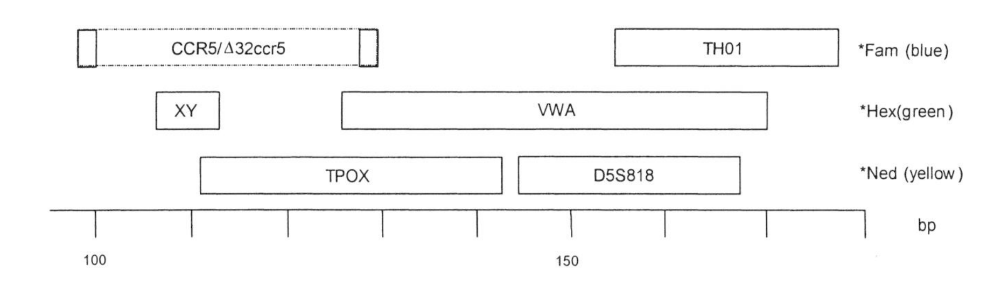

Fig. 2.26. Fragment lengths and dye labels of a multiplex PCR consisting of the CCR5/Δ32ccr5, the autosomal STRs TPOX, D5S818, VWA, TH01 and the amelogenin locus (*XY*)

2.2.7 ΔF508

About 30 different mutational events are known to be responsible for cystic fibrosis. Of those, the so-called ΔF508 mutation is the one that shows the highest frequency, ranging from 50–70% throughout Europe (e.g., Tsui 1992). The deletion leads to the absence of a phenylalanine in position 508 of the amino acid chain of the cystic fibrosis transmembrane conductance regulator. Cystic fibrosis (CF) is the most common lethal autosomal recessive disorder affecting people of European origin in a homozygous status with an incidence of 1 in 2,500 live births and a heterozygote frequency between 1:20 and 1:25. The disease is characterized by chronic obstructive lung disease, pancreatic enzyme insufficiency and an elevated sweat electrolyte concentration because of an incorrect regulation of the chloride ion channels. The location of ΔF508 is on the long arm of chromosome 7 (locus q31.2).

The defect observed in the regulation of chloride ion channels in homozygotes favors the hypothesis that heterozygote carriers show a higher resistance against electrolyte-wasting diarrheas, a major cause of infant mortality in the past (Romeo et al. 1989; Morral et al. 1994; Betranpetit and Calafell 1996) and against deaths resulting from cholera infections (Ryan et al. 2000). Another hypothesis postulates a resistance to tuberculosis proceeding from the observation that carriers for CF produce excessive amounts of hyaluronic acid, which serves as the host's primary defence against such diseases as pulmonary tuberculosis (Meindl 1987).

In order to enable a reliable screening for ΔF508 mutations in ancient populations, the primer design was made to generate a very short amplification product (95 bp for the wild-type / 92 bp in case of ΔF508) and to make possible the building in of the primer set to the AmpFlSTRProfilerPlus multiplex assay (Bramanti et al. 2000), which is applied to ancient samples for kinship analysis and population genetics purposes. In this way, the CF data are collected in addition to the STR data; this at the same time ensures the validation of the data. An example for the application on ancient skeletal material is given in section 7.4.5. The arrangement of the primers is given in Fig 2.27.

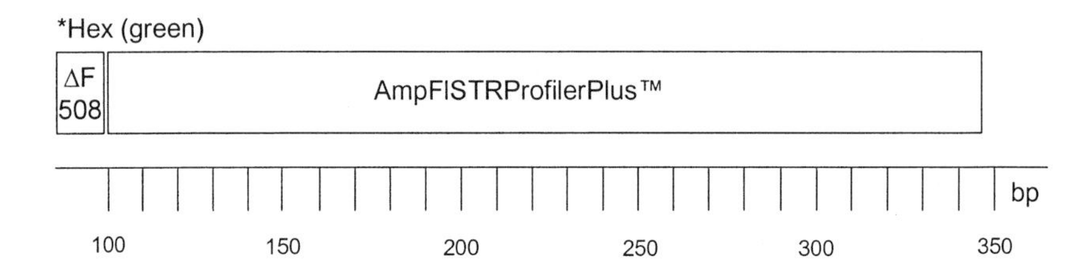

Fig. 2.27. Arrangement of ΔF508 to the AmpFlSTR-ProfilerPlus (also cf. Fig. 2.19)

References

Anderson S, Bankier A, Barrell B, De Bruijn M, Coulson A, Drouin J, Eperon I, Nierlich D, Roe B, Sanger F, Schreier P, Smith A, Staden R, Young I (1981) Sequence and organization of the human mitochondrial genome. Nature 290:457–465

Bacher J, Schumm JW (1998) Development of highly polymorphic pentanucleotide tandem repeat loci with low stutter. Profiles in DNA, Gene Print 1998:3–6

Bailey JF, Richards MB, Macauley VA, Colson IB, James IT, Bradley DG, Hedges REM, Sykes BC (1996) Ancient DNA suggests a recent expansion of European cattle from a diverse wild progentior species. Proc R Soc Lond B 263:1467–1473

Bär W, Brinkmann B, Budowle B, Carracedo A, Gill P, Holland M, Lincoln PJ, Mayr W, Morling N, Olaisen B, Schneider PM, Tully G, Wilson M (2000) DNA Commission of the International Society for Forensic Genetics: guidelines for mitochondrial DNA typing. Int J Legal Med 113:193–196

Bertranpetit J, Calafell F (1996) Genetic and geographical variability in cystic fibrosis: evolutionary considerations. Ciba Foundation Symposium 197:97–114

Blanpain C, Lee B, Tackoen M, Puffer B, Boom A, Libert F, Sharron M, Wittamer V, Vassart G, Doms RW, Parmentier M (2000) Multiple nonfunctional alleles of CCR5 are frequent in various human populations. Blood 96:1638–1645

Boerwinkle E, Xiong W, Fourest E, Chan L (1989) Rapid typing of tandemly repeated hypervariable loci by the polymerase chain reaction: application to the apolipoprotein B 3' hypervariable. Proc Natl Acad Sci USA 86:212–216

Boles TC, Snow CC, Stover E (1995) Forensic DNA testing on skeletal remains from mass graves: a pilot project in Guatemala. J Forensic Sci 40:349–355

Botstein D, White RL, Skolnick M, Davis RW (1980) Construction of a genetic-linkage map in man using restriction fragment length polymorphisms. Am J Hum Genet 32: 314–331

Bramanti B, Sineo L, Vianello M, Caramelli D, Hummel S, Chiarelli B, Herrmann B (2000) The selective advantage of cystic fibrosis heterocygotes tested by aDNA analysis. Int J Anthropol 15:255–262

Brenner CH (1993) A note on paternity computation in cases lacking a mother. Transfusion 33:51–54

Brenner CH, Morris JW (1990) Paternity index calculations in single locus hypervariable DNA probes: validation and other studies. Proc Int Sympos Human Ident 1989, Promega Corporation. pp 21–53

Brinkmann B, Klintschar M, Neuhuber F, Hühne J, Rolf B (1998) Mutation rate in human microsatellites: influence of the structure and length of the tandem repeat. Am J Hum Genet 62:1408–1415

Burger J, Hummel S, Pfeiffer I, Herrmann B (2000) Palaeogenetic analysis of (pre)historic artifacts and its significance for anthropology. Anthropol Anz 58:69–76

Butler JM (2001) Forensic DNA typing. Biology and technology behind STR markers. Academic Press, San Diego

Butler JM, Levin BC (1998) Forensic applications of mitochondrial DNA. Focus 16:158–162

Calì F, Forster P, Kersting C, Mirisola MG, D'Anna R, De Leo G, Romano V (2002) DXYS156: a multi-purpose short tandem repeat locus for determination of sex, paternal and maternal geographic origins and DNA fingerprinting. Int J Legal Med 116:133-138

Cann RL, Brown WM, Wilson AC (1984) Polymorphic sites and the mechanism of evolution in human mitochondrial DNA. Genetics 106:479–499

Carvalho-Silva DR, Santos FR, Hutz MH, Salzano FM, Pena SDJ (1999) Divergent human Y-chromosome microsatellite evolution rates. J Mol Evol 49:204–214

Cavalli-Sforza LL (1998) The DNA revolution in population genetics. Trends Genet 14:60–65

Chakraborty R, Jin L (1993) Determination of relatedness between individuals using DNA fingerprinting. Hum Biol 65:875–895

Chakraborty R, Stivers DN, Su B, Zhong Y, Budowle B (1999) The utility of short tandem repeat loci beyond human identification: implications for development of new DNA typing systems. Electrophoresis 20:1682–1696

Deka R, Shriver MD, Yu LM, Ferrell RE, Chakraborty R (1995) Intra- and inter-population diversity at short tandem repeat loci in diverse populations of the world. Electrophoresis 16:1659–1664

Deka R, Jin L, Shriver MD, Yu LM, Saha N, Barrantes R, Chakraborty R, Ferrell RE (1996) Dispersion of human Y chromosome haplotypes based on five microsatellites in global populations. Genome Res 6:1177–1184

DeKnijff P, Kayser M, Caglia A, Corach D, Fretwell N, Gehrig C, Graziosi G, Heidorn F, Herrmann S, Herzog B, Hidding M, Honda K, Jobling M, Krawczak M, Leim K, Meuser S, Meyer E, Oesterreich W, Pandya A, Parson W, Penacino G, Perez-Lezaun A, Piccinini A (1997) Chromosome Y microsatellites: population genetic and evolutionary aspects. Int J Legal Med 110:134–140

Edelmann J, Hering S, Michael M, Lessig R, Deichsel D, Meier-Sundhausen G, Roewer L, Plate I, Szibor R (2001) Sixteen X-chromosome STR loci frequency data from a German population. Forensic Sci Int 3192:1–4

Edelmann J, Szibor R (2001) DXS101: A highly polymorphic X-linked STR. Int J Legal Med 114:301–304

Edwards A, Hammond HA, Jin L, Caskey CT, Chakraborty R (1992) Genetic variation at five trimeric and tetrameric tandem repeat loci in four human populatuion groups. Genomics 12:241–253

Eizirik E, Bonatto SL, Johnson WE, Crawshaw PGJ, Vie JC, Brousset DM, O'Brien SJ, Salzano FM (1998) Phylogeographic patterns and evolution of the mitochondrial DNA control region in two neotropical cats (Mammalia, felidae). J Mol Evol 47:613–624

Elliott JC, Fourney RM, Budowle B, Aubin RA (1993) Quantitative reproduction of DNA typing minisatellites resolved on ultrathin silver-stained polyacrylamide gels with X-ray duplicating film. BioTechniques 14:702–704

Encalada SE, Bjorndal KA, Bolten AB, Zurita JC, Schroeder B, Possardt E, Sears CJ, Bowen BW (1998) Population structure of loggerhead turtle (*Caretta caretta*) nesting in the Atlantic and Mediterranean as inferred from mitochondrial DNA control region sequences. Marine Biol 130:567–575

Faruque A, Mahalanabis D, Hoque S, Albert M (1994) The relationship between ABO blood groups and susceptibility to diarrhea due to *Vibrio Cholerae* 0139. Clin Infect Dis 18: 827–828

Foster EA, Jobling MA, Taylor PG, Donelly P, de Knijff P, Mieremet R, Zerjal T, Tyler-Smith C (1998) Jefferson fathered slave's last child. Nature 396:27–28

Gerstenberger J, Hummel S, Schultes T, Hack B, Herrmann B (1999) Reconstruction of a historical genealogy by means of STR analysis and Y-haplotyping of ancient DNA. Eur J Hum Genet 7:469–477

Gill P, Jeffreys AJ, Werrett DJ (1985) Forensic application of DNA 'fingerprints'. Nature 318:577–579

Gill P, Ivanov PL, Kimpton C, Piercy R, Benson N, Tully G, Evett I, Hagelberg E, Sullivan K (1994) Identification of the remains of the Romanov family by DNA analysis. Nat Genet 6:130–135

Gill P, Brenner C, Brinkmann B, Budowle B, Carracedo A, Jobling MA, de Knijff P, Kayser M, Krawczak M, Mayr WR, Morling N, Olaisen B, Pascali V, Prinz M, Roewer L, Schneider PM, Sajantila A, Tyler-Smith C (2001) DNA commission of the International Society of Forensic Genetics: recommendations on forensic analysis using Y-chromosome STRs. Int J Legal Med 114:305–309

Ginther C, Issel-Tarver L, King MC (1992) Identifying individuals by sequencing mitochondrial DNA from teeth. Nat Genet 2:135–138

Greenwood AD, Pääbo S (1999) Nuclear insertion sequences of mitochondrial DNA predominate in hair but not in blood of elephants. Mol Ecol 8:133–137

Grunnet N, Steffensen R, Bennett E, Clausen H (1994) Evaluation of histo-blood group ABO genotyping in a Danish population: frequency of a novel O allele defined as O2. Vox Sang 67:210–215

Hadly EA, Kohn MH, Leonard JA, Wayne RK (1998) A genetic record of population isolation in pocket gophers during Holocene climatic change. Proc Natl Acad Sci USA 95:6893–6896

Hagelberg E, Clegg JB (1993) Genetic polymorphisms in prehistoric Pacific islanders determined by analysis of ancient bone DNA. Proc R Soc Lond B 252:163–170

Hamada H, Petrino MG, Kakunaga T (1982) A novel repeated element with Z-DNA forming potential is widely found in evolutionary diverse eukaryote genomes. Proc Natl Acad Sci USA 79:6465–6469

Hammond HA, Jin L, Zhong Y, Caskey CT, Chakraborty R (1994) Evaluation of 13 short tandem repeat loci for use in personal identification applications. Am J Hum Genet 55:175–189

Han GR, Lee YW, Lee HL, Kim SM, Ku TW, Kang IH, Lee HS, Hwang JJ (2000) A Korean population study of the nine STR loci FGA, VWA, D3S1358, D18S51, D21S11, D8S1179, D7S820, D13S317 and D5S818. Int J Legal Med 114:41–44

Häntschel M, Hausmann R, Lederer T, Martus P, Betz P (1999) Population genetics of nine short tandem repeat (STR) loci – DNA typing using the AmpFlSTR profiler PCR amplification kit. Int J Legal Med 112:393–395

Hardy C, Callou C, Vinge JD, Casane D, Dennebouy N, Mounolou JC, Monnerot M (1995) Rabbit mitochondrial DNA diversity from prehistoric to modern times. J Mol Evol 40:227–237

Harihara S, Hirai M, Suutou Y, Shimizu K, Omoto K (1992) Frequency of a 9-bp deletion in the mitochondrial DNA among Asian populations. Hum Biol 2:160–166

Harihara S, Lee SB, Suryobroto B, Omoto K, Kawamoto Y, Takenaka O (1995) Mitochondrial DNA type of Sulawesi macaques in the borderland between *M. hecki* and *M. tonkeana*. Studies on non-human primates 9:37–40

Hauge XY, Litt M (1993) A study of the origin of "shadow bands" seen when typing dinucleotide repeat polymorphisms by the PCR. Hum Mol Genet 2:411–415

Holland MM, Fisher DL, Mitchell LG, Rodriquez WC, Canik JJ, Merril CR, Weedn VW (1993) Mitochondrial DNA sequence analysis of human skeletal remains: identification of remains from the Vietnam war. J Forensic Sci 38:542–553

Horn GT, Richards B, Klinger KW (1989) Amplification of a highly polymorphic VNTR segment by the polymerase chain reaction. Nucleic Acids Res 17:2140–2140

Huckenbeck W, Kuntze K, Scheil HG (1997) The distribution of the human DNA-PCR polymorphisms. Köster, Berlin

Hühne J, Pfeiffer H, Brinkmann B (1999) Heteroplasmic substitutions in the mitochondrial DNA control region in mother and child samples. Int J Legal Med 112:27–30

Irwin DM, Kocher TD, Wilson AC (1991) Evolution of the cytochrome b gene of mammals. J Mol Evol 32:128–144

Ivanov P, Parsons, T, Wadhams M, Holland M, Rhoby R, and Weedn V (1996) Mitochondrial DNA variations in the Hessian lineage: heteroplasmy found in Grand Duke of Russia Georgij Romanov ends disputes over authenticity ot the remains of Tzar Nicholas II. In: Proceedings of the ISFH, Hakone Symposium on DNA polymorphisms. Hakone, Japan.

Jeffreys AJ, Wilson V, Thein SL (1985) Individual-specific "fingerprints" of human DNA. Nature 316:76–79

Jeffreys AJ, Wilson V, Neumann R, Keyte J (1988) Amplification of human minisatallites by polymerase chain reaction: towards DNA fingerprinting of single cells. Nucleic Acids Res 16:10953–10971

Jeffreys AJ, Allen MJ, Armour JAL, Collick A, Dubrova Y, Fretwell N, Guram T, Jobling M, May CA, Neil DL, Neumann R (1995) Mutation processes at human minisatellites. Electrophoresis 16:1577–1585

Jehaes E, Decorte R, Peneau A, Petrie JH, Boiry PA, Gilissen A, Moisan JP, Van den Berghe H, Pascal O, Cassiman JJ (1998) Mitochondrial DNA analysis on remains of a Putative son of Louis XVII, King of France and Marie-Antoinette. Eur J Hum Genet 6:383–395

Jobling MA, Pandya A, Tyler-Smith C (1997) The Y chromosome in forensic analysis and paternity testing. Int J Legal Med 110:118–124

Jones DA (1972) Blood samples: probability of discrimination. J Forensic Sci Soc 12: 355–359

Jörgensen G (1980) ABO blood groups and life expectancy. Dtsch Med Wochenschr 105:103–106

Kayser M, Caglia A, Corach D, Fretwell N, Gehrig C, Graziosi G, Heidorn F, Herrmann S, Herzog B, Hidding M, Honda K, Jobling M, Krawczak M, Leim K, Meuser S, Meyer E, Oesterreich W, Pandya A, Parson W, Penacino G, Perez-Lezaun A, Piccinini A (1997) Evaluation of Y-chromosomal STRs: a multicenter study. Int J Legal Med 110:125–133

Kimpton C, Fisher D, Watson S, Adams M, Urquhart A, Lygo J, Gill P (1994) Evaluation of an automated DNA profiling system employing multiplex amplification of four tetrameric STR loci. Int J Legal Med 106:302–311

Kimpton CP, Gill P, Walton A, Urquhart A, Millican ES, Adams M (1993) Automated DNA profiling employing multiplex amplification of short tandem repeat loci. PCR Meth Appl 3:13–22

Krajewski C, Driskell AC, Baverstock PR, Braun MJ (1992) Phylogenetic relationships of the thylacine (Mammalia: Thylacinidae) among dasyuroid marsupials: evidence from cytochrome b DNA. Proc R Soc Lond B 250:19–27

Krings M, Stone A, Schmitz RW, Krainitzki H, Stoneking M, Pääbo S (1997) Neandertal DNA sequences and the origin of modern humans. Cell 90:19–30

Krüger J, Fuhrmann W, Lichte KH, Steffens C (1968) Zur Verwendung der sauren Erythrozytenphosphatase bei der Verwandtschaftsbegutachtung. Dtsch Z Gerichtl Med 64:127–146

LaFountain MJ, Schwartz MB, Svete PA, Walkinshaw MA, Buel E (2001) TWGDAM validation of the AmpFlSTRProfiler Plus and AmpFlSTR COfiler STR multiplex systems using capillary electrophoresis. J Forensic Sci 46:1191–1198

Lander RS, Ellis JJ (1998) Founding father. Nature 396:13–14

Launhardt K, Epplen C, Epplen JT, Winkler P (1998) Amplification of microsatellites adapted from human systems in faecal DNA of wild Hanuman langurs (*Presbytis entellus*). Electrophoresis 19:1356–1361

Lee J, Chang J (1992) ABO genotyping by polymerase chain reaction. J Forensic Sci 37:1269–1275

Li R, el-Mallakh RS (1997) Triplet repeat gene sequences in neuropsychatric diseases. Harv Rev Psychiatry 5:66–74

Limborska SA, Balanovsky OP, Balanovskaya EV, Slominsky PA, Schadrina MI, Livshits LA, Kravchenko SA, Pampuha VM, Khusnutdinova EK, Spitsyn VA (2002) Analysis of CCR5Delta32 geographic distribution and its correlation with some climatic and geographic factors. Hum Hered 53:49–54

Litt M, Hauge X, Sharma V (1993) Shadow bands seen when typing polymorphic dinucleotide repeats: some causes and cures. BioTechniques 15:280–283

Liu R, Paxton WA, Choe S, Ceradini D, Martin SR, Horuk R, MacDonald ME, Stuhlmann H, Koup RA, Landau NR (1996) Homozygous defect in HIV-1 coreceptor accounts for resistance of some multiply-exposed individuals to HIV-1 infection. Cell 86:367–377

Loreille O, Vigne JD, Hard C, Callou C, Treinen-Claustre F, Dennebouy N, Monnerot M (1997) First distinction of sheep and goat archaeological bones by the means of their fossil mtDNA. J Archaeol Sci 24:33–37

Mannucci A, Sullivan KM, Ivanov PL, Gill P (1994) Forensic application of a rapid and quantitative DNA sex test by amplification of the X-Y homologous gene amelogenin. Int J Legal Med 106:190–193

Meindl RS (1987) Hypothesis: a selective advantage for cystic fibrosis heterozygotes. Am J Phys Anthropol 74:39–45

Merriwether DA (1999) Freezer anthropology: new uses for old blood. Philos Trans R Soc Lond B Biol Sci 354:121–129

Moretti TR, Baumstark AL, Defenbaugh DA, Keys KM, Smerick JB, Budowle B (2001) Validation of short tandem repeats (STRs) for forensic usage: performance testing of fluorescent multiplex STR systems and analysis of authentic and simulated forensic samples. J Forensic Sci 46:647–660

Morral N, Bertranpetit J, Estivill X, Nunes V, Casals T, Giménez J, Reis A, Varon-Mateeva R, Macek M, Kalaydjieva L, Angelicheva D, Dancheva R, Romeo G, Russo MP, Garnerone S, Restagno G, Ferrari M, Magnani C, Claustres M, Desgeorges M (1994) The origin of the major cystic fibrosis mutation (deltaF508) in European populations. Nat Genet 7:169–175

Nasir J, Goldberg YP, Hayden MR (1996) Huntington disease: new insights into the relationship between CAG expansion and disease. Hum Mol Genet 5:1431–1435

Nebel A, Filon D, Weiss DA, Weale M, Faerman M, Oppenheim A, Thomas MG (2000) High-resolution Y chromosome haplotypes of Israeli and Palestinian Arabs reveal geographic substructure and substantial overlap with haplotypes of Jews. Hum Genet107:630-641

Nei M, Tajima F, Tateno Y (1983) Accuracy of estimated phylogenetic trees from molecular data. II. Gene frequency data. J Mol Evol 19:153–170

Nishimukai H, Fukumori Y, Okiura T, Yuasa I, Shinomiya T, Ohnoki S, Shibata H, Vogt U (1996) Genotyping of the ABO blood group system: Analysis of nucleotide position 802 by PCR-RFLP and the distribution of ABO genotypes in a German population. Int J Legal Med 109:90–93

Pascal O, Aubert D, Gilbert E, Moisan JP (1991) Sexing of forensic samples using PCR. Int J Legal Med 104:205–207

Parr RL, Carlyle SW, O'Rourke DH (1996) Ancient DNA analysis of Fremont American Indians of the Great Salt Lake Wetlands. Am J Phys Anthropol 99:507–518

Parson W, Pegoraro K, Niederstatter H, Foger M, Steinlechner M (2000) Species identification by means of the cytochrome b gene. Int J Legal Med 114:23–28

Perez-Lezaun A, Calafell F, Mateu E, Comas D, Ruiz-Pacheco R, Bertranpetit J (1997a) Microsatellite variation and the differentiation of modern humans. Hum Genet 99:1–7

Perez-Lezaun A, Calafell F, Seielstad M, Mateu E, Comas D, Bosch E, Bertranpetit J (1997b) Population genetics of Y-Chromosome short tandem repeats in humans. J Mol Evol 45:265–270

Pfeiffer H, Hühne J, Ortmann C, Waterkamp K, Brinkmann B (1999) Mitochondrial DNA typing from human axillary, pubic and head hair shafts – success rates and sequence comparisons. Int J Legal Med 112:287–290

Queller DC, Strassmann JE (1993) Microsatellites and kinship. Trends Ecol Evol 8:285–288

Richards M, Corte-Real H, Forster P, Macaulay V, Wilkinson H, Demaine A, Papiha S, Hedges R, Bandelt HJ, Sykes B (1996) Paleolithic and Neolithic lineages in the European mitochondrial gene pool. Am J Hum Genet 59:185–203

Roewer L, Arnemann J, Spurr NK, Grzeschik KH, Epplen JT (1992) Simple repeat sequences on the human Y chromosome are equally polymorphic as their autosomal counterparts. Hum Genet 89:389–394

Romeo G, Devoto M, Galietta LJ (1989) Why is the cystic fibrosis gene so frequent? Hum Genet 84:1–5

Ryan ET, Dhar U, Khan WA, Salam MA, Faruque AS, Fuchs GJ, Calderwood SB, Bennish ML (2000) Mortality, morbidity, and microbiology of endemic cholera among hospitalized patients in Dhaka, Bangladesh. Am J Trop Med Hyg 63:12–20

Samson M, Labbe O, Mollereau C, Vassart G, Parmentier M (1996a) Molecular cloning and functional expression of a new human CC-chemokine receptor gene. Biochemistry 35:3362–3367

Samson M, Libert F, Doranz BJ, Rucker J, Liesnard C, Farber CM, Saragosti S, Lapoumeroulie C, Cognaux J, Forceille C, Muyldermans G, Verhofstede C, Burtonboy G, Georges M, Imai T, Rana S, Yi Y, Smyth RJ, Collman RG, Doms RW, Vassart G, Parmentier M (1996b) Resistance to HIV-1 infection in Caucasian individuals bearing mutant alleles of the CCR-5 chemokine receptor gene. Nature 382:722–725

Santos FR, Pandya A, Tyler-Smith C (1998) Reliability of DNA-based sex tests. Nat Genet 18:103

Schneider PM, Seo Y, Rittner C (1999) Forensic mtDNA hair analysis excludes a dog from having caused a traffic accident. Int J Legal Med 112:315–316

Schultes T (2000) Typisierung alter DNA zur Rekonstruktion von Verwandtschaft in einem bronzezeitlichen Skelettkollektiv. Dissertation, Georg August-Universität, Göttingen. Cuvellier, Göttingen

Seielstad M, Bekele E, Ibrahim M, Toure A, Traore M (1999) A view of modern human origins from Y chromosome microsatellite variation. Genome Res 9:558–567

Skorecki K, Selig S, Blazer S, Bradman R, Bradman N, Waburton PJ, Ismajlowicz M, Hammer MF (1997) Y chromosomes of Jewish priests. Nature 385:32

Soodyall H, Vigilant L, Hill AV, Stoneking M, Jenkins T (1996) mtDNA control-region sequence variation suggests multiple independent origins of an "Asian-specific" 9-bp deletion in sub-Saharan Africans. Am J Hum Genet 58:595–608

Stone AC, Stoneking M (1998) mtDNA analysis of a prehistoric Oneota population: implications for the peopling of the New World. Am J Hum Genet 62:1153–1170

Stoneking M (2000) Hypervariable sites in the mtDNA control region are mutational hotspots. Am J Hum Genet 67:1029-1032

Stoneking M, Soodyall H (1996) Human evolution and the mitochondrial genome. Curr Opin Genet Dev 6:731–736

Sullivan KM, Mannucci A, Kimpton CP, Gill P (1993) A rapid and quantitative DNA gender test: fluorescence-based PCR-analysis of X-Y homologous gene amelogenin. BioTechniques 15:636–641

Sykes B, Irven C (2000) Surnames and the Y chromosome. Am J Hum Genet 66:1417–1419

Torroni A, Lott MT, Cabell MF, Chen YS, Lavergne L, Wallace DC (1994) mtDNA and the origin of Caucasians: identification of ancient Caucasian-specific haplogroups, one of which is prone to a recurrent somatic duplication in the D-loop region. Am J Hum Genet 55:760–776

Tsai L, Kao L, Chang J, Lee H, Linacre A, Lee J (2000) Rapid identification of the ABO genotypes by their single-strand confirmation polymorphism. Electrophoresis 21:537–540

Tsui LC (1992) The spectrum of cystic fibrosis mutations. Trends Genet 8:392–398

Tun Z, Honda K, Nakatome M, Nakamura M, Shimada S, Ogura Y, Kuroki H, Yamazaki M, Terada M, Matoba R (1999) Simultaneous detection of multiple STR loci on sex chromosomes for forensic testing of sex and identity. J Forensic Sci 44:772–777

Turbon D, Lalueza A, Perez-Perez A, Prats E, Moreno P, Pons P (1994) Absence of the 9 bp mt DNA region V deletion in ancient remains of aborigines from Tierra del Fuego. Ancient DNA Newsletter 2/1:24–26

Urquhart A, Oldroyd NJ, Kimpton CP, Gill P (1995) Highly discriminating heptaplex short tandem repeat PCR system for forensic identification. BioTechniques 18:116–121

Van Loon F, Clemens J, Sack D, Rao M, Faruque A, Chowdhury S, Harris J, Ali M, Chakraborty J, Khan M, Neogy P, Svennerholm A, Holmgren J (1991)

ABO blood groups and the risk of diarrhea due to enterotoxigenic *Escherichia coli*. J Infect Dis 163:1243–1246
Vogel F, Motulski AG (1997) Human genetics. Problems and approachs (3rd edn) Springer, Heidelberg
Wallace D, Torroni A (1992) American Indian prehistory as written in the mitochondrial DNA: A review. Hum Biol 64:403–416
Walsh PS, Fildes NJ, Reynolds R (1996) Sequence analysis and characterisation of stutter products at the tetranucleotide repeat locus vWA. Nucleic Acids Res 24:2807–2812
Weiss ML, Turner TR (1992) Hypervariable minisatellites and VNTRs. In: Devor EJ (ed) Molecular applications in biological anthropology. Cambridge University Press, Cambridge. pp 76–102
Wilson JF, Weiss DA, Richards M, Thomas MG, Bradman N, Goldstein DB (2001) Genetic evidence for different male and female roles during cultural transitions in the British Isles. Proc Natl Acad Sci USA 98:5078–5083
Wrischnik LA, Higuchi RG, Stoneking M, Erlich HA, Arnheim N, Wilson AC (1987) Length mutations in human mitochondrial DNA: direct sequencing of enzymatically amplified DNA. Nucleic Acids Res 15:529–541
Yamamoto F, Clausen H, White T, Marken J, Hakomori S (1990) Molecular basis of the histo-blood group ABO system. Nature 345:229–233
Yao YG, Watkins WS, Zhang YP (2000) Evolutionary history of the mtDNA 9-bp deletion in Chinese populations and its relevance to the peopling of east and southeast Asia. Hum Genet 107:504–512
Yip S (2000) Single-tube multiplex PCR-SSCP analysis distinguishes 7 common ABO alleles and readily identifies new alleles. Blood 95: 1487–1492
Zischler H (1999) Mitochondrial DNA: diversity analysis and possible pitfalls. In: Epplen J, Lubjuhn T (eds) DNA profiling and DNA fingerprinting. Birkhäuser, Basel
Zischler H, Höss M, Handt O, von Haeseler A, van der Kuyl AC, Goudsmit J, Pääbo S (1995) Detecting dinosaur DNA. Science 268:1192–1193

3 aDNA extraction

DNA extraction is probably the most crucial step in any ancient DNA analysis. At this initial phase of the investigation, a wrong decision can reduce or in fact destroy all the potential information that may have been preserved through time and retained in the sample. Despite the importance of the extraction techniques, there has been very little debate about the most appropriate method (Fisher et al. 1993; Lalu et al. 1994; Cattaneo et al. 1997; Vandenberg et al. 1997; Vince et al. 1998). This is difficult to understand, given the controversies over later stages of analysis and the fact that all subsequent steps in the analysis are dependent on the correct decision concerning the suitable extraction method.

A brief perusal through the materials-and-methods sections of manuscripts on ancient DNA reveals almost as many extraction techniques as there are publications. Even within the same laboratory, extraction protocols vary. This is generally not because of differences in the basic approach used, but rather because of the details about how it should be carried out. It is possible that in most cases there are quite valid reasons for this. The best reason would be that the variations in extraction protocols reflect an evolutionary process in a comparatively young research field where, from the beginning, people have continued to try their best to optimize their methods. Another reason would be that in contrast to modern DNA samples, the ancient samples are truly individual, with each one requiring its own procedure. However, this point is not reflected by the publications. On the one hand, strongly deviating protocols are used for the same types of samples, while on the other hand, basically the same protocol is used for samples of completely different qualities, such as bone, soft tissue or keratinous material.

Of course, for many reasons it is neither possible nor recommendable to perform an individual protocol on each sample. But in light of the considerable number of protocols that have accumulated during the last decade, it seems timely to compare and evaluate basic strategies of DNA extraction.

3.1 Comparison of extraction methods

Extraction protocols are almost as numerous as publications, making it difficult to test each of them. Still, there are many variants if the experiment is reduced to the extraction of DNA from archaeological skeletal material. The protocols that claim to successfully recover DNA from ancient bone can basically be divided up into four major strategies. These strategies are best characterized by calling them the phenol, silica, boiling and chloroform methods. However, the latter two are outsiders in the sense that they are rarely represented in the literature (Meijer et al. 1992; Yang et al. 1997); the phenol and the silica methods (Baron et al. 1996; Höss and Pääbo 1993) are used frequently, each with more or less numerous variations. In both cases the founder protocols, to which all later users refer in their applications, were included in an experiment that aimed to compare the effectiveness of extraction protocols claiming to recover aDNA from ancient skeletal material[5] (cf. also Müller 2000). Since there are some extraction kits available that also claim to be suitable for DNA extraction from bone, also a representative of this approach, which promises easy handling and good recovery, was chosen. An overview of the characteristic steps of each protocol is given in Table 3.1.

In order not to bias any of the protocols, all methods were tested on the same skeletal element, a human Bronze Age tibia from the Lichtenstein cave (see section 7.3.3). Since archaeological bone is known to show considerable differences in DNA preservation throughout its entirety (e.g., Schultes et al. 1997), the tibia was cut in slices and assigned to the five protocols as indicated in Fig. 3.1. Each piece of bone assigned to a certain method was pretreated as indicated by the respective protocol. This included, for example, washing the surface with various solutions, removing the surface using a scalpel or administering UV irradiation for a certain time span. Protocol demands, such as using the spongy parts of the bone only or the suggested way of crushing the sample material to pieces, were followed as indicated in the manuscript. Before starting the DNA extraction procedures, the pretreated bone slices were pulverized in order to ensure homogenous aliquots for the repeated extractions that were carried out for each method.

To examine the extraction success and to enable comparability, the final water and buffer volumes to solute the extracted DNA were modified in

[5] The experiments and results were presented at the 5th International Ancient DNA Conference in Manchester, UK, July 2000

such a way that the ratio of bone material (g) to extract volume (µl) was identical for all protocols.

Table 3.1. Basic steps of the bone extraction protocols

Baron et al. 1996	Yang et al. 1997	Höss and Pääbo 1993[6]	Meijer et al. 1992	InViSorb forensic kit	Chelex 100
Phenol	Chloroform	Silica	Boil	InViSorb	Chelex
0.5 M EDTA ⇩ supernatant is used ⇩ proteinase K ⇩ phenol:chloroform:isoamylalcohol ⇩ 2-propanol + silica ⇩ wash by ethanol ⇩ solve DNA/silica in water	20 mM EDTA ⇩ supernatant is discarded ⇩ CTAB ⇩ chloroform:iso amylalcohol ⇩ ethanol ⇩ wash by ethanol ⇩ solve DNA in TE-buffer	GuSCN-buff 1 ⇩ GuSCN-buff 1 + silica ⇩ wash by GuSCN-buff 2 ⇩ wash by ethanol ⇩ wash by acetone ⇩ solve DNA in water or TE-buffer	0.1 M EDTA TritonX-100 Dithiotreitol ⇩ boil ⇩ centrifuge ⇩ supernatant ready for PCR	buffer for cell lysis ⇩ carrier substance ⇩ wash solution ⇩ elution of DNA to TE-buffer	Chelex 100 ⇩ boil ⇩ centrifuge ⇩ supernatant ready for PCR

Following this, all extracts were amplified according to a standardized procedure. All amplifications were carried out after the DNA extracts had been stored at -20°C for 24 h and a second time after a 6 weeks storage at -20°C. The amplifications employed a commercially available autosomal STR amplification kit (cf. section 2.2.2) and a multiplex approach for the mitochondrial hypervariable regions I and II (cf. section 2.1.1), following

[6] Although many publications refer to this protocol, it cannot be carried out because 10 M GuSCN is in a solid state that can neither be pipetted nor mixed with the sample. We took instead the protocol as given in the Ph.D. thesis of Höss (1995), which basically follows the same procedure except that 5 M GuSCN is used, and the 0.1 M Tris/HCl has a pH of 7.4 instead of 6.4, as in the publication of Höss and Pääbo (1993).

the amplification protocols given in chapter 8. Both the STR and the mtDNA amplifications were carried out for 35 cycles. The results that were to be expected from the STR typing were known from an earlier application of the STR amplification kit to extracts derived from another skeletal element belonging to the same individual (cf. section 7.2.1).

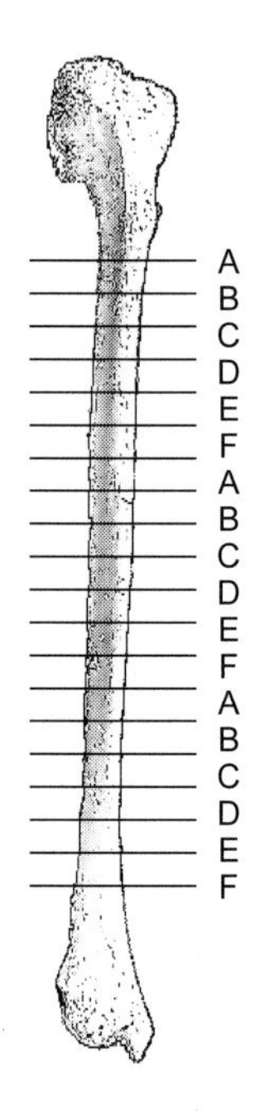

Fig. 3.1. A Bronze Age tibia was cut in slices and assigned to one of the six protocols. This way, possible effects on the extraction results because of differences in DNA preservation that may occur along the diaphysis were minimized. The sample preparations and DNA extractions were carried out precisely as described by the respective manuscript

The amplification results for the mitochondrial regions and the autosomal STRs differed remarkably. Autosomal STR amplifications were only achieved from the DNA extracts of three protocols: Phenol (Baron et al. 1996), Silica (Höss and Pääbo 1993) and InVisorb (InVisorb forensic kit, Invitek GmbH), while the other protocols apparently failed completely to enable the extraction of chromosomal DNA.

But even within the three protocols that provided chromosomal DNA, major differences in the success of the amplification were observed (Fig. 3.2). Only the protocol based on a phenol extraction provided the amount of intact targets needed for a full STR typing. However, the allele determinations of the STR typing that are listed in Table 3.2 show that the isolated alleles that were amplified from the extracts generated following the silica method and the forensic kit are also most probably authentic ones.

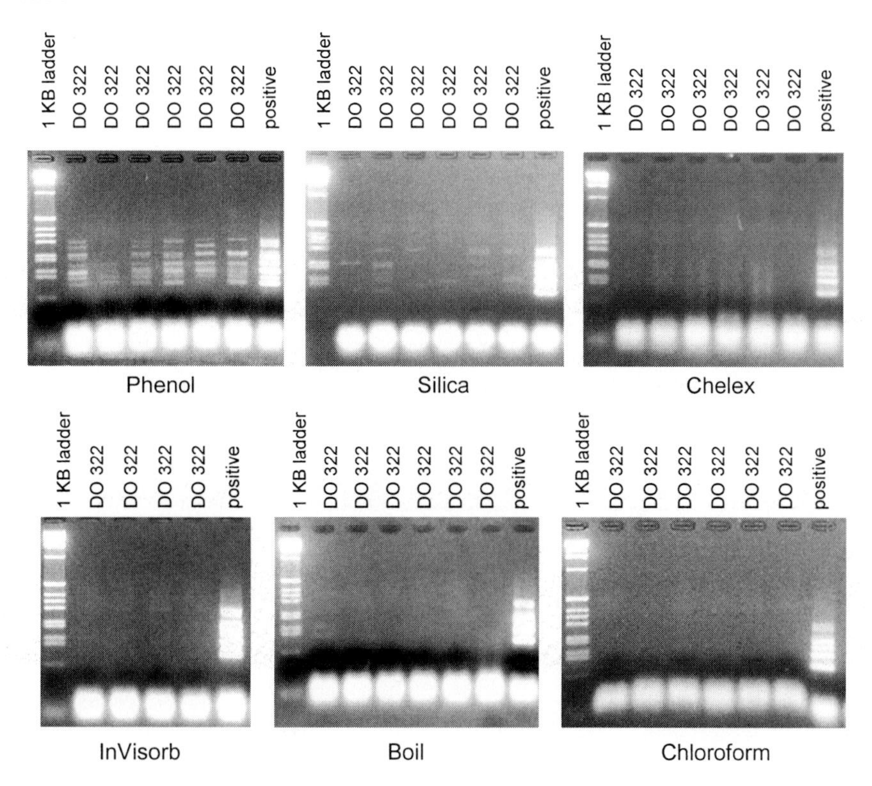

Fig. 3.2. Chromosomal DNA was extracted only by the phenol (Baron et al. 1996, also cf. section 8.3), the silica (Höss and Pääbo 1993; Höss 1995) and the InVisorb (InVisorb) extraction protocols. This was proven by 35-cycle amplifications of autosomal STRs (AmpFlSTRProfilerPlus). The other protocols obviously failed (Chelex; Meijer et al. 1992; Yang et al. 1997) to extract chromosomal DNA from the bone samples, since no amplification products were detectable

However, all extraction protocols except one succeeded to provide mitochondrial DNA although differing with respect to amount and quality (Fig. 3.3). From this it can be deduced that also possible differences in the ability of the different protocols to remove impurities from the DNA extract vary substantially and have an enormous impact on generating data.

Table 3.2. STR allele determinations derived from fragment length analysis[7]

	Amel	D3S1358	D8S1179	D5S818	VWA	D21S11	D13S317	FGA	D18S51	D7S820
Phenol	XY	15/16	12/14/15	10	15/16	29/33.2	11/12	20/21	18	7/9
	XY	15/16	15	9/10/12	16	29/33.2	11	20	14/18	7
Silica	Y	-	-	-	16	-	11	-	14	9
	XY	15	12/13/15	9/12	-	-	-	-	13/14	10
InVisorb	X	16	-	-	16	-	-	-	-	-
	-	-	12	-	-	33.2	-	-	14	-

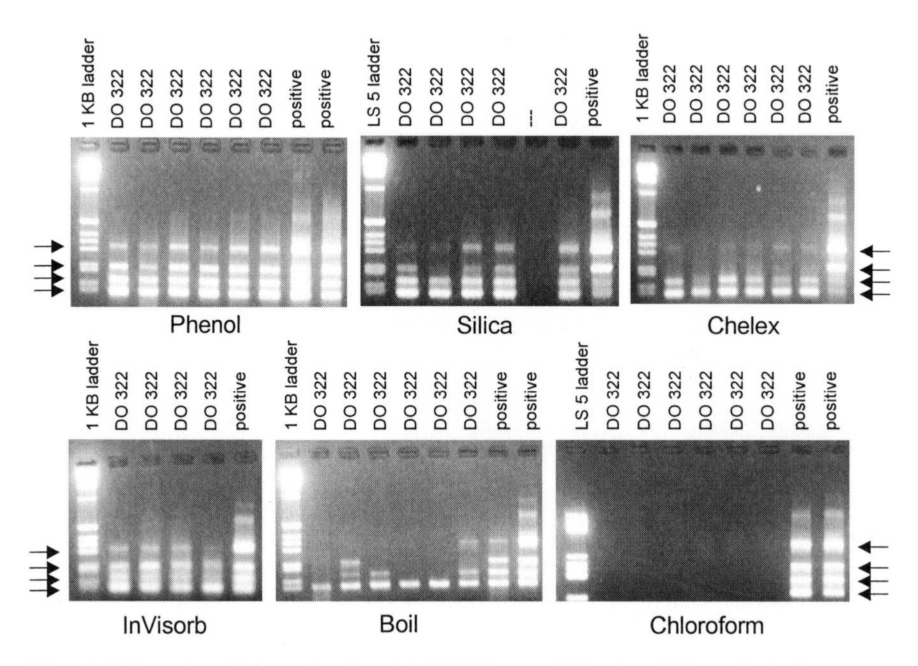

Fig. 3.3. Results of the mitochondrial DNA amplification (35 cycles) that were possible from extracts of five protocols (Baron et al. 1996; Höss and Pääbo 1993; Höss 1995; Chelex; InVisorb; Meijer et al. 1992). One protocol (Yang et al. 1997) failed to extract DNA at all

The results suggest that in addition to proteinase K, a strong protein denaturing agent such as phenol is necessary to break down the histones, which form a stabile cover that effectively protects the nucleus from decay over time. Although this was not shown by these experiments, it could be

[7] The fragment-length analysis was carried out on a capillary sequencing machine (Type 310, software 672 Gene Scan Analysis, Applied Biosystems).

plausible to assume that this may apply to tissues other than bone as well (e.g., Fridez and Coquoz 1996). In general, the findings are supported by empirical evidence, which indicates that the majority of publications reporting successful amplification of chromosomal DNA used phenol-based extraction methods. In contrast, those that succeeded in mtDNA typing but explicitly declare to have failed in chromosomal DNA analysis used silica-based methods.

3.2 aDNA yield

In modern DNA analysis, the DNA yield is usually determined in order not to overload the reaction mix, which would cause an amplification failure (cf. section 4.6.3). In contrast, in ancient DNA analysis the determination of the DNA yield appears to be of secondary interest. Due to the degradation process, any DNA that is present is usually rare, and in most cases, an amplification failure will be due to inhibition rather than a DNA overload (cf. section 4.6). In such cases a further clean-up of the sample (cf. section 8.3) will generally lead to amplification success. However, even in ancient samples that for some fortunate circumstances reveal better-than-average DNA preservation (cf. section 3.3), DNA overload is obviously an underestimated cause for amplification failure. Probably, further clean-up would also lead to amplification success, but primarily because any clean-up procedure is accompanied by DNA loss.

There is a reliable indication that indeed DNA overload may be the reason for amplification failures even in ancient DNA investigations. In such cases a smear is visible after agarose gel electrophesis throughout the lane from low to high molecular-weight regions. Smears contraindicate too much remaining inhibitor in an extract, because they demonstrate that an amplification has taken place. Although the amplification is unspecific, it means the Taq polymerase was not inhibited. Alternatively, smears may also occur if a product is overamplified (Fig. 3.4).

Nonetheless, in both cases the mechanism for the generation of the smear is the same: the reaction runs out of sufficient amounts of primers, and the amplification products begin to prime themselves. In addition, if there is still sufficient dNTPs present, unspecific, high molecular-weight products will be generated in the subsequent elongation phases.

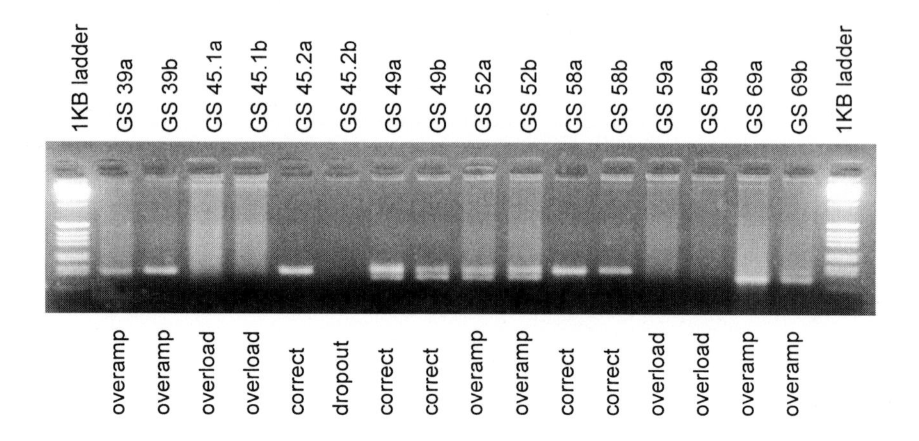

Fig. 3.4. Smears in aDNA amplifications can either result from DNA overload or overcycling. The singleplex amplification on CCR5 (cf. also section 2.2.6) was carried out in double assays (*a* and *b*) for 45 cycles using 5-µl aDNA extract in each reaction. In two cases (GS 45.1 and GS 59), this was too much DNA; in three cases (GS 39, GS 52 and GS 69), the reaction was slightly overamplified

In case of overloading aDNA, the reaction will never reach an effective amplification phase of the short product. The primers are exhausted by generating the initial asymmetric product (section 4.1) from the excessive supply of targets. In case a smear occurs as a result of overcycling, the sample would show specific products some cycles earlier (section 4.6.4). Therefore, the trial-and-error method to discriminate among the possible causes for an amplification failure with more or less intense smears visible on the agarose gel would be to run fewer cycles. A reliable way is to determine routinely the DNA yield of extracts that originate from series of samples that are assumed to contain very well preserved, indigenous DNA by UV spectrophotometry. If these values are taken into account to determine the volume of aDNA extract that is subjected to the PCR, the amplification success can be reached without the need to waste Taq polymerase. In our experience, if not too much microorganism DNA is present in the sample (which can be checked by microsections of the sample; cf. section 3.3.2), approximately 100–400 ng of aDNA results in specific amplification products.

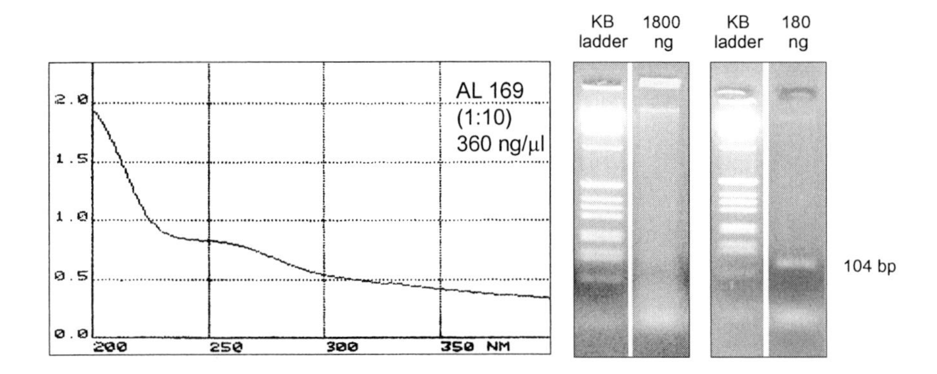

Fig. 3.5. UV-spectrophotometry shows that 360 ng/µl DNA (calculated from OD_{260}) are present in the extract of a femur-bone sample of AL 169 (19th century, Sicily). If 5 µl of the aDNA extract (1,800 ng) were subjected to the PCR reaction, the amplification resulted in a smear because of DNA overload. Reducing the aDNA amount to 180 ng revealed a specific amplification product. The amplification reactions targeted a 104-bp fragment on exon 6 of chromosome 9 (ABO blood group typing; cf. sections 2.2.5 and 7.4.3) and were run for 45 cycles

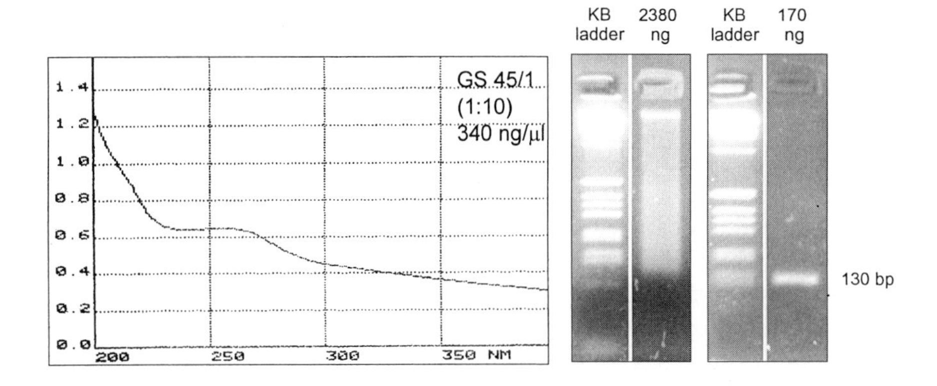

Fig. 3.6. UV spectrophotometry shows that 340 ng/µl DNA (calculated from OD_{260}) are present in the extract of a femur bone sample of GS 45/1 (17–18th century, northern Germany). If 7 µl of the aDNA extract (2,380 ng) were subjected to the PCR reaction, the amplification resulted in an intense smear because of the considerable DNA overload. Reducing the aDNA amount to 170 ng revealed a specific amplification product. The amplification reactions targeted a 130-bp fragment of the CCR 5 chemokine receptor gene on chromosome 3 (CCR 5 typing; cf. sections 2.2.6 and 7.4.4) and were run for 45 cycles

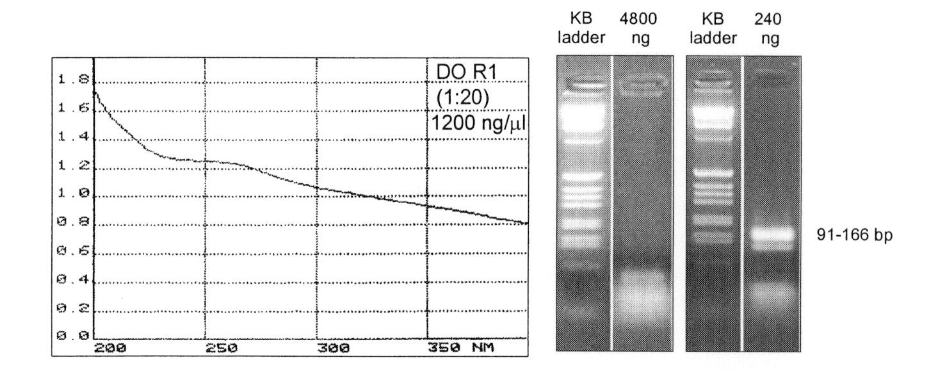

Fig. 3.7. UV spectrophotometry shows that 1,200 ng/µl DNA (calculated from OD_{260}) are present in the extract of a tooth sample of DO R1 (Bronze Age cave find, northern Germany). When 4 µl of the aDNA extract (4,800 ng) were subjected to the PCR reaction, the amplification resulted in a faint smear. Although the smear indicates DNA overload, its weakness also points to a minor inhibition. This is supported by the spectrophotometric diagram showing a less pronounced DNA peak at 260 nm. Reducing the aDNA amount to 240 ng, and thereby also the possible inhibitor, resulted in a specific amplification product. The multiplex amplification reactions targeted multiple fragments with a range of 91–166 bp on the X and the Y chromosomes (sexing by multiplex PCR; cf. sections 2.2.4 and 7.1.2) and were run for 50 cycles

3.3 aDNA preservation

The preservation of ancient DNA involves various aspects, beginning with the positive proof of the existence of authentic aDNA. Although PCR can prove that a result is authentic with a high level of plausibility (chapter 6), a positive proof can only be ascertained by a technique that does not use a DNA amplification step. Further aspects of DNA preservation concern the patterns of degradation, the environmental conditions that favor DNA preservation, the length of the preserved sequence fragments and, finally, questions about the storage of aDNA extracts for further analysis even years later.

3.3.1 aDNA degradation patterns

PCR-based analysis of ancient DNA may reveal false-positive results because of contamination with contemporary DNA (chapter 6). Some of the published results of mitochondrial and chloroplast DNA sequences of the early 1990s were discovered to be the results of contamination. This led to discussion about whether ancient DNA existed at all. However, more contemporary experiments using comparative genome hybridization (CGH) showed that the doubts were without reason. Unlike PCR, CGH (e.g., Kallioniemi et al. 1992) requires no amplification steps. Therefore, false results because of accidental contamination can practically be excluded. The experiments proved the authenticity of ancient DNA for an approximately 250-year-old, ethanol-fixed embryo (Fig. 3.8) as well as for a 3,000-year-old bone sample (Tönnies et al. 1998; Hummel et al. 1999).

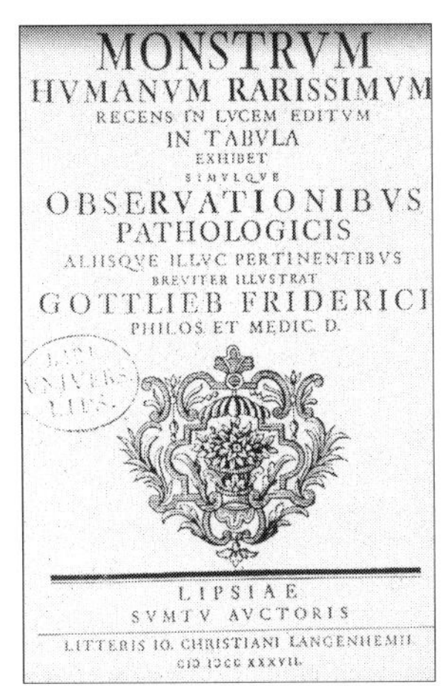
MONSTRVM
HVMANVM RARISSIMVM
RECENS IN LVCEM EDITVM
IN TABVLA
EXHIBET
SIMVLQVE
OBSERVATIONIBVS
PATHOLOGICIS
ALIISQVE ILLVC PERTINENTIBVS
BREVITER ILLVSTRAT
GOTTLIEB FRIDERICI
PHILOS. ET MEDIC. D.

LIPSIAE
SVMTV AVCTORIS
LITTERIS IO. CHRISTIANI LANGENHEMII
CIϽ IϽCC XXXVII.

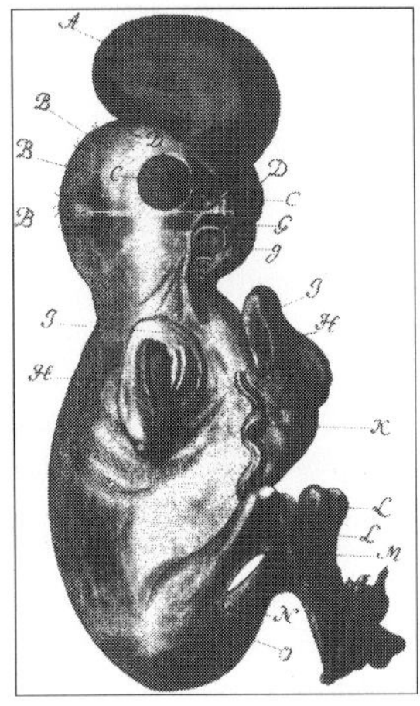

Fig. 3.8. Historical drawing of the embryo showing multiple malformations. The embryo was formalin-fixed following a miscarriage

The experiments also revealed a rare chromosomal aberration, a partial monosomy of chromosome 17 (Fig. 3.9). This defect causes certain death in an embryo usually already in the early stages of pregnancy. The CGH findings were fully concordant with the strongly deviating morphological aspect of the 250-year-old embryo.[8]

Unfortunately, it is not possible to investigate the preservation of mitochondrial DNA using this technique. However, the CGH technique offers not only the chance to prove the existence of ancient chromosomal DNA, but also enables an understanding of the patterns of chromosomal DNA degradation. The results of the experiments carried out on different specimens clearly show that the telomeric regions of ancient chromosomal DNA suffer the most from degradation, while the preservation of the

[8] The experiments were carried out in cooperation with Dr. Holger Tönnies and Prof. Dr. Karl Sperling of the Human Genetics Department of the Charité, Berlin.

centromeric regions is better than average. This is clearly shown by the relative amounts of aDNA compared to the modern DNA competitor at the different chromosomal regions.

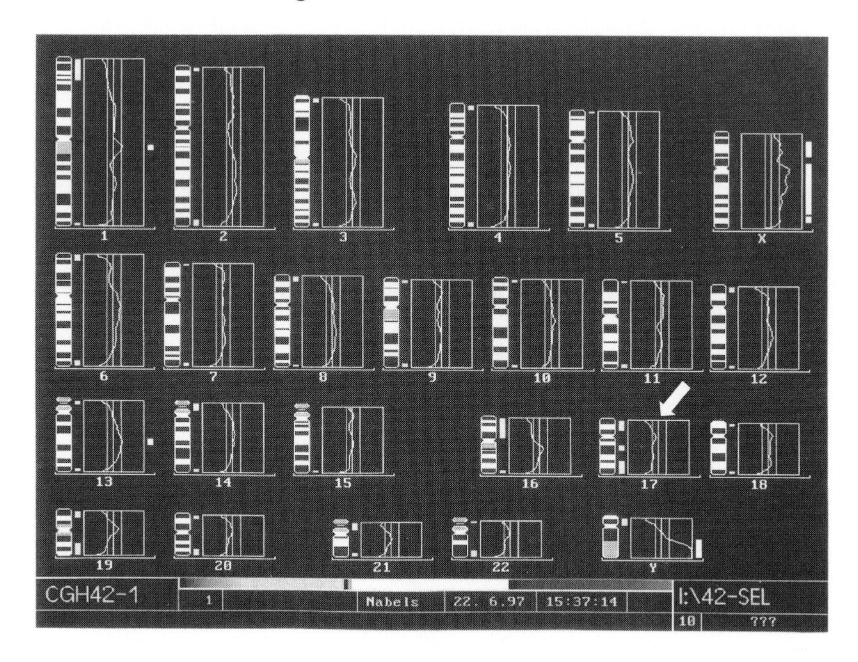

Fig. 3.9. CGH analysis reveals a rare chromosomal aberration of chromosome 17, which is positively lethal at the embryonic stage. Additionally, it clearly shows telomeric regions suffering most from degradation (signals exceeding the threshold to the *left*), while the centromeric regions show better-than-average preservation (signals exceeding the threshold to the *right*)

3.3.2 The age of aDNA

The preservation of DNA and the possible damage to the nucleic acid sequences have been discussed from the onset of aDNA analysis (e.g., Bär et al. 1988; Eglington and Logan 1991; Tuross 1994). From theoretical predictions ancient DNA would only to be expected in comparatively young specimens, not exceeding a few thousand years of age (Lindahl 1993). Nonetheless, the same theoretical predictions also would not suggest that any skeletal element would remain for longer than about the same few thousand years, which has proved wrong, as many skeletons have been found that are much older. The reason is the many assumptions in the theoretical approach, such as relatively low pH values, that generally are not found in real settings.

Although it is certainly true that all chemical reactions in bone and DNA result from decay occurring over time, it is the environmental conditions that determine the speed and extent of chemically induced and biogenic decay. This has already been supposed by many researchers because of their empiric findings on aDNA that has been preserved in specimens of various ages (Höss et al. 1994; Poinar 1994; Colson et al. 1997; Schultes et al. 1997).

The first systematic and empirical study to address the question about which conditions favor DNA preservation was carried out through experiments on skeletal material of about the same age (approximately 3,000 years BP) but from different environmental in situ conditions (Burger et al. 1999). Furthermore, the experiments investigated teeth samples that were derived from the same site but had different histories of post-excavation storage. The ability to reproduce the amplification of chromosomal STR sequences was taken as a criterion for DNA preservation. In relation to the skeletal material from different environmental conditions, the findings of the experiments showed that aDNA preservation was strongly dependent on a particular coincidence of factors (Table 3.3), in which the most important factors seemed to be the occurrence and amount of microorganisms.

Table 3.3. DNA preservation and environmental conditions of the burial site

Site	Conditions	PCRs	Chromosomal loci amplified			
			Amelo	VWA	TH01	FES/FPS
Lichtenstein cave, Germany	Cold (8°C), humid, no microorg., stored at -20°C	1	X	16	6/9.3	11/12
		2	X	16	6/9.3	11/12
		3	X	16	6/9.3	11/12
		4	X	16	6/9.3	11/12
		5	X	16	6/9.3	11/12
Shimal desert, UAE	Hot (27°C), dry, no microorg., stored 10 years at RT	1	XY	18/19	9	-
		2	-	19	7/9	-
		3	X	19	-	-
		4	-	18	-	-
		5	X	-	6	-
Mustang cave, Nepal	Cold (12°C), dry-humid, high microorg., stored 1 year at RT	1	-	-	-	-
		2	-	-	-	-
		3	-	-	-	-
		4	-	-	-	-
		5	-	-	-	-

The destructive power of microorganisms that may colonize in bone (Fig. 3.10) are well known from investigations on the micromorphology of skeletal elements (Piepenbrink 1989; Herrmann et al. 1990). The results thus confirm at the DNA level what is indeed presumed by archaeologists and anthropologists who diagnose skeletons concerning the macro- and micromorphologic appearance of skeletal material.

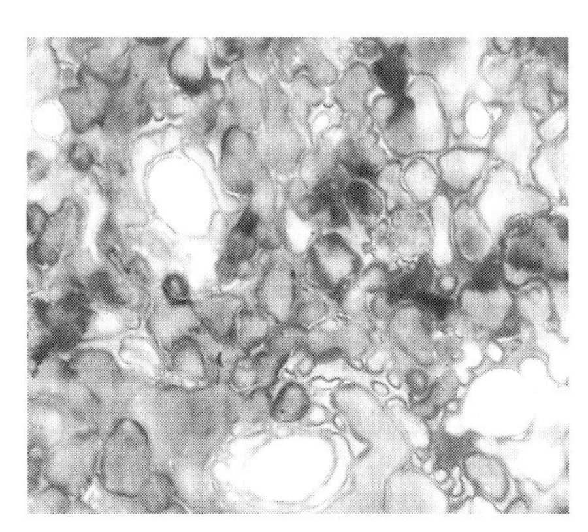
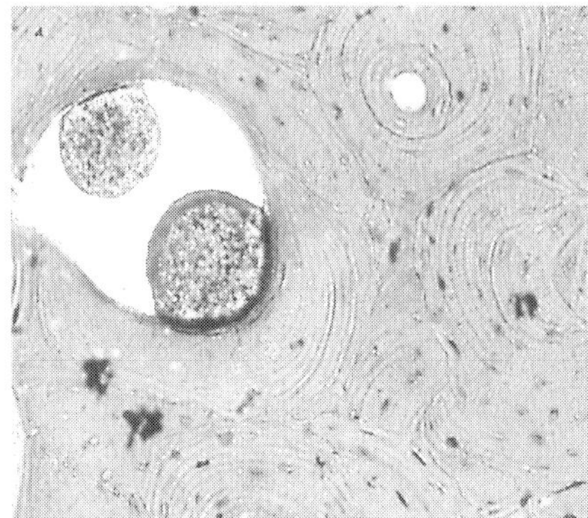

Fig. 3.10. Thin sections of two medieval femoral bones: the specimen on the left reveals large areas of destruction by microorganisms colonizing the bone, producing bore channels. The specimen at the right is almost intact. (magnifications ×400)

Slightly acidic pH conditions induce chemical changes in the bone, transforming the mineral hydroxy apatite into brushite (Fig. 3.11, Fig. 3.12 and Table 3.4) (Herrmann and Newesely 1982; Newesely 1987). This process eventually leads to soil silhouettes as the only remains of a skeleton (Fig. 3.13).

Table 3.4. Bone preservation and pH values of the soil

pH	7.5–7.0	7.0–6.0	6.0–4.5	<4.5
Mineral	Hydroxy apatite	Octo-calcium phosphate	Brushite	(Only soil silhouettes remain)
Formula	$Ca_5(PO_4)_3OH$	$Ca_4H(PO_4)_3x2H_2O$	$CaHPO_4x2H_2O$	-

This makes clear that even slight changes towards acidic pH values are detrimental to both the preservation of DNA and of the source material itself. In other words, if skeletal material has been preserved in macroscopically good condition in general, then the pH value of the soil cannot be the reason for a considerable reduction in DNA preservation. The only exceptions are the few burial sites where extreme alkaline conditions resulting from covering layers of animal dung are observed, particularly because dung contains urea that further destabilizes the DNA, but does not affect the overall preservation of the bone morphology.

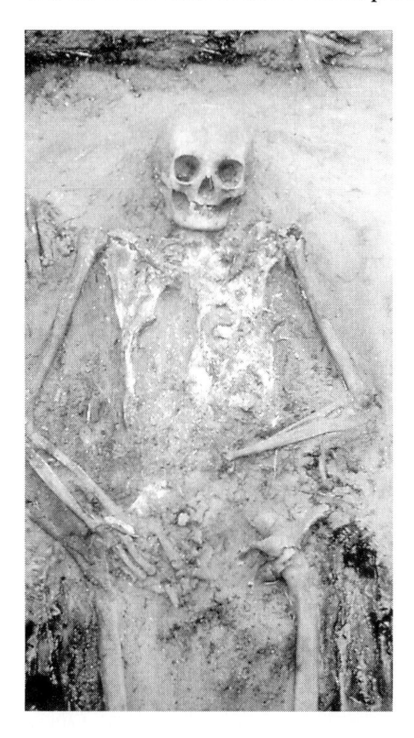

Fig. 3.11. Macromorphological aspect of brushite formation (white mineral crystals) in the area of the chest that was caused by slightly acidic pH values

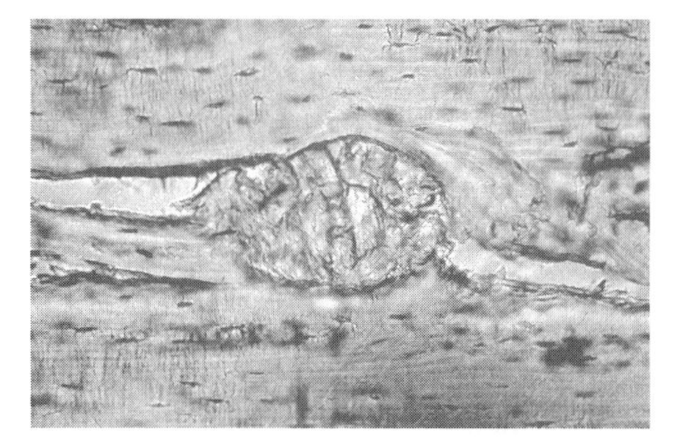

Fig. 3.12. Micromorphology of brushite formation in the thin section of a historic bone. It is clearly visible that the brushite crystals cause both chemical and physical damage by cracking the bone material. (magnification ×800)

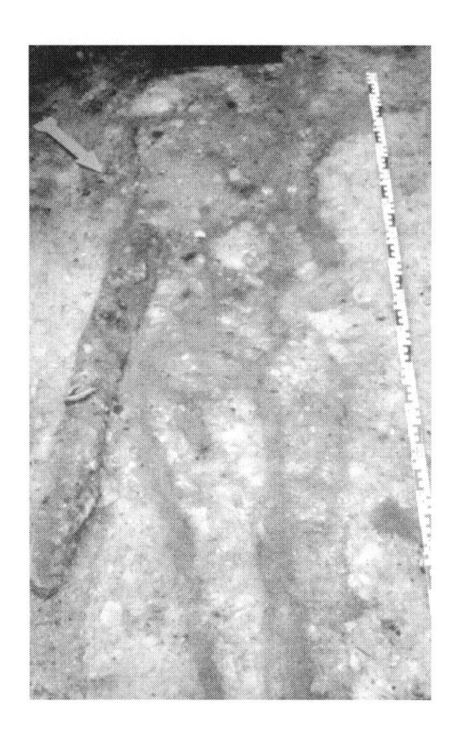

Fig. 3.13. A soil silhouette is all that is left of a skeleton buried in acidic soil

However, another of the study's findings does not have its evident counterpart in the morphologic aspect: the storage of skeletal material after excavation. Skeletal material available from the Lichtenstein cave (section 7.3.3) provided different histories of storage at the time of the investigations. This was more than apparent in its effect on DNA preservation (Table 3.5).

Table 3.5. Storage of samples after excavation and DNA preservation

Lichtenstein cave	Storage after excavation	0–2 years at RT	16 years at RT
(8°C)	Amplifiable STRs	72–90%	68%
(teeth samples)	Reproducibility	86–91%	40%

These results strongly indicated that samples of excavated skeletal material that are intended for molecular investigations should be stored frozen as soon as possible after excavation, since long-term storage at room temperature, which is common for anthropological and zoological material, destroys DNA that has already been preserved for thousands of years.

3.3.3 Fragment lengths of aDNA

Fragment lengths of aDNA have been a matter of debate above all in the context of the authenticity of results from ancient DNA. It has been suggested by many researchers working with ancient DNA that authentic fragments longer than 200–300 bp cannot be amplified from aDNA extracts (e.g., Handt et al. 1994; Hardy et al. 1994; Krings et al. 1997). These statements may certainly be true in the very generalized context about average aDNA preservation, in particular if mitochondrial DNA is concerned with its inferior protection against decay resulting from a lack of histones (cf. sections 2.1 and 2.2).

However, it becomes critical if the fragment length of amplified DNA is taken as a criterion of authenticity, claiming that short fragments' ability to be amplified exclusively validates the results (e.g., DeSalle 1994; Handt et al. 1994; Hardy et al. 1994; Braun et al. 1998). This argument does not consider the possibility that contaminating DNA may also be severely degraded (chapter 6), which must be assumed in particular for famous specimens that have been passed through many hands for decades before coming to rest in a museum collection. Alternatively, it cannot be assumed that a long fragment amplified from an ancient specimen is necessarily a result of modern contamination (as stated by e.g., Pääbo in 1989).

The following experiments show that it is possible to amplify authentic, long aDNA fragments from skeletal material that comes from burial conditions that obviously favored better-than-average aDNA preservation (Haack et al. 2000). By generating individual-specific STR alleles (section 2.2.2), the design of the study ensured the authenticity of the results (Fig. 3.14).

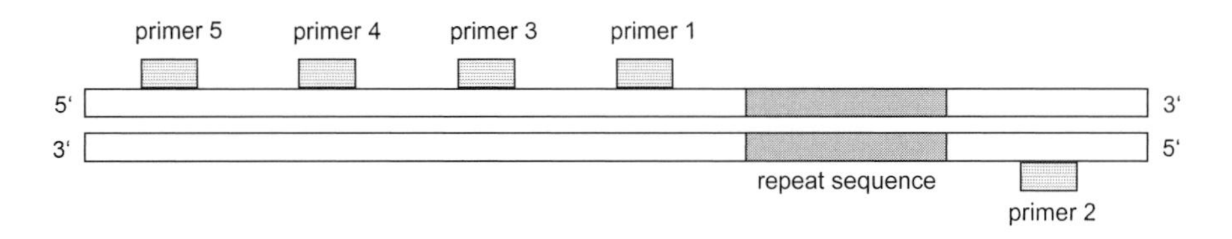

Fig. 3.14. Experimental design for the amplification of long STR fragments. The different primer sets enabled incremental increases of fragment lengths for the amplification of the VWA locus

From the precise locations of the primers in the region of the repeat structure of the VWA locus, it was possible to calculate the fragment lengths for each specimen as they were to predict for all primer pair combinations. The eight specimens included in the study all showed fully reproducible results in earlier studies employing the AmpFlSTR-ProfilerPlus amplification kit (Applied Biosystems) (Fig. 3.15).

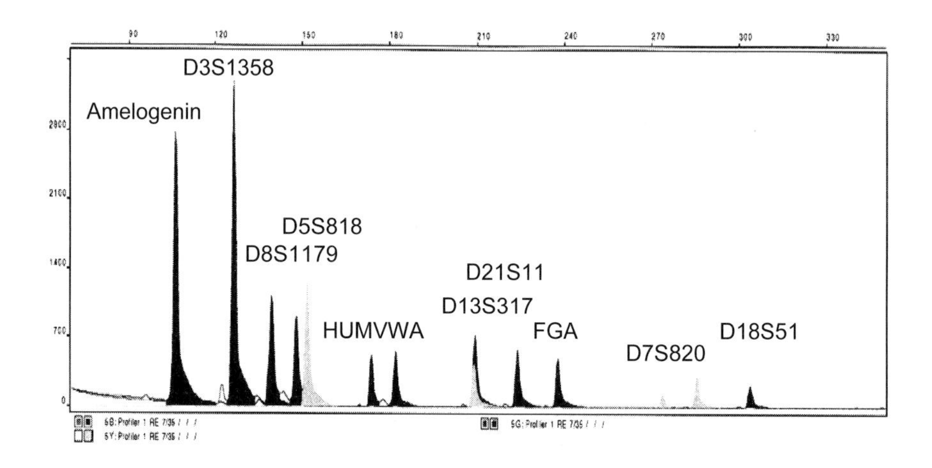

Fig. 3.15. Electropherogram of a medieval bone (RE 8) revealing a full autosomal STR fingerprint through amplification by AmpFlSTRProfilerPlus

The results of the study show (Table 3.6) that it is possible to amplify fragments longer than 300 bp if the specimen has remained in a good state of preservation (also cf. Lambert et al. 2002). From seven of the eight specimen fragments, about 400 bp were amplifiable. From three of the eight specimens, 600 bp could be amplified. All of the specimens were derived from the 250-year-old Goslar burial site in Lower Saxony (Bramanti et al. 2000a; Bramanti et al. 2000b). In one case (GS 32), it was even possible to amplify 800-bp fragments, which may be considered to be authentic because of the matching of the calculated expected bp-values. Nevertheless, if results are to be obtained from as many specimens as possible, it is much more desirable to design primers (section 4.4) that amplify the targeted sequence information by giving a PCR product that is as short as possible.

Table 3.6. Amplification of long STR fragments from ancient DNA extracts

Sample	Genotype and MP		VWA primer set 2/1	VWA primer set 2/3*	VWA primer set 2/4*	VWA primer set 2/5*
DO 183	14/19	Expected	139/159	382/402	618/638	786/806
	P_m=0.009	Analyzed	139.0 139.3 139.1/159.6	-/-	-/-	-/-
DO 1044	15/17	Expected	143/151	286/394	622/630	790/798
	P_m=0.026	Analyzed	142.7/150.9 143.6/151.6 142.6/150.6	388.6 389.2 389.2	-/-	-/-
DO 1500	17/18	Expected	151/155	394/398	630/634	798/802
	P_m=0.068	Analyzed	151.0 151.4/155.6 151.1/155.2	397.3	-/-	-/-
RE 3	14/16	Expected	139/147	382/390	618/626	786/794
	P_m=0.017	Analyzed	139.0/147.0 139.0/147.0	386.5	-/-	-/-
RE 8	15/17	Expected	143/151	386/394	622/630	790/798
	P_m=0.021	Analyzed	142.9/151.1 143.1/151.2	396.6	-/-	-/-
GS 32	16/19	Expected	147/159	390/402	626/638	794/806
	P_m=0.021	Analyzed	147.1/159.3 147.0/159.2 146.7/159.2	192.5 393.0/405.7	627.6/641.5 625.8 625.9 640.9	791.6 807.6
GS 34	17/20	Expected	151/163	394/406	630/642	798/810
	P_m=0.004	Analyzed	151.1/163.5 150.9/163.3	397.3/409.5 397.5/409.8	631.3/644.2 632.5/644.7 631.9 630.9	-/-
GS 35	17/20	Expected	151/163	394/406	630/642	798/810
	P_m=0.004	Analyzed	152.7 151.0 151.2/163.4	409.5	645.4	-/-

* all results are systematically approximately 2–3 bp longer than the expected values. This may be because of either a sequence polymorphism (deletion) of the reference sequence the primers were designed to or a deviation because of the algorithm used by the 672 Genescan software program (Applied Biosystems) that was used for fragment-length determination. P_m=matching probability (cf. section 2.2)

3.3.4 Storage of aDNA extracts

Often a multiple analysis from an aDNA extract needs to be repeated or, until new methods become available, the extracts need to be stored for some time: weeks, months or even years. When a phenol-based DNA extraction is carried out, the final steps of DNA extraction employ a silica-particle solution (e.g., Glassmilk, QBiogene) in order to reduce the volume and purify the extract (cf. section 8.3). The suppliers of silica solutions recommend DNA storage in pure water or TE-buffered solutions after removing the glass particles. This is done by incubating the DNA-silica solution at 56°C for 10–15 min, followed by a short centrifugation and transfer of the eluated DNA to a new Eppendorf tube. The aDNA extracts that were prepared this way worked well if they were subject to amplification within a few days or weeks. However, as experiments revealed, aDNA extracts, which originally revealed sufficient intact targets for successful amplifications of single-copy sequences, started to degrade to a remarkable degree already after a period of only 6 to 14 weeks if the DNA was stored at normal refrigerator temperatures. If the DNA was eluated from the silica particles, no intact targets were found at all, but when it was attached to the silica particles, the DNA degradation was reduced (Fig. 3.16).

In contrast, the stability of aDNA was clearly improved at a storage temperature of -20°C when the silica beads were not removed from the aqueous extracts. Then, the amplification results showed no distinguishable differences after a long-term storage compared to the freshly extracted samples (Fig. 3.17). Even after 3 years, aqueous aDNA extracts that were stored with the silica particles were unchanged and showed perfectly reproducible amplification results (cf. section 7.4.3 and 7.4.4).

The best long-term results, independent of the storage temperature, were reached if the aDNA was allowed to remain attached to the silica particles in the presence of ethanol, i.e., the final extraction steps, drying the pellet and dissolving them in water, should only be done prior to the use of the aDNA in an amplification reaction. Analysis carried out after a 3-month period did not show any differences compared to results obtained from freshly extracted specimens. Therefore, this method of storage is recommended especially when shipping aDNA extracts.

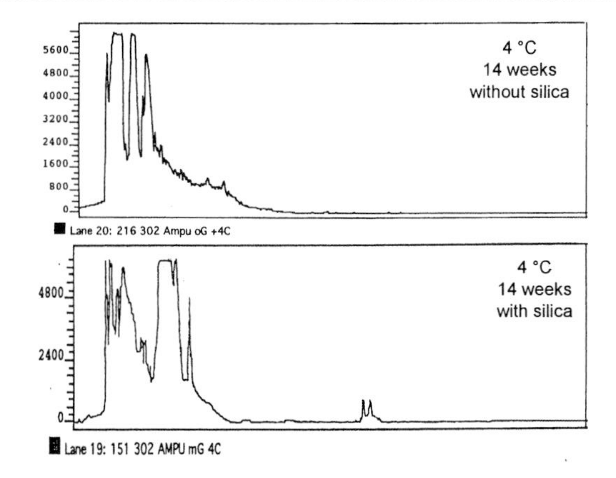

Fig. 3.16. Amplification results for aDNA that was stored at +4°C for 14 weeks. Both of the extracts stored attached to the silica particles, and the extracts stored in the eluated form suffer considerably from being stored at refrigerator temperatures. However, the DNA attached to silica still shows signals, i.e., intact targets are present, while the purified extract contains no more intact targets

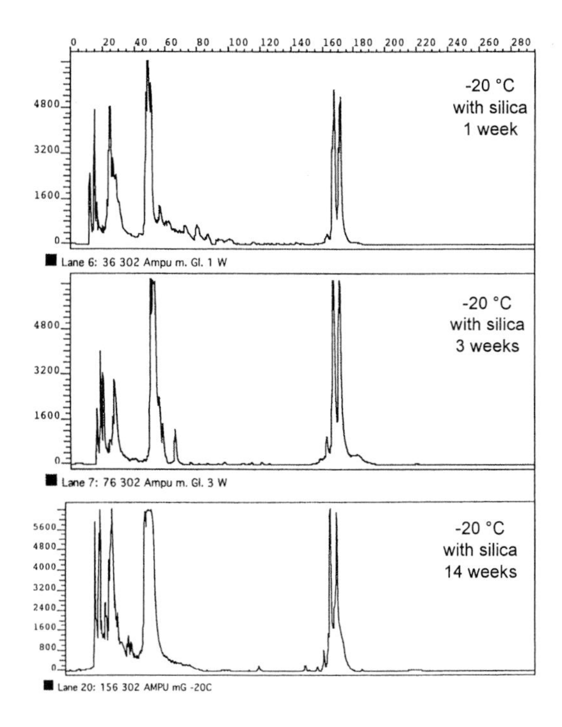

Fig. 3.17. Amplification results for aDNA that was stored attached to silica particles at -20°C. These conditions revealed the best results with apparently unchanged numbers of intact targets in the extracts

References

Bär W, Kratzer A, Mächler M, Schmid W (1988) Postmortem stability of DNA. Forensic Sci Int 39:59-70

Baron H, Hummel S, Herrmann B (1996) *Mycobacterium tuberculosis* complex DNA in ancient human bones. J Archaeol Sci 23:667–671

Bramanti B, Hummel S, Schultes T, Herrmann B (2000a) STR allelic frequencies in a German skeleton collection. Anthrop Anz 58:45–49

Bramanti B, Hummel S, Schultes T, Herrmann B (2000b) Genetic characterization of a historical human society by means of aDNA analysis of autosomal STRs. Bienniel Books of EAA 1:147–163

Braun M, Collins-Cook D (1998) DNA from *Mycobacterium tuberculosis* complex identified in North American, pre-Columbian human Skeletal remains. J Archaeol Sci 25:271–277

Burger J, Hummel S, Herrmann B, Henke W (1999) DNA preservation: A microsatellite-DNA study on ancient skeletal remains. Electrophoresis 20:1722–1728

Cattaneo C, Craig OE, James NT, Sokol RJ (1997) Comparison of three DNA extraction methods on bone and blood stains up to 43 years old and amplification of three different gene sequences. J Forensic Sci 42:1126–1135

Colson IB, Bailey JF, Vercauteren M, Sykes B (1997) The preservation of ancient DNA and bone diagenesis. Ancient Biomol 1:109–117

DeSalle R (1994) Implications of ancient DNA for phylogenetic studies. Experientia 50:543–550

Eglington G, Logan GA (1991) Molecular preservation. Phil Trans R Soc Lond B 333:315–328

Fisher DL, Holland MM, Mitchell L, Sledzik PS, Wilcox AW, Wadhams M, Weedn VW (1993) Extraction, evaluation, and amplification of DNA from decalcified and undecalcified United States Civil War bone. J Forensic Sci 38:60–68

Fridez F, Coquoz R (1996) PCR DNA typing of stamps: evaluation of the DNA extraction. Forensic Sci Int 78:103-110

Haack K, Hummel S, Herrmann B (2000) Ancient DNA fragments longer than 300 bp. Anthrop Anz 58:51–56

Handt O, Richards M, Trommsdorf M, Kilger C, Simanainen J, Georgiev O, Bauer K, Stone A, Hedges R, Schaffner W, Utermann G, Sykes B, Pääbo S (1994) Molecular genetic analyses of the Tyrolean Ice Man. Science 264:1775–1778

Hardy C, Casane D, Vigne JD, Callou C, Dennebouy N, Mounolou JC, Monnerot M (1994) Ancient DNA from Bronze Age bones of European rabbit (*Oryctolagus cuniculus*). Experientia 50:564–570

Herrmann B, Grupe G, Hummel S, Piepenbrink H, Schutkowski H (1990) Prähistorische Anthropologie. Leitfaden der Feld- und Labormethoden. Springer, Heidelberg

Herrmann B, Newesely H (1982) Dekompositionsvorgänge des Knochens unter langer Liegezeit. Die mineralische Phase. Anthrop Anz 40:19–31

Höss M (1995) Extraktion und Analyse von DNA aus archäologischem Knochenmaterial. Dissertation, Ludwig Maximilians-Universität, München
Höss M, Pääbo S (1993) DNA extraction from Pleistocene bones by a silica-based purification method. Nucleic Acids Res 21:3913–3914
Höss M, Pääbo S, Vereshchagin NK (1994) Mammoth DNA sequences. Nature 370:333
Hummel S, Herrmann B, Rameckers J, Müller D, Sperling K, Neitzel H, Tönnies H (1999) Proving the authenticity of ancient DNA (aDNA) by comparative genomic hybridization (CGH). Naturwissenschaften 86:500–503
Kallioniemi A, Kallioniemi OP, Sudar D, Rutovitz D, Gray JW, Waldman F, Pinkel D (1992) Comparative genomic hybridization for cytogenetic analysis of solid tumors. Science 258:818–820
Krings M, Stone A, Schmitz RW, Krainitzki H, Stoneking M, Pääbo S (1997) Neanderthal DNA sequences and the origin of modern humans. Cell 90:19–30
Lalu K, Karhunen PJ, Sajantila A (1994) Comparison of DNA-extraction methods from compact bone tissue. Advances Forensic Haematogen 5:160-163
Lambert DM, Ritchie PA, Millar CD, Holland B, Drummond AJ, Baroni C (2002) Rates of evolution in ancient DNA from Adelie penguins. Science 295:2270-2273
Lindahl T (1993) Instability and decay of the primary structure of DNA. Nature 362:709–715
Meijer H, Perizonius WRK, Geraedts JPM (1992) Recovery and identification of DNA sequences harboured in preserved ancient human bones. Biochem Biophys Res Commun 183:367–374
Müller A (2000) aDNA–Extraktion aus Knochen – ein Methodenvergleich. Diplomarbeit, Georg August-Universität, Göttingen
Newesely H (1987) Chemical stability of hydroxyapatite under different conditions. In: Grupe G, Herrmann B (eds) Trace elements in environmental history. Springer, Berlin, Heidelberg, New York
Pääbo S (1989) Ancient DNA: Extraction, characterization, molecular cloning, and enzymatic amplification. Proc Natl Acad Sci USA 86:1939–1943
Piepenbrink H (1989) Examples of chemical changes during fossilisation. Appl Geochem 4:273–280
Poinar GO (1994) The range of life in amber: significance and implications in DNA studies. Experientia 50:536–542
Schultes T, Hummel S, Herrmann B (1997) Recognizing and overcoming inconsistencies in microsatellite typing of ancient DNA samples. Ancient Biomol 1:227–233
Tönnies H, Müller D, Hummel S, Herrmann B, Sperling K, Neitzel H (1998) Chromosome analysis of a 262-years preserved fetus with multiple congenital malformations; first application of comparative genomic hybridization to ancient DNA. Europ J Hum Genet 6:86 ff
Tuross N (1994) The biochemistry of ancient DNA in bone. Experientia 50:530–535
Vandenberg N, van Oorschot RA, Mitchell RJ (1997) An evaluation of selected DNA extraction strategies for short tandem repeat typing. Electrophoresis 18:1624–1626

Vince A, Poljak M, Seme K (1998) DNA extraction from archival Giemsa-stained bone-marrow slides: comparison of six rapid methods. Br J Haematol 101:349–351

Yang H, Golenberg EM, Shoshani J (1997) Proboscidean DNA from museum and fossil specimens: an assessment of ancient DNA extraction and amplification techniques. Biochem Genet 35:165–179

4 PCR

Polymerase chain reaction (PCR) is a well known and established technique in many scientific, medical and molecular genetics service disciplines. It enables an investigator to detect trace amounts of DNA by amplifying the sequences of interest by exposing the DNA repeatedly to different temperatures (e.g., Erlich 1989; Innis et al. 1990; Mullis et al. 1994; Innis et al. 1995).

In general, the technique is straightforward and easy to carry out, and its application is supported by a broad variety of commercially available kits, ranging from the preparation and extraction of the DNA samples, to the PCR set up itself to the analysis of the amplification product. If a simple presence/absence result is sought from the analysis, or, if fragment length polymorphisms with inter-product deviations longer than about five to six base pairs are analysed, then even the technical equipment that is necessary for carrying out the experiments and the analysis (cf. chapter 5) is comparatively simple and inexpensive. Overall, it has become a reliable technique that can simply be used as a methodological tool whose parameters need no further attention. This may apply in routine analysis using standardized DNA templates, such as those using fresh tissues that are processed with the help of one of the many DNA extraction kits. However, for any application examining ancient and degraded DNA, there is no "standardized" DNA template. Although the quality and quantity of the aDNA extract depends on the actual method applied for sample preparation and DNA extraction (cf. chapter 3), it is the understanding of and the potential capability of the PCR technique, its parameters and its components that are important for any aDNA investigator. Understanding PCR as a flexible mechanism and at the same time being aware of its pitfalls allow investigations using degraded DNA to utilize its full potential and to avoid false positives or otherwise misleading results.

4.1 Basic PCR mechanism

The PCR technique is an in vitro amplification driven by enzymes and is usually applied to relatively short DNA sequences. Only very few components are essential for the amplification process, including a pH-buffered solution consisting of $MgCl_2$, KCl and Tris in addition to the primers, nucleotide bases, a heat-stabile DNA polymerase and the actual DNA sample (e.g., Landre et al. 1995). The reaction itself is based on three steps: the denaturation, annealing and elongation phases. All of these are repeatedly passed in sequence, through "cycling" (Fig. 4.1).

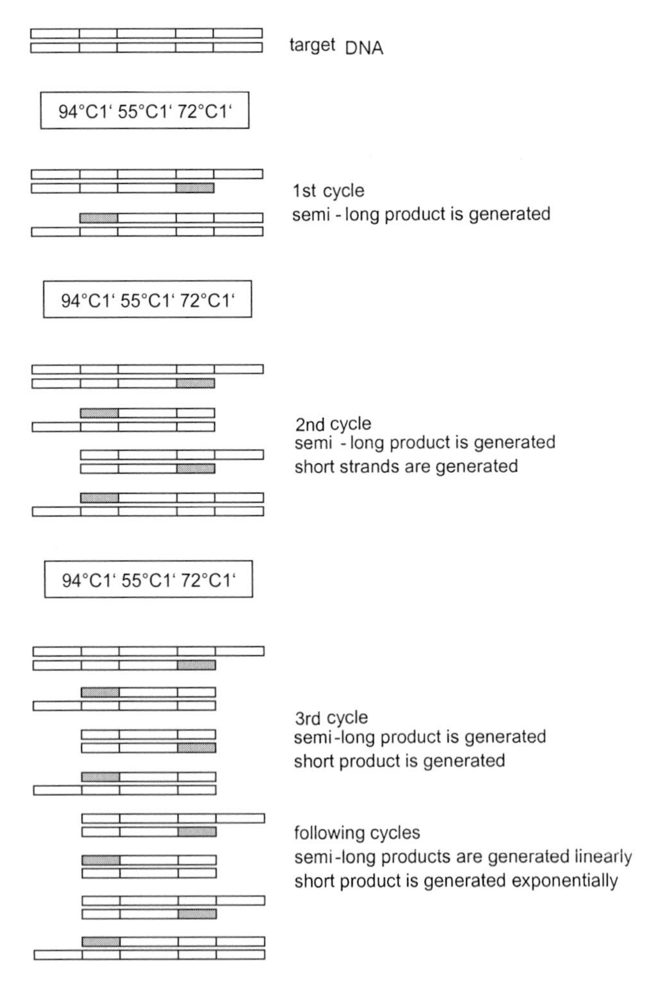

Fig. 4.1. PCR mechanism

During the denaturation phase, which is typically carried out for about 60 s at 94°C, the double-stranded DNA molecules "fall apart" (denature) into single strands. The hydrogen bonds of the complementary bases are forced open through the high energy input at this elevated temperature. From this follows the annealing phase, which also typically lasts for 60 s. This stage enables the primers or oligonucleotides (synthetic single-stranded nucleotide chains that are 20–25 base pairs long) to hybridize to their recognition sites at considerably lower temperatures lying between 50–60°C; the exact value will depend on the actual primer sequence (cf. section 4.4). During the final elongation phase, which is usually carried out for 60 s at 72°C, a heat-stabile DNA polymerase begins to build in nucleotide bases starting from the 3'-end of the primers. Theoretically, at the end of this first cycle consisting of the three essential PCR phases, already a doubling of the targeted sequence and part of its neighboring sequences has taken place. Any further cycle will lead to further doublings, causing an exponential amplification of the targeted sequence that follows the equation:

$$Y=A\ (1+F)^n$$

where Y= product yield [ng DNA], A= target concentration [ng DNA], F= reaction efficiency and *n*= number of amplification cycles.

In a practical situation, a true exponential amplification (F=1, or a reaction efficiency of 100%) cannot be reached in the reaction even with fully intact DNA targets that are in a pure extraction medium. Given these optimal conditions, there still remain a number of factors that prevent the efficiency (F) of a PCR reaction from approaching 1.0 (e.g., Cha and Thilly 1993; Sardelli 1993). Typically, an optimal PCR that has been set up for the amplification of intact DNA targets from highly purified extracts reaches F-values of between 0.8 and 0.9. With sub-optimal energy profiles, the primers will generate even lower efficiencies, falling to about F=0.7. In the case of amplifying ancient and degraded DNA, those factors are accompanied by additional ones (e.g., impurities in the extract), which further lower the F value to 0.4–0.7 and therefore further inhibit the PCR reaction (Table 4.1). The effect of a decreasing Y value (i.e. a non-detectable PCR product) can only be compensated for by increasing the number of amplification cycles.

Table 4.1. Factors affecting the efficiency (F) of a PCR-based DNA amplification

Factor	Cause
• Changing/suboptimal component balances (DNA-primers-dNTPs) • Unstable pH value due to heat lability of buffer substances • "Overshooting" DNA polymerase activity in first cycles (leads to unspecific products)* • Weak DNA-polymerase activity at last cycles*	• Is PCR immanent (cannot be avoided)
• Suboptimal elongation temperature of 72°C must be used (Taq DNA polymerases has its maximum activity at 80°C)	• Short hybridized double strands (primer-matrix) start to denature at higher temperatures
• Suboptimal hybridization of primers	• Suboptimal primer design • Sequence polymorphism at primer site
• Unspecific amplification by-products	• Suboptimal primer design (this is generally unforeseeable due to incomplete knowledge of respective genomes and/or mistakes in databases)
• Primer dimer formation	• Suboptimal primer design
• Impurities within the extract **	• Inefficient purification (this is generally unavoidable in aDNA preparation due to excessive DNA loss in any purification procedure)
• Excess of strongly fragmented DNA compared to intact targets**	• May lead to PCR jumping effects (i.e., semi-long products generated from fragmented targets hybridize incorrectly and build sequence chimerae)

*effect is strongly reduced by the use of inhibited Taq DNA polymerases that need heat-induced activation (e.g., AmpliTaq Gold, Applied Biosystems)
**generally applies to aDNA extracts only

4.2 Number of amplification cycles

The number of cycles that are used for an amplification reaction are crucial for the success of the analysis. Although this appears fairly obvious, the decisions about how many amplification cycles are necessary often seem to be guided by a mixture of unfounded conclusions and the endeavour to avoid false results because of contamination. Of course, every effort should be made to avoid contamination throughout the entire process of examining ancient DNA. However, false results because of contamination cannot be avoided by choosing to carry out only a few amplification cycles that are barely above the detection limit, because this strategy presumes that the amount of intact authentic targets from the sample outnumber the amount of possibly contaminating targets. This may be a pragmatic and successful strategy in the analysis of modern DNA samples; however, it does not need much imagination to recognize that this implicit prerequisite may not necessarily apply to degraded DNA analysis. Therefore, following the common PCR-rule, "don't trust more than 30 cycles," leads to mistaken confidence in aDNA analysis. On the other hand, running no more than 30 cycles would also conceal many authentic results that can be gained from aDNA by applying a suitable number of additional amplification cycles (Wages and Fowler 1993; Rameckers et al. 1997). To escape this dilemma, the avoidance of false-positive results should not be controlled by the number of amplification cycles but rather must be done by minimizing contamination risks and by elaborated ways of monitoring for false-positive results (cf. chapter 6).

The number of cycles suitable to detect even small amounts of target sequences (<100) can be calculated by transforming the equation describing the PCR process (cf. section 2.4.1) to:

$$n = \frac{\ln(Y / A)}{\ln(1 + F)}$$

In order to calculate *n*-values (= number of required cycles) following this equation, it is necessary to determine an appropriate Y-value (= DNA yield). The detection limit on an ethidium-bromide-stained agarose gel is 2 ng of DNA, given that a 200 bp fragment is amplified. If a standard 10-µl aliquot of a 50-µl total reaction volume is electrophoresed, the total amount of amplified product Y should be fixed at 10 ng. Values for A (= amount of initial target sequences) and F (= efficiency) have been varied discontinuously for the calculation of cycle numbers (Table 4.2). Thereby, the values for A, based on an average of 3×10^5 templates of a single copy target, are represented in 1 µg of total mammalian genomic DNA, and the length of the targeted sequence is 200 bp.

Following the values given in Table 4.2, which were calculated from the prior equation, attempts to derive amplification products from only a few intact target sequences with substantially fewer than 45–50 cycles must be unsuccessful. This particularly applies to aDNA extracts where there is a considerable reduction in the reaction efficiency (<0.7), presumably as a result of remnants of inhibiting impurities.

Table 4.2. Number of cycles that yield 10 ng of a 200-bp amplification product

	Number of targets (A)					
	1	10	100	1,000	10,000	100,000
Efficiency (F)	Number of amplification cycles (*n*)					
1.00	34	30	27	24	20	17
0.90	36	33	29	26	22	18
0.80	40	36	32	28	24	20
0.70	44	40	35	31	27	22
0.60	49	45	40	35	30	25
0.50	57	52	46	40	35	29
0.40	69	62	55	48	42	35

4.3 Cycling parameters

One of the most important factors for the success of a PCR amplification reaction is the temperature for the different cycling phases. In contrast, the durations of the actual amplification stages influence the result to a lesser extent.

4.3.1 Annealing temperature

It is common knowledge that the annealing temperature is a fundamental PCR parameter. It needs to be carefully adjusted for each set of primers. Depending on the length and the individual base-pair composition of the primers, the annealing temperature will range between 50–60°C in most cases. The optimal annealing temperature is a compromise between the hybridization specificity of the actual primer and a full reaction efficiency. In order to determine the optimal temperature that accounts for both of these parameters, it is necessary to carry out experiments over a series of different temperatures for any new primer set (cf. section 4.4). This should also be done if primers and protocols are taken directly from publications on modern DNA analysis. Often, these protocols succeed at very low annealing temperatures, which is disadvantageous when specific requirements, such as high cycle numbers, are needed.

4.3.2 Denaturation temperature

In the vast majority of PCR protocols, little attention has been paid to the denaturation temperature, which is usually 94°C. This temperature will work well in most cases, because 90–92°C will fully denature any DNA sequence that has a balanced GC/AT content (Saiki 1989). However, analysis failures are predictable in any PCR machine that uses a glue to transmit temperature from the heating appliance to the metal block where the reaction tubes are placed. This hardware problem is a feature of many PCR machines. Although these glues are heat resistant, significant degradation occurs, leading to considerably lower temperatures in the metal block than the ones displayed. In such cases, GC-rich sequences will no longer be completely denatured, and the amplification will fail at least partially (Fig. 4.2).

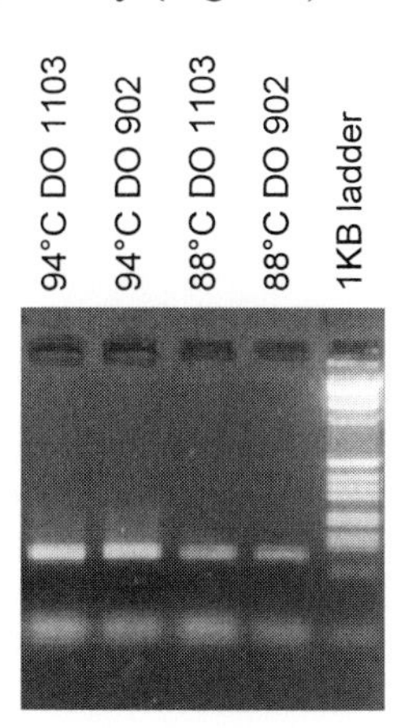

Fig. 4.2. Partial amplification failure due to a denaturation temperature of only 88°C. If the same amplification is carried out at 94°C for denaturation but with otherwise constant parameters, there are remarkably more amplification products. The lesser amount of products at a denaturation temperaure of 88°C indicates an ineffective amplification resulting from incomplete denaturation of the GT-rich product

What makes this particular reason for an amplification failure so difficult to determine is the fact that other reactions that amplify sequences with lower GC content may still work perfectly well. Commonly, one is not aware of the GC content of the sequence amplified. Therefore, and since there are so many other reasons why an aDNA amplification might fail, this hardware problem is often overlooked as a possible source of amplification failure, although it will certainly occur at some stage. In order to control both the denaturation as well as the annealing/elongation temperatures, it is advisable to check the profiles of the PCR cycler regularly using an external high-precision thermometer.

4.3.3 Elongation temperature

Just as little focus has been given to the elongation temperature as to the temperature in the annealing and denaturing steps. This is probably understandable in most cases, except in the amplification of certain short tandem repeat (STR) sequences (cf. section 2.2.2). There is an indication that lowering the elongation temperature reduces stutter bands, a common STR artifact[9] (Fig. 4.3).

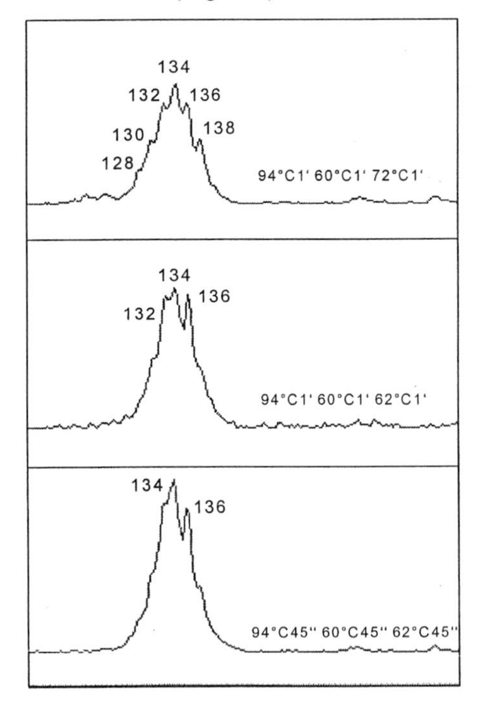

Fig. 4.3. Reduction of stutter bands in a dinucleotide amplification by lowering the elongation temperature to 62°C

[9] The practical experiments were carried out by Dr. Odile Loreille and were presented at the 5th International Ancient DNA Conference in Manchester, UK, July 2000.

The explanation for this lies in the short motif itself. As a result of the different energy profile of the repeat structure (cf. sections 2.2.2 and 4.4), it tends to early denaturation ("waving") during the elongation phase at 72°C. This phenomenon increases stutter products because of an incorrect hybridization in the renaturing phase (section 2.2.2). The aim of exposing the amplification reaction to a lower elongation temperature is that early denaturing is reduced in the repeat motif of short-tandem repeat sequences (Su et al. 1996).

4.3.4 Durations of the cycling steps

The majority of PCR protocols that amplify fragments of less than 2 kb use a 1-min time span for each of the three amplification phases. However, so-called "booster-PCR" and "two-step" protocols have demonstrated that already much shorter time spans of down to 1 s at least for the annealing and the elongation steps are sufficient to generate products (e.g., Ruano et al. 1989; Wittwer et al. 1994). Our own experiments amplifying a single-copy sequence from aDNA extracts with time intervals of all temperature steps reduced to 15 s revealed only very slight differences in the amount of product that is generated (Fig. 4.4). Positive effects of the reduction of the time intervals are possibly increased reaction specificity (e.g., Odelberg and White 1993; Swerdlow et al. 1993; Wittwer et al. 1994), the time-saving aspects and the elongated Taq-polymerase activity. It is therefore possible to run more cycles on the same amount of Taq polymerase, or, alternatively, fewer units of Taq polymerase can be used without reducing the overall enzymatic activity. Of course, this latter advantage can also be achieved by reducing the duration of the denaturing time only, which does not affect the efficiency of the reaction as soon as the exponential phase is reached when the majority of target sequences are already short products (cf. Fig. 4.1).

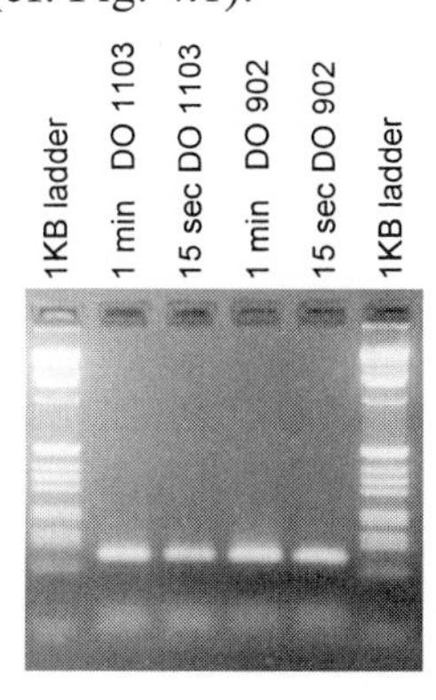

Fig. 4.4. Amplification of an exon-6 sequence relevant for ABO genotyping from Bronze Age aDNA extracts. Compared are amplification products that were generated by 45 cycles but with different durations of the cycling steps (*1min* all cycling steps carried out for 1 min; *15 sec* all cycling steps carried out for 15 sec). As in this case, often only very few differences in the amount of amplification product can be observed if the amplifications are carried out using short cycles. This enables savings of Taq polymerase and time

4.4 Primer design

The primer sequences determine which particular sequence of a human, animal, plant or microorganism genome will be targeted in a PCR run. Most important for a high quality PCR is the specificity of the reaction. This is possible if the primers target a particular locus exclusively throughout the entire genome, which may be present either as a single-copy or multi-copy sequence. Primarily, the specificity of the reaction is ensured when the primer sequences are a complete match with the intended hybridization site and are unique within the genome in focus. This should be checked by a "blast search" in GenBank.

A closer examination reveals that there are additional factors that may have a strong impact on the efficiency and specificity of the reaction (cf. section 4.1) (e.g., Dieffenbach et al. 1993; Brownie et al. 1997) and that are therefore also essential for a good primer design. The most important demands for a good primer design are listed stepwise in Table 4.5, while the subsequent sections of this chapter provide the background information.

Table 4.5. Requirements of good primer design

Requirement	Aim
• Compare multiple sequences of a targeted amplification region by alignment, if possible	• Avoids accidentally building in mismatches because of possible sequence polymorphisms
• Avoid any mismatch site	• Optimizes reaction efficiency
• If a mismatch is unavoidable, place it as close as possible to the 5'-end	• Avoids amplification failure, although reaction efficiency may be decreased
• Intended mismatches must be placed at the 3'-end	• Effective discrimination against unwanted sequence amplification (e.g., contaminating DNA)
• 5'-end binding should be strong, 3'-end binding weak	• Optimizes reaction specificity and thereby also efficiency
• Annealing temperatures of primers should not differ significantly (<2°C)	• Optimizes reaction efficiency
• Avoid primer dimers and hairpins	• Optimizes reaction efficiency

4.4.1 Matching primers

It does not need to be explained further that ideal primers should fully match the intended hybridization sites (Fig. 4.5). However, this may not be quite so simple to achieve, since there are a number of other factors that need to be observed for good primer design (cf. sections 4.4.2 and 4.4.3) and that may complicate the process. In particular, this applies when primers are to be designed in sequence regions that reveal many sequence polymorphisms, e.g., when designing primers for species identification using the mitochondrial cytochrome-b sequence (cf. section 2.1.2, Fig. 2.7). However, even when primers are designed for more conserved sequences, such as nuclear DNA, there is a possibility of accidentally building mismatches into the primers. This may be the situation when some individuals reveal polymorphic sites within the hybridization site. Apparently, this occurs more often than expected, as empiric observations show, for example, in the STR amplification of intact modern DNA with validated PCR kits (e.g., LaFountain et al. 2001). In spite of all standardized conditions, the relative amplification product yields within the multiplex assays may vary considerably among different individuals. This phenomenon is commonly explained by polymorphic sites neighboring the repeat sequences which affects the effective hybridization of the primers.

Fig. 4.5. A fully matching primer and a 5'-end mismatch primer. Mismatches at the 5'-end do not affect the reaction efficiency

In order to minimize the risk of accidentally building in mismatches to a primer, one strategy is to download a couple of matrix sequences from GeneBank that originate from different individuals and to check for possible sequence polymorphisms in the intended primer design areas through alignment. The other possible strategy is to carry out the primer design first and then later align the primer to several other sequences. If the directions for internal primer energy profiles (sections 4.4.2 and 7.4.3) are followed and if a mismatch site is unavoidable, it should be located as closely as possible to the 5'-end in order not to affect the reaction efficiency (Fig. 4.5). In contrast, mismatch locations at the 3'-end may even lead to complete amplification failure (sections 4.4.4 and 7.2.7).

4.4.2 Primer energy profile

The major impact of the internal primer energy profile on the reaction specificity lies in the hybridization stabilities of both the 5'- and the 3'-end of a primer at a given annealing temperature. If the hybridization stability of the 3'-end substantially exceeds that of the 5'-end, the probability increases that the primer binds with its 3'-end at a site other than the targeted sequence and initiates elongation from this wrong site. Once this has happened, the unspecific product will serve as a perfect template in further cycles. Therefore, as a rule of thumb and different from earlier recommendations for primer design (Lowe et al. 1990; Dopazo and Sobrino 1993), the binding capacity of the 3'-end should be weak and that of the 5'-end should be strong (Fig. 4.6). This ensures that elongation can only be initiated if the entire primer sequence matches the hybridization site.

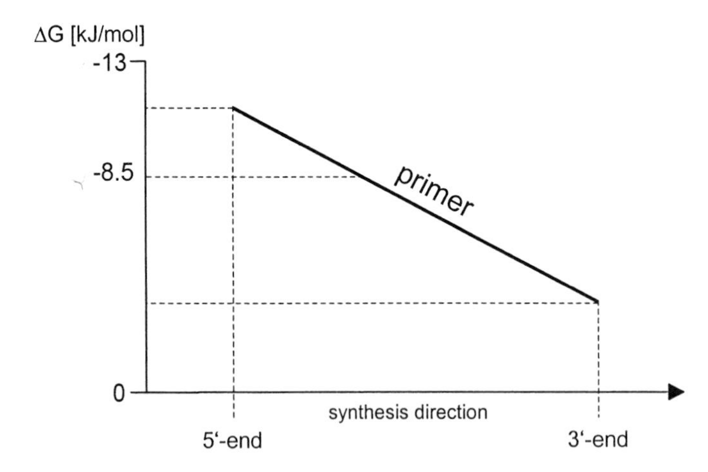

Fig. 4.6. Energy profiles of primers should enable stable binding of the 5'-end and weaker binding of the 3'-end. This avoids generation of unspecific amplification products. ΔG represents the energy that must be supplied to denature the primer from the hybridization site. In an optimal primer, ΔG should be large for the 5'-end and small for the 3'-end

In modern intact DNA amplification reactions, the internal energy profiles of the primers do not appear to play such an important role because there are a greater number of target sequences in the beginning of the reaction. The correct elongations simply outnumber the events of mismatch elongations. Therefore, if primers from modern DNA protocols are used, it is advisable to make sure they reveal a suitable internal energy profile. As a by-product of this effort, in most cases it is possible also to reduce the amplified product length at the same time, which always increases the success rate of aDNA amplifications (Fig. 4.7).

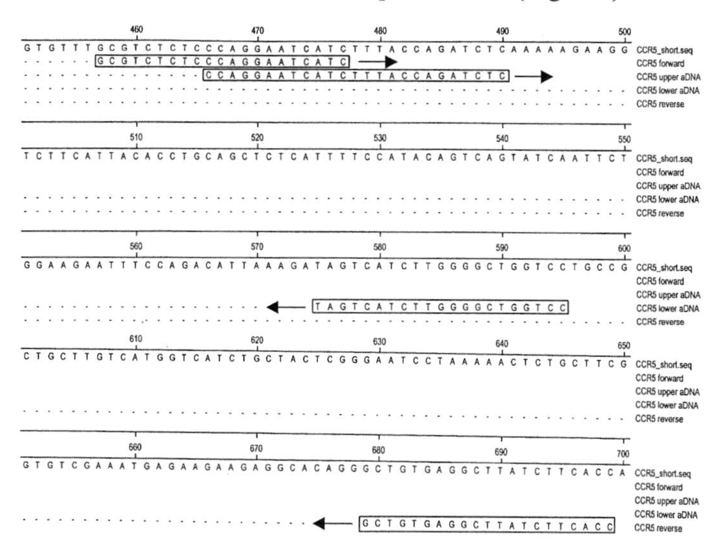

Fig. 4.7. The originally published primers of the CCR5 locus (*CCR5 forward* and *CCR5 reverse*) generated a 248/206-bp fragment and co-amplified unspecific products as a result of suboptimal ΔG values of the primers. Their redesign (*CCR5 upper aDNA* and *CCR5 lower aDNA*) enabled the optimization of the energy profile and at the same time cut down the product length to 130/98 bp, respectively (section 2.2.6). This measurement considerably increased the amplification specificity and efficiency in aDNA applications (section 7.4.4)

4.4.3 Annealing temperature, primer dimer and hairpins

Another advantage of redesigning primers is the possibility also to check the calculated annealing temperatures of the primers. Again, this function is often suboptimal in modern DNA protocols, i.e., the annealing temperatures differ by up to 4–5°C within a primer set. As a result, only the primer with the lowest annealing temperature can run under high stringency conditions, which may also severely affect the specificity of a reaction. Additionally, greatly differing annealing temperatures may lead to an asymmetric assay, i.e., one primer is capable of binding elsewhere due to a suboptimal annealing temperature and will therefore not only generate unspecific, smear-like products, but will also be exhausted earlier.

The early consumption of a primer usually becomes a major problem, particularly in the case of extensive primer dimer formation, self-dimerization and the so-called hairpin formation, which is a particular type of self-dimerization. These reactions basically compete with the specific target sequence amplification. In the cases of dimer and hairpin formations that allow elongations (Fig. 4.8), the impact on the reaction efficiency may be very substantial because of the drastic reduction in the reaction component resources (dNTPs, primers, etc.). In the worst case, dimer formations may totally overwhelm the intended reaction. If primer dimers and hairpins do not enable elongation (Fig. 4.9), they generally do not affect the reaction unless their binding energy is not too high. Therefore, any extensive primer dimer formations and hairpins that allow elongation should be avoided (Dieffenbach et al. 1993; Brownie et al. 1997), although a moderate primer dimer formation may also serve as an indicator if an amplification failure is due to inhibition (cf. section 4.6.2).

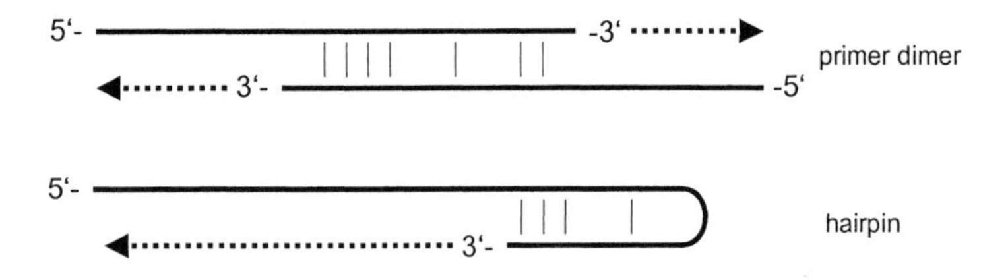

Fig. 4.8. Scheme of primer dimer and hairpin formation. Due to free 5'-ends in the sequence of primer formations, elongations are enabled starting at the 3'-ends. These reactions compete with the specific target elongations, exhausting the reaction component resources

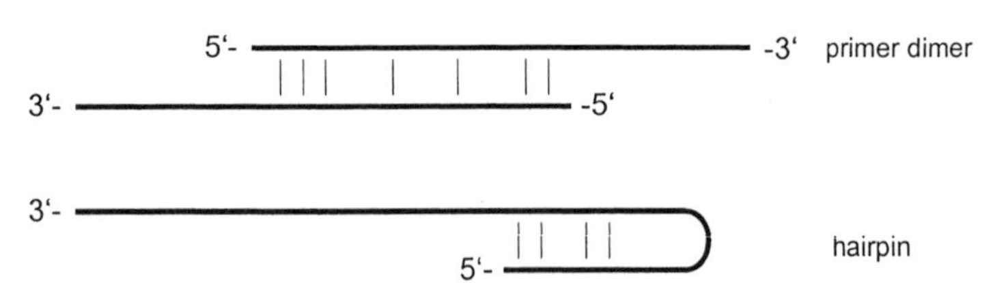

Fig. 4.9. Scheme of primer dimer and hairpin formation. Due to the free 3'-ends of the primer formations, elongations are prevented and therefore do not lead to an early exhaustion of the amplification reaction

4.4.4 Mismatch primers

In general, primers are designed to match perfectly with the complementary position of the target sequence (cf. section 4.4.1). However, an intended inclusion of sequence mismatch sites can be used as a tool to discriminate between the intended target sequence and contaminating or other unwanted DNA (e.g., Kwok et al. 1995). This strategy has proved to be particularly valuable in the amplification of semi-conserved regions of the cytochrome-b genes that are employed to identify species.

The mismatch primer strategy takes advantage of the fact that the elongation procedure will only take place if the 3'-end of a primer hybridizes to the DNA sequence. The insertion of only a few mismatched bases at the 3'-end on either one or both primers next to the target sequence is sufficient to effectively suppress the amplification of unwanted DNA (Fig. 4.10).

Fig. 4.10. Mismatches at the 3'-end effectively suppress the amplification of non-target DNA. This is valuable, e.g., in the amplification of cytochrome-b sequences from animals where the co-amplification of contaminating human DNA can be avoided by this strategy

The amplification of the targeted sequences will not yet be affected and will take place on a level of high specificity. This was shown by a primer design that intended to amplify a cytochrome-b region that enables the discrimination of several mammals by sequence analysis. In order not to co-amplify possible contaminating human sequences that might be present in the reaction tubes (cf. section 6.1.2), the primers were designed with 3'-end mismatches to the respective regions of the human sequence (cf. chapter 6 and section 7.2.7).

4.4.5 Software packages

Primer design software has been a great help and includes checks for balanced annealing temperatures and for possible dimers and hairpins. Thus, the software is extremely valuable for creating singleplex assays, and it is also invaluable for the design of multiplex PCR (cf. section 4.5). Software packages that support all facets of primer design are numerous, and some can even be downloaded from the internet. Most of the primer sequences published in this book (cf. section 8.5) have been designed with the assistance of *PrimerSelect*, *EditSeq* and *MegAlign*, which are components of the *DNAStar* software package (Lasergene). *PrimerSelect* supports a large number of criteria for primer design, including a multiplex primer design, checks for internal primer stability, energy profile, primer dimer and hairpin formation and a mismatch primer design. The other parts of the package are necessary to enable downloads of sequences from databases such as GenBank and carry out alignments that must be performed when, for example, sequence polymorphisms are the focus of interest (http://www.ncbi.nlm.nih.gov/Genbank/index.html).

Information on software packages that support primer design and alignment tasks can be found at: http://gsu.med.ohio-state.edu/ppt/primer_design/index.htm.

4.4.6 Primer design – step by step

The purpose of this chapter is to show an illustrated example of how to amalgamate all of the previous sections from theory to practice.

(1) The chromosome, chromosomal region or mitochondrial region of the intended sequence should be known. This information is used as the keywords for searching a database, e.g., GenBank. (In this example, we knew that we were looking for the X-chromosomal STR DXS6789). Furthermore, it may be very helpful if nucleotide positions of possible sequence polymorphism are known, or, as in our example, the primer sequences of modern DNA amplifications.

(2) Begin the GenBank search with the respective keywords and try to check as many of the accessions numbers as possible that claim to refer to the intended sequences.[10] This check can only be done by downloading a number of sequences and aligning them in order to discover a suitable

[10] Checking a number of sequences should be done whenever possible, since anyone can place sequences at GenBank, and there is no review-like type of control. This means that incorrect sequences are found in the database as well.

sequence, which will serve as the reference sequence for primer design. Figure 4.11 illustrates an accession site in GenBank.

```
1: G08105. human STS CHLC.GA...[gi:938655]
LOCUS       G08105      330 bp    DNA          STS       05-FEB-1997
DEFINITION  human STS CHLC.GATA31F01.P14964 clone GATA31F01, sequence tagged
            site.
ACCESSION   G08105
VERSION     G08105.1  GI:938655
KEYWORDS    STS; STS sequence; primer; sequence tagged site.
SOURCE      human vector=pJCP1 host=E.coli dut+ung+ (DH10B) Marker Selected
            genomic DNA prepared from XY individual of French nationality.
  ORGANISM  Homo sapiens
            Eukaryota; Metazoa; Chordata; Craniata; Vertebrata; Euteleostomi;
            Mammalia; Eutheria; Primates; Catarrhini; Hominidae; Homo.
REFERENCE   1  (bases 1 to 330)
  AUTHORS   Murray,J., Sheffield,V, Weber,J.L., Duyk,G. and Buetow,K.H.
  TITLE     Cooperative Human Linkage Center
  JOURNAL   Unpublished
COMMENT     Synonyms: GATA31F01, CHLC.GATA31F01.#T14877
            GDB: G00-364-256
            Contact: Dr. Jeffrey C. Murray
            UofI
            The University of Iowa
            Department of Pediatrics, Iowa City, IA 52242, USA
            Tel: (319) 356-3508
            Fax: (319) 356-3347
            Email: jeff-murray@uiowa.edu

            Primer A: TTGGTACTTAATAAACCCTCTTTT
            Primer B: CTAGAGGGACAGAACCAATAGG
            STS size: 149
            PCR Profile:
                    denature:   30 seconds at 94 degrees C
                    annealing:  75 seconds at 55 degrees C
                    extension:  15 seconds at 72 degrees C
                    PCR cycles: 27
                    extension:  6 minutes at 72 degress C
            Protocol:
                    Template:      30ng genomic DNA
                    Primer:        each 1.5 pmole
                    dNTPs:         each 200 uM
                    Taq Polymerase:  0.3 units
                    Total Vol:     10 ul

            Buffer:
                    MgCl2:  1.5mM
                    KCl:    50mM
                    Tris:   10mM
                    pH:     8.3.
FEATURES             Location/Qualifiers
     source          1..330
                     /organism="Homo sapiens"
                     /db_xref="taxon:9606"
     STS             3..151
     primer_bind     3..26
     primer_bind     complement(130..151)
BASE COUNT      97 a     62 c     43 g    126 t      2 others
ORIGIN
        1 agttggtact taataaaccc tcttttatct atgtatgtat gtatgtatgt atgtatgtat
       61 gtatgtatgt atctatctat ctatctatct atctatctat ctatctatct atctatctac
      121 catctatctc ctattggttc tgtccctcta gggatccctg actaatacaa taggacatca
      181 aataacttct tagatctcct aatttttttg ccaacaacca cagtgattgg ctatcttttc
      241 gatttttttt accccagcac ataatgaagt ttgcaatcat ttagtaaaag ttaaataant
      301 gtaaaagggg aagnaaaacc accttggaat
//

Revised: October 24, 2001.
```

Fig. 4.11. Accession site of GenBank (http://www.ncbi.nlm.nih.gov/Genbank/index.html)

(3) Downloading of sequences, which can be found at the bottom of GenBank accession sites, is done by a simple copy-and-paste procedure. Through this procedure, the sequence, which is a (.txt)-format, is imported

directly into *EditSeq*, where it will become a (.seq)-format when saved there.

(4) With a search/find option, you can now locate your region of interest more precisely. For example, let the program find your primer sequences or the repeat structure your are interested in. Delete those parts of the sequence you are sure not to need for your primer design. Save the remaining sequence in order to generate the (.seq)-format.

(5) Import the sequence to *PrimerSelect*. Enter both original primer sequences that are known from the modern DNA publication. If no original primer sequences are known, just take some sequence parts flanking your region of interest as if these were the primers. This will initiate the primer design options of the program.

(6) Match the primer sequences (or the makeshift primer sequences) to the matrix sequence, which is visualized by the program as shown in Fig. 4.12.

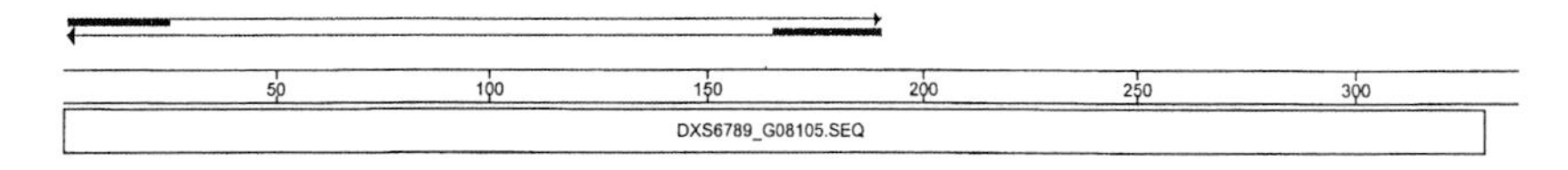

Fig. 4.12. The program recognizes the primer sequences as part of the entire sequence and visualizes this by marking the hybridization sites

(7) Check the energy profile of the original primers or the randomly chosen sequences that just served as makeshift primers. If the situation looks like the one demonstrated for the lower primer in Fig. 4.13, the primer placement needs to be improved.

(8) The lower primer in Fig. 4.13 shows exactly the reverse energy profile (weak 5'-end and strong 3'-end) of what a well-designed primer should have (cf. chapter 4.4.2). From the profiles of the neighboring sequences, the region about 35 bp to the left (3' direction) appears more suitable.

(9) Moving the primer to another position is done in a sub-program, which is displayed on the screen as shown in Fig. 4.14. This sub-program already shows the most stable self-dimer and hairpin formation. Additionally, the respective endonucleases for possibly intended restriction sites are displayed.

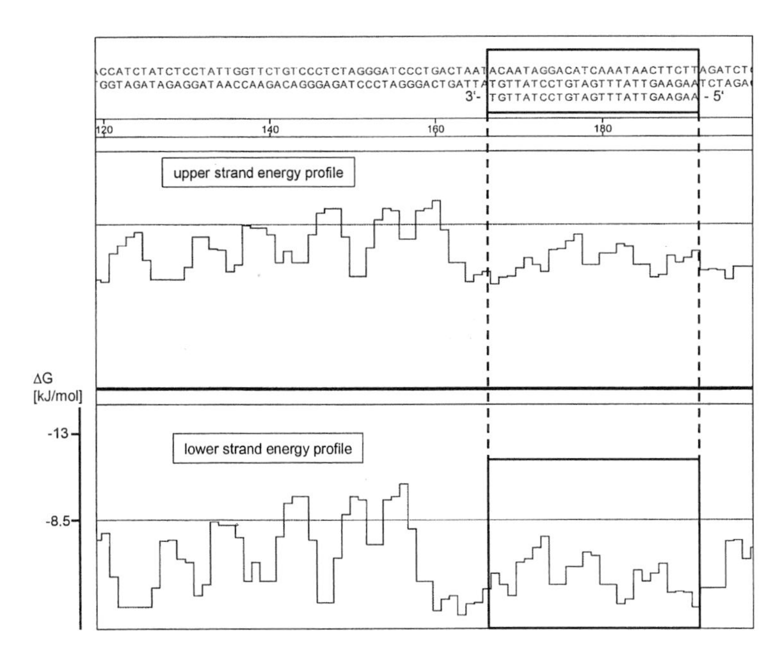

Fig. 4.13. The lower primer has almost the reverse energy profile to that recommended for good primer design. This primer would most likely create unspecific products beside the targeted product sequence

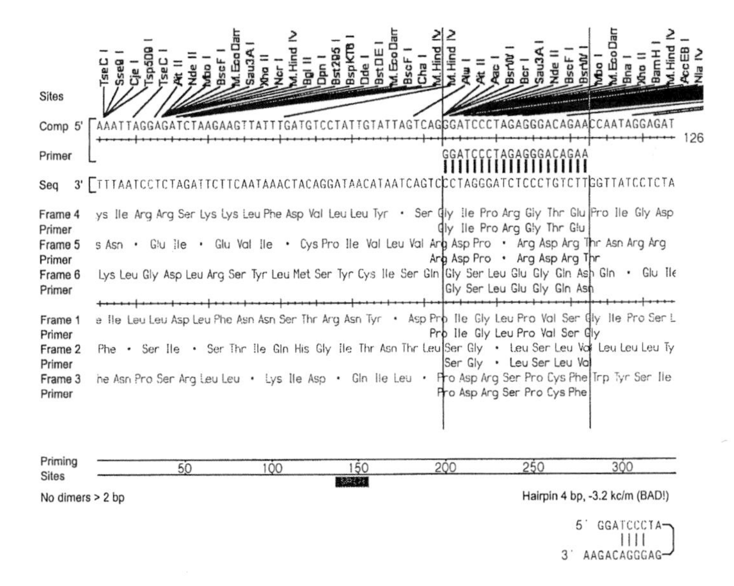

Fig. 4.14. Subroutine enabling the movement of primers. The possible hairpins and self dimers of the primers are given at the bottom of the display

(10) In this case, the hairpin and the most stable self primer should not interfere with good amplification results, since the 3'-ends are free, i.e., the primer cannot anneal to itself as a template for elongation. At the same time, the energy profile of this primer should enable highly specific products (Fig. 4.15).

(11) If this procedure was carried out for both primers with satisfying results concerning the amplification product length, the energy profiles of the primers and the annealing temperatures, the primers should be saved. A separate sub-routine of the program now enables the user to check for possible primer dimers involving both primers, a task that may be laborious in the case of a multiplex PCR design.

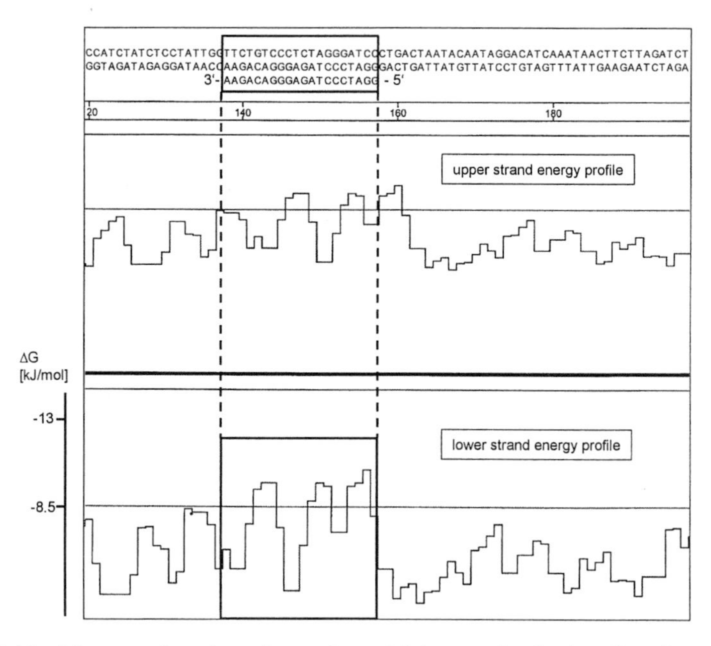

Fig. 4.15. After moving the primer about 35 base pairs in the direction 5'>3', the energy profile promises a stable primer (strong 5'-end and a weaker 3'-end) that does not create unspecific products

(12) Finally, an amplification summary that reveals important data, such as primer sequences, primer locations, product lengths and the minimum annealing temperature for each primer, is produced (Fig. 4.16).

(13) In case of a multiplex primer design, the next sequence can now be entered from Genbank. If all sequences are loaded to the same file, a final cross-check for possible primer dimers for all the possible primer sets is possible.

Upper Primer: DXS6789-upper 25-mer 5' GTTGGTACTTAATAAACCCTCTTTT 3' Lower Primer: DXS6789-lower 20-mer 5' GGATCCCTAGAGGGACAGAA 3'		
DNA 250 pM, Salt 50 mM	Upper Primer	Lower Primer
Primer Tm Primer Overall Stability Primer Location	49.1 °C -42.5 kc/m 2..26	49.6 °C -37.0 kc/m 157..138
Product Tm - Primer Tm Primers Tm Difference Optimal Annealing Temperature	19.6 °C 0.5 °C 47.9 °C	
Product Length Product Tm (%GC Method) Product GC Content Product Tm at 6xSSC	156 bp 68.7 °C 32.1% 90.3 °C	

Fig. 4.16. Amplification summary listing the most important PCR parameters

4.5 Multiplex PCR

Multiplex amplification assays strongly improve data validation in aDNA analysis. This has been used for many years in forensic autosomal STR typing of evidence materials from crime scenes or mass disasters by various commercially available certified and validated amplification kits (e.g., Kimpton et al. 1996; LaFountain et al. 2001; Moretti et al. 2001). The more information (i.e., polymorphism) that is obtained simultaneously from a multiplex assay, the less likely it is that this can happen only by chance (i.e., contamination) in a repeated analysis (cf. section 6.2). Any multiplex analysis is a step towards the individualization of the data sets (Chamberlain and Chamberlain 1994; Edwards and Gibbs 1995; Hummel and Schultes 2000; Hummel et al. 2000).

Using the multiplex approach, it is even possible to fully individualize data that are primarily non-individual, such as sequences that only reveal mutations in some individuals (cf. sections 2.2.6 and 2.2.7). In singleplex assays, this type of analysis would be of no value because of the many risks of contamination entering an aDNA analysis (section 6.1).

Another great advantage of multiplex assays is the saving of sample material. From as little as 1.0 g of bone material, an enormous amount of information can be collected. This includes autosomal and Y-chromosomal STR typing, sex determination and mitochondrial haplotyping, and, if desired, a collection of data about comparatively rare mutational events can be carried out. Furthermore, also sequences that will be the subject of direct sequencing may be amplified and analyzed successfully in a multiplex assay (cf. sections 2.1 and 7.3.3).

A prerequisite for successful multiplex analysis is a good primer design that does not enable unspecific product generation, extensive primer dimer formation and preferential amplification as a result of unbalanced primer characteristics (Chamberlain and Chamberlain 1994; Edwards and Gibbs 1995; Weissensteiner and Lanchbury 1996) (cf. section 4.4). Numerous suggestions, applications and protocols for multiplex assays are found throughout this book.

4.6 PCR failures

PCR failure can be attributed to many reasons in aDNA analysis. Those most likely include DNA degradation or too many remnants of inhibiting substances in the aDNA extract. Much rarer events in aDNA analysis, but also possible reasons for PCR failures, are DNA overcycling and DNA overload. These are discussed below.

4.6.1 DNA degradation

DNA degradation is presumably the most common reason for PCR failure. Most aDNA extracts do not reveal significant amounts of intact targets exceeding 200 bp. One of the reasons is that the internal structure of the DNA molecule reveals about 190 bp within a single helical loop, which obviously is a more stabile subunit than any longer molecule fragments.

The very stabile formation of the 190-bp helical subunit is reflected perfectly in particular by those extracts that contain longer fragments. If they are amplified in a multiplex assay consisting of targets throughout the entire range from about 100–350 bp, a clear decrease of product size can be recognized for all those fragments exceeding 200 bp (Fig. 4.17). For the aDNA extract, this indicates that shorter intact targets were represented numerously, while the longer ones were rare. This effect is also reflected by allelic dropouts (section 2.2.2), which occur much more commonly in long amplification products than in the shorter ones.

Unfortunately, the fragment lengths of the indigenous sequences in an aDNA extract cannot be determined prior to amplification. UV-spectrophotometry only measures the nucleic acid content and does not refer to fragment lengths. However, the lengths can be roughly estimated by electrophoresis of aliquots of the extract (Fig. 4.18), although DNA originating from microorganisms may severely bias the evaluation. Therefore, often there is no way, except by the PCR amplification itself, to find an answer to this question. In general, primer design should try to enable products that are as short as possible and do not exceed 200 bp.

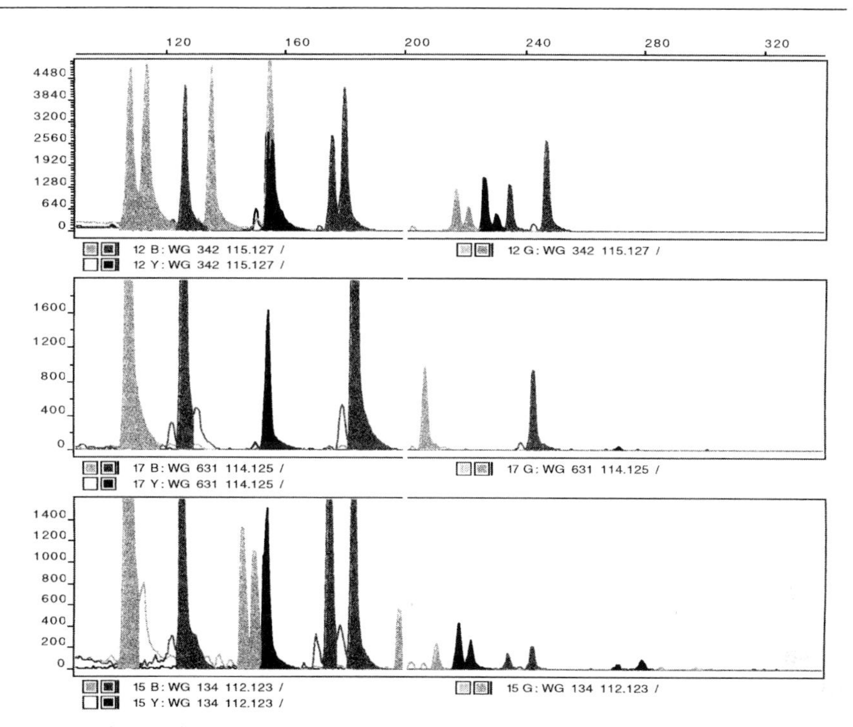

Fig. 4.17. Electropherograms of multiplex autosomal STR amplifications of three early medieval specimens (WG 342, WG 631 and WG 134), revealing significantly lower peak heights for fragments exceeding 200 bp. This indicates numerous short aDNA fragments (<200 bp) and comparatively few longer intact aDNA targets (>200 bp)

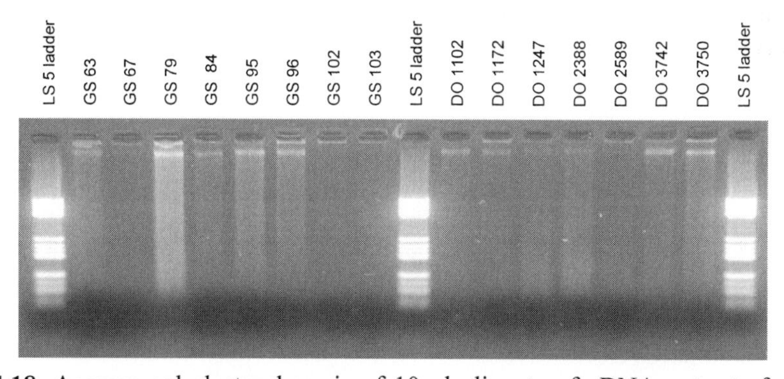

Fig. 4.18. Agarose gel electrophoresis of 10-µl aliquots of aDNA extracts from specimens that are approximately 200 years old (GS) and 3,000 years old (DO). From this analysis, it cannot be concluded whether the DNA is indigenous or of microbial origin (cf. also section 3.3.2). In general, DNA with a high molecular weight is most likely a result of microorganisms. However, the samples shown here revealed excellent results in autosomal STR multiplex assays (cf. Fig. 4.17), suggesting that even a smaller portion of the larger fragments are indigenous DNA

4.6.2 Inhibition

Probably the second most important reason for amplification failure in aDNA analysis is remaining inhibitors. If a PCR failure is suspected to be due to an excess of inhibitors, this can be clearly recognized by the lack of any primer dimers, lack of smears or any other unspecific amplification products. If any one of those is present, it is not possible that major inhibition caused the amplification failure.

Most commonly, inhibiting substances come from soil, which contains humic acids, tannins, maillard products and fulvic acids to various extents. All of these components derive mainly from decomposing plant material and inhibit the Taq polymerase activity. The actual amounts of each of the chemically heterogenic groups in a particular environment depend on such various factors as the specific plants growing in an area, pH values, temperatures and moisture. In an extracted aDNA sample, the inhibitors are recognized by yellow to dark brown stainings of the sample materials (e.g., Pääbo 1990; Cooper 1992) and by specific characteristics, such as in UV spectrophotometric diagrams (Tuross 1994). They fluoresce in a bright turquoise color on agarose gel when illuminated by 254-nm UV light. It is not even necessary to stain the gel with ethidium bromide (Fig. 4.19) (cf. also Hänni et al. 1995a). In contrast, if the extract also contains comparatively large amounts of strongly degraded DNA, it could be recommendable not to stain the gel in order not to cover the fluorescence of the inhibiting substances that usually migrate through the gel along with 20–60 bp DNA fragments. Another possible potent inhibitor, collagene I, may stem from the bone sample material itself as was shown by Scholz et al. (1998) in experimental assays. Collagen makes up a major portion of the organic fraction of the bone material and is recognized on unstained agarose gels as an autofluorescent smear migrating along with DNA of about 500–1,000 bp. However, since collagen is not soluble in aqueous solutions, it will only be co-extracted if an extraction protocol is used that includes the bone powder fraction for further processing after the initial EDTA extraction step (e.g., Hagelberg and Clegg 1991; Hänni et al. 1995b; Scholz et al. 1998). The protocol of Baron et al. (1996) (cf. sections 3.1 and 8.3) discards the bone powder fraction and thereby the still insoluble collagen.

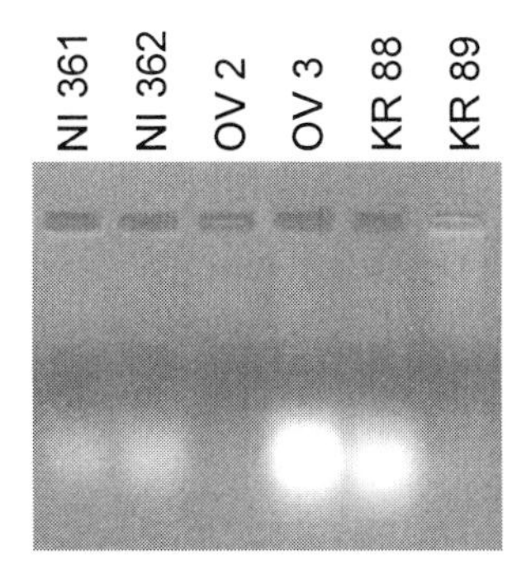

Fig. 4.19. An unstained agarose gel clearly shows the inhibiting substances that remained in the aDNA extracts from ancient bones from three different burial sites. These fluoresce at a turquoise color at 254-nm UV light. If the gel had been stained with ethidium bromide, the DNA fluorescence might cover the fluroscence of the inhibitors

In order to enable amplification, it may be sufficient just to dilute the aDNA extract either by adding some more water or buffer or by using smaller aliquots in the actual amplification reaction.

In case of larger amounts of soil-derived inhibiting substances, an additional clean-up of the extract should be carried out. This can be done either by a new precipitation, e.g., with another type of alcohol than the one used in the original precipitation and including silica particles or by spin columns (e.g., Wizard Prep, Promega Inc.).

Alternatively, there are most promising attempts to quantitatively remove inhibitors from aDNA extracts by polyvinylpolypyrrolidone (PVPP), which is a water- insoluble polymer that complexes polyphenols and tannins. The use of PVPP in order to remove soil-derived inhibitors from DNA extracts proved to be successful in environmental microbiology (Berthelet et al. 1996; Zhou et al. 1996; Brunner et al. 2001; Chiou et al. 2001). Our own initial experiments showed that humic acids that were intentionally subjected to positive controls indeed were removed quantitatively by a PPVP treatment, as was shown by gaining back the full amplification efficiency (Krebs 2000).

4.6.3 DNA overload and overcycling

Another possible cause of amplification failure, which is often not even considered in aDNA analysis, is DNA overload in the reaction mix. Examples of this are given in section 3.2. This also explains why it is recommended to determine the DNA yield in case of better-than-average aDNA preservation in a series of samples.

In contrast to inhibition, DNA overload usually results in smears throughout the gel, often not looking very much different from the agarose gel electrophoresis of an aDNA extract. Since DNA overload can also occur because of aDNA extracts that contain only fragmented sequences of less than 100 bp, for example, it does not necessarily mean that the target is present intact in case of a DNA overload in the amplification reaction, although this will usually be thought to be the case. If overload was the reason for the amplification failure, then the reaction did not reach the exponential phase, i.e., the phase in which the short product is amplified very effectively. This is due to the fact that the primers had already been exhausted by generating the semi-long product. If primers are in short supply, the semi-long products will start to prime themselves, which results in very long amplification products, creating the smear. In such cases, the specific product can than be amplified by much less extract in the reaction (cf. section 3.2, Figs. 3.5 to 3.7) or, alternatively, the amount of primer used in the reaction can be increased.

Another phenomenon that may be mixed up with an amplification failure because of DNA overload is overcycling of the reaction. A reaction that has been run for too many cycles also will show smears, but often the original products can still be seen. In this case, running fewer cycles will help (Fig. 4.20). For both phenomena, also refer to Fig. 3.4 in section 3.2.

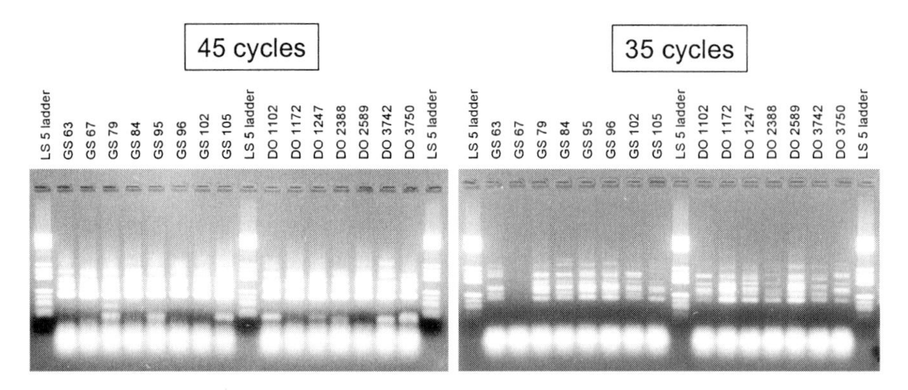

Fig. 4.20. If too many cycles were carried out, the amplification products will show intensive smears (*left*). A reduction of the cycle number considerably improves the results and the following fragment length detection. In the presented case, autosomal STR multiplex amplifications were carried out using 45 and 35 cycles, respectively. The aDNA extracts came from early modern bone samples (*GS*) and aDNA extracts of Bronze Age bone samples (*DO*)

References

Baron H, Hummel S, Herrmann B (1996) *Mycobacterium tuberculosis* complex DNA in ancient human bones. J Archaeol Sci 23:667–671

Berthelet M, Whyte LG, Greer CW (1996) Rapid, direct extraction of DNA from soils for PCR analysis using polyvinylpolypyrrolidone spin columns. Microbiol Lett 138:17–22

Brownie J, Shawcross S, Theaker J, Whitcomb D, Ferrie R, Newton C, Little S (1997) The elimination of primer-dimer accumulation in PCR. Nucleic Acids Res 25:3235–3241

Brunner I, Brodbeck S, Buchler U, Sperisen C (2001) Molecular identification of fine roots of trees from the Alps: reliable and fast DNA extraction and PCR-RFLP analyses of plastid DNA. Mol Ecol 10:2079–2087

Cha R, Thilly WG (1993) Specifity, efficiency, and fidelity of PCR. PCR Meth Appl 3:S18–S29

Chamberlain JS, Chamberlain JR (1994) Optimization of multiplex PCRs. In: Mullis KB, Ferré F, Gibbs RA (eds) PCR. Polymerase chain reaction. Birkhäuser, Boston

Chiou FS, Hsu YP, Tsai CW, Yang CH (2001) Extraction of human DNA for PCR from chewed residues of betel quid using a novel "PVP/CTAB" method. J Forensic Sci 46:1174–1179

Cooper A (1992) Removal of colourings, inhibitors of PCR, and the carrier effect of PCR contamination from ancient DNA samples. Ancient DNA Newsletter 1/2:31–32

Dieffenbach CW, Lowe TMJ, Dveksler GS (1993) General concepts for PCR primer design. PCR Meth Appl 3:S30–S37

Dopazo J, Sobrino F (1993) A computer program for the design of PCR primers for diagnosis of highly variable genomes. J Virol Meth 41:157–166

Edwards MC, Gibbs RA (1995) Multiplex PCR. In: Dieffenbach CW, Dveksler GA (eds) PCR primer. A laboratory manual. Cold Spring Harbour Laboratory Press, New York

Erlich HA (ed) (1989) PCR technology. Principles and applications for DNA amplification. Stockton Press, New York

Hagelberg E, Clegg JB (1991) Isolation and characterization of DNA from archaeological bone. Proc R Soc Lond B 244:45–50

Hänni C, Brousseau T, Laudet V, Stehelin D (1995a) Isopropanol precipitation removes PCR inhibitors from ancient bone extracts. Nucleic Acids Res 23:881–882

Hänni C, Begue A, Laudet V, Stéhelin D (1995b) Molecular typing of Neolithic human bones. J Archaeol Sci 22:649–658

Hummel S, Bramanti B, Schultes T, Kahle M, Haffner S, Herrmann B (2000) Megaplex DNA typing can provide a strong indication of the authenticity of ancient DNA amplifications by clearly recognizing any possible type of modern contamination. Anthrop Anz 58:15–21

Hummel S, Schultes T (2000) From skeletons to fingerprints – STR typing of ancient DNA. Ancient Biomol 3:103–116

Innis MA, Gelfand DH, Sninsky JJ, White TJ (eds) (1990) PCR protocols. Academic Press, San Diego

Innis MA, Gelfand DH, Sninsky JJ (eds) (1995) PCR Strategies. Academic Press, San Diego

Kimpton CP, Oldroyd NJ, Watson SK, Frazier RR, Johnson PE, Millican ES, Urquhart A, Sparkes BL, Gill P (1996) Validation of highly discriminating multiplex short tandem repeat amplification systems for individual identification. Electrophoresis 17:1283–1293

Krebs O (2000) Extraktion von aDNA aus stark inhibierten Proben durch HPLC. Diploma thesis, University of Göttingen

Kwok S, Chang SY, Sninsky JJ, Wang A (1995) Design and use of mismatch and degenerate primers. In: Dieffenbach CW, Dveksler GS (eds) PCR primer. A laboratory manual. Cold Spring Harbour Laboratory Press, New York

LaFountain MJ, Schwartz MB, Svete PA, Walkinshaw MA, Buel E (2001) TWGDAM validation of the AmpFlSTR Profiler Plus and AmpFlSTR COfiler STR multiplex systems using capillary electrophoresis. J Forensic Sci 46:1191–1198

Landre PA, Gelfand DH, Watson RM (1995) The use of cosolvents to enhance amplification by the polymerase chain reaction. In: Innis MA, Gelfand DH, Sninski JJ (eds) PCR Strategies. Academic Press, San Diego

Lowe T, Sharefkin J, Yang SQ, Dieffenbach CW (1990) A computer program for selection of oligonucleotide primers for polymerase chain reactions. Nucleic Acids Res 18:1757–1761

Moretti TR, Baumstark AL, Defenbaugh DA, Keys KM, Smerick JB, Budowle B (2001) Validation of short tandem repeats (STRs) for forensic usage: performance testing of fluorescent multiplex STR systems and analysis of authentic and simulated forensic samples. J Forensic Sci 46:647–660

Mullis KB, Ferré F, Gibbs RA (eds) (1994) PCR. Polymerase chain reaction. Birkhäuser, Boston

Odelberg SJ, White R (1993) A method for accurate amplification of polymorphic CA-repeat sequences. PCR Meth Appl 3:7–12

Pääbo S (1990) Amplifying ancient DNA. In: Innis MA, Gelfand DH, Sninsky JJ, White TJ (eds) PCR protocols. Academic Press, San Diego

Rameckers, J., Hummel, S., Herrmann, B. (1997) How many cycles does a PCR need? Determination of cycle numbers depending on the number of targets and the reaction efficiency factor. Naturwissenschaften 84:259–262

Ruano G, Fenton W, Kidd KK (1989) Biphasic amplification of very dilute DNA samples via "booster" PCR. Nucleic Acids Res 17:5407

Saiki RK (1989) The design and optimization of the PCR. In: Erlich HA (ed) PCR Technology. Principles and Applications for DNA amplification. Stockton Press, New York

Sardelli AD (1993) Plateau effect – understanding PCR limitations. Amplifications 9:1–5

Scholz M, Giddings I, Pusch CM (1998) A polymerase chain reaction inhibitor of ancient hard and soft tissue DNA extracts is determined as human collagen type I. Anal Biochem 259:283–286

Su XZ, Wu Y, Sifri CD, Wellems TE (1996) Reduced extension temperatures required for PCR amplification of extremely A+T-rich DNA. Nucleic Acids Res 24:1574–1575

Swerdlow H, Dew-Jager K, Gesteland RF (1993) Rapid cycle sequencing in an air thermal cycler. BioTechniques 15:512–519

Tuross N (1994) The biochemistry of ancient DNA in bone. Experientia 50:530–535

Wages JM, Fowler AK (1993) Amplification of low copy number sequences. Amplifications 11:1–3

Weissensteiner T, Lanchbury JS (1996) Strategy for controlling preferential amplification and avoiding false negatives in PCR typing. BioTechniques 21:1102–1108

Wittwer CT, Reed GB, Ririe KM (1994) Rapid cycle DNA amplification. In: Mullis KB, Ferré F, Gibbs RA (eds) PCR. Polymerase chain reaction. Birkhäuser, Boston

Zhou J, Bruns MA, Tiedje JM (1996) DNA recovery from soils of diverse composition. Appl Environ Microbiol 62:316–322

5 Analysis of PCR products

The basic technology required for analyzing PCR products is electrophoresis (Sambrook et al. 1989; Rickwood and Hames 1990). Electrophoresis, which separates DNA fragments according to their size, may either be carried out on slab gels or through capillaries. If slab gels are used, two different carrier substances are available, agarose or polyacrylamide. The method of electrophoresis that is chosen depends on the type of analysis and resolution with which the scientific question is being asked or on the type of answer that is expected (Fig. 5.1).

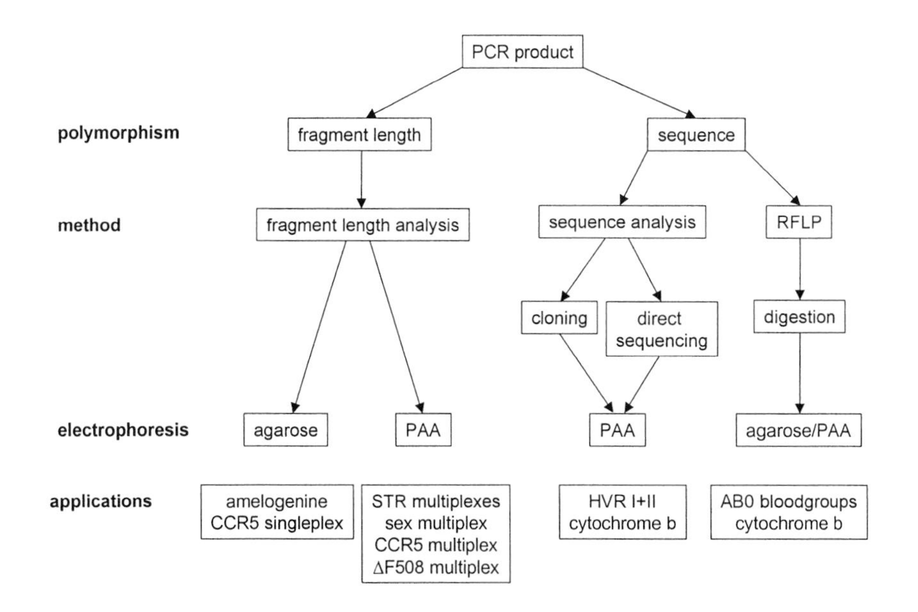

Fig. 5.1. Principles of analysis strategies

The most straightforward case is a binary PCR result, in other words, one in which a particular sequence is either present or absent, for example, screenings for sequences originating from pathogens. In such cases, agarose gel electrophoresis is usually sufficient to detect, interpret and present the results. Agarose gels are also appropriate in situations where the amplification products reveal sufficient length polymorphisms (cf. sections 2.2.1 and 2.2.6). Agarose gel electrophoresis is also suitable for RFLP analysis of amplification products if a sufficient length difference between the restriction products and undigested product is guaranteed.

If multiple-sequence polymorphisms or short-length polymorphisms of 2–6 bp (see sections 2.6 and chapter 7, respectively) are investigated, it is necessary to analyze the amplification products on high-resolution electrophoresis units. To obtain enough resolution to discriminate these fragments, polyacrylamide is used as the separating medium. Both sequence analysis and fragment length analysis can be carried out using automated machines that are supported by a respective analysis software, making the task of interpreting the results considerably easier. In contrast to standardized modern DNA amplifications, aDNA amplifications require agarose electrophoresis prior to further analysis in order to evaluate the amount of amplification products. Based on this pre-check, parameters such as the amount of PCR product that is subjected to further analysis steps can be established.

5.1 Fragment length analysis

A fragment length analysis, or the separation of DNA molecules according to their size, is the most basic concept in all electrophoresis techniques. However, the results will differ in respect to their resolution depending on the separation medium that is used. In case of sequence analysis, introducing additional analysis steps prior to electrophoresis will enable the transformation of fragment length data into base-by-base sequence data.

If the generation of fragment length data is the intended outcome, it will be necessary to run DNA markers along with the samples in the electrophoresis. These DNA markers consist of standardized DNA fragments of known length. For this reason, they serve as reference lengths for comparison with those of the sample. Therefore, in automated fragment-length detection on high-precision electrophoresis units, the DNA markers enable the user to calculate sample fragment lengths with a resolution of one base pair. In contrast to this, in agarose electrophoresis, the estimation of the sample size is more of an elaborated approximation. Although many DNA markers for agarose electrophoresis are available on the market, only a few are suitable for aDNA analysis, providing sufficient fragments between 50–200 bp.

5.1.1 Agarose gel electrophoresis

Agarose gel electrophoresis is a comparatively easy technique (cf. section 8.6). It requires an electrophoresis unit that is horizontally positioned and basically looks like the illustration in Fig. 5.2. The DNA fragments that are to be separated possess a negative charge and therefore migrate towards the anode as soon as a voltage is placed over the unit.

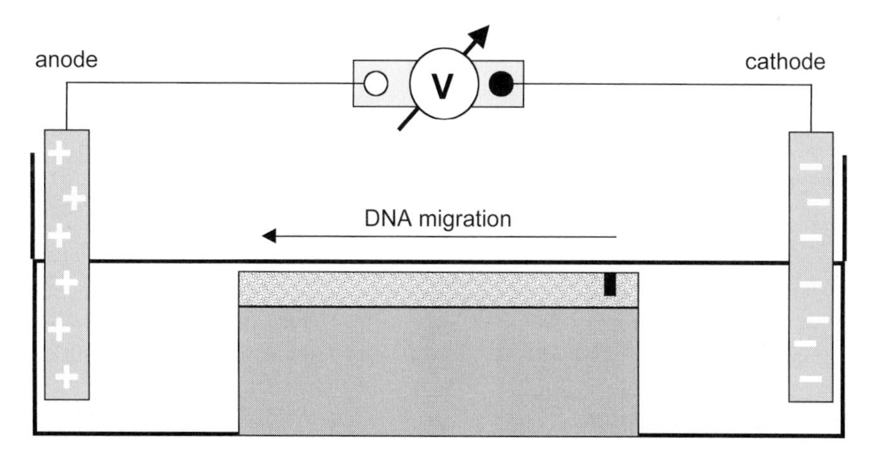

Fig. 5.2. Horizontal electrophoresis unit

Primarily, the relative rates at which molecules migrate through the gel is proportional to their size, i.e., smaller molecules move faster than the larger ones. How much faster the small molecules move compared to the larger molecules can be influenced by several factors. Firstly, there is the percentage of the gel. Lower-percentage agarose gels (e.g., 1%) will cause a similar rate of movement of the small molecules (50–500 bp) through the gel. This results in poor fragment separation of exactly those fragment sizes that are of interest in a DNA analysis. Only high molecular-weight fragments (>500 bp) are separated successfully using low-percentage agarose gels. In contrast, high-percentage agarose gels (3.75–4.4%) almost totally prevent any high molecular-weight DNA from migrating through the gel; however, they separate fragments within the size range of 50–500 bp perfectly. Even such small differences in length as about 5 bp can be resolved on high-percentage agarose gels if the fragments are within the size range of 50–100 bp.

Further factors that influence the fragment separation are the distance the molecules are moved through and the thickness of the gel. Longer distances separate short fragments better than short ones, and thin gels separate better than thick gels (Fig. 5.3). While the first factor is not surprising, the latter factor enables quite a nice "trick." If the gel is poured on a slope, the maximal loading capacity of the wells can be combined with a maximum separation capacity for the portion of small molecules in the distant part of the gel (Fig 5.4).

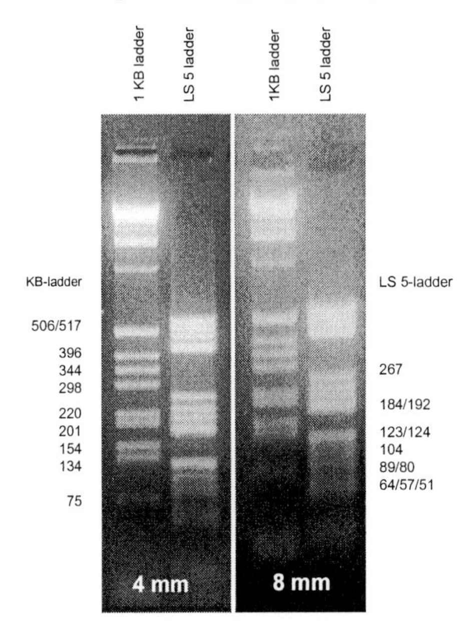

Fig. 5.3. Separation results on 2% agarose gels for two DNA standards. The separation is much increased by a thin gel (*left*, 4 mm) compared to a thick gel (*right*, 8 mm)

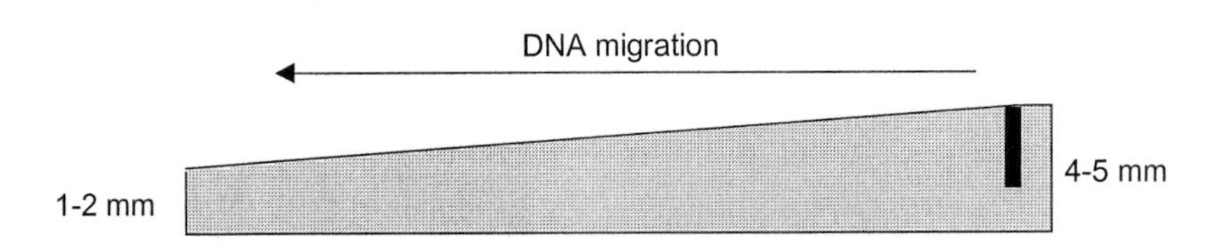

Fig. 5.4. A sloped gel enables the combination of a maximum loading capacity with optimum separation of smaller fragments

Another factor that can enhance the quality of an agarose gel is the layer of buffer covering the gel. The gel should just be submerged to preventing it from drying out. If excess buffer is used, not only does the electrophoresis time increase, but the buffer will also impair the separation of DNA fragments.

If agrose gel electrophoresis is carried out under optimal conditions (Table 5.1), it can be used successfully to separate even down to 5-bp differences (cf. chapter 7), within the size range of interest in aDNA analysis.

Table 5.1. Optimal agarose electrophoresis conditions

Condition	Aim
~1% agarose gel	Not suitable for aDNA analysis
>2% agarose gel	Separates about 30-bp differences in a 50–500 bp range
>4% agarose gel	Separates down to 5–6 bp in the 50–120 bp range
Thin gels, long distance	Enhances fragment separation
No across-gel slopes	Straight molecule moving fronts
Minimum buffer layer	Short electrophoresis time, enhances fragment separation

5.1.2 Automated fragment length determination

Software-based fragment length detection systems are comparatively easy to manage and enable the user to reduce the many chances of producing false data as a result of incorrect use of the technique or the software. They enable very precise size determination, down to differences of 1 bp throughout the entire detection range. The detection range is determined by which DNA marker is used. Due to the number of fluorescent dyes that are available for DNA markers and PCR samples, it is possible to run a standard within each lane to ensures that lane-to-lane differences in the migration rate do not affect the precision of the size determination (Ziegele et al. 1992; Mertes et al. 1997; Butler 2001). However, if very short length differences are expected to occur within and between samples, like those used in the analysis of STR alleles (cf. section 7.2), it is also recommended to use "allelic ladders" (Fig. 5.5). If STR amplification kits are used, the allelic ladders are usually supplied with the kit. In cases where amplifications are performed with STR primers that are either self-designed or adapted from publications, allelic ladders should also be generated. These ladders prevent misleading determinations and considerably speed up allelic determination. Allelic ladders are generated by combining PCR products that consist of the desired alleles. Very small aliquots of these mixes can be amplified further, which does mean that it is not necessary to produce new allelic ladders each time they are required.

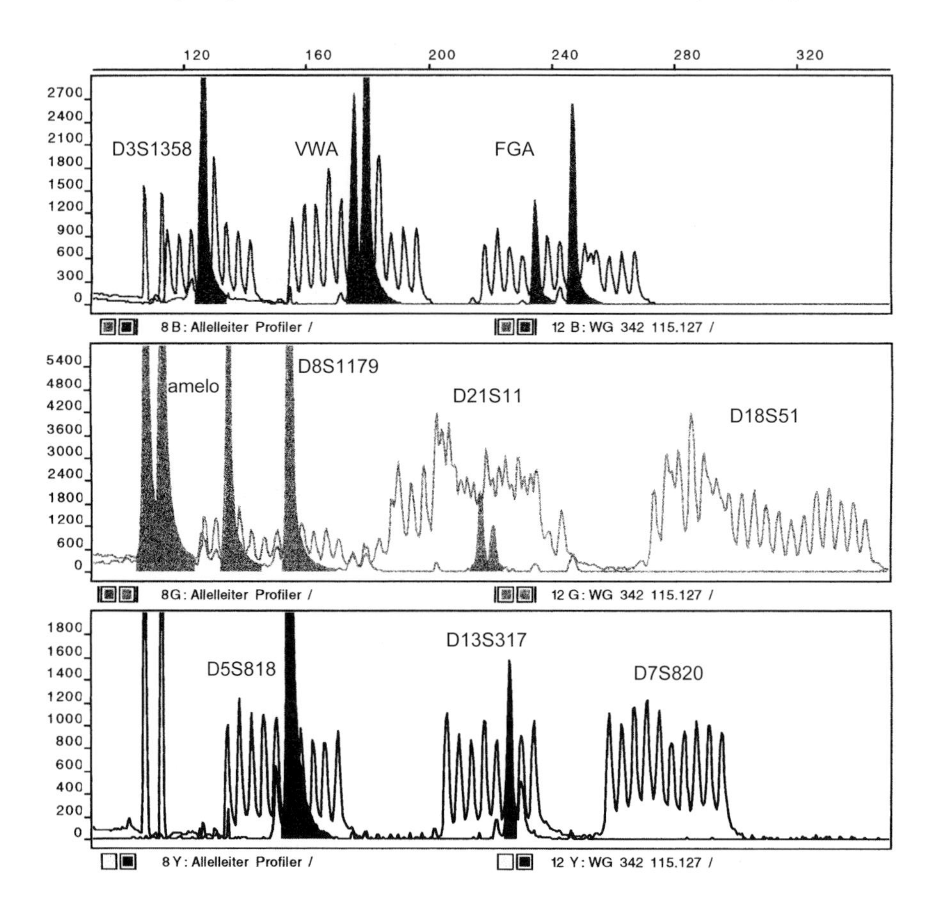

Fig. 5.5. The electropherogram shows the analysis results of a medieval sample (lane 12: WG342, solid peaks) amplified with the AmpFlSTRProfilerPlus kit (Applied Biosystems) and the signals of the allelic ladders of the kit (lane 8: Allelleiter, silhouette peaks). The analysis was carried out on a 373 stretch automated sequencer using the 672 GeneScan analysis software that enables the separate analysis of the three different dye labels (B = blue, G = green and Y = yellow) for each lane, indicating the values for the peak heights (= amount of amplification product) on the vertical scale and the fragment length (= base pairs) at the horizontal scale. The superimposition of the sample analysis results (= lane 8) on the allelic ladder (= lane 12) enables an easy and sound allele determination

Basically, there are only two events that may significantly affect the size-calculation algorithm that leads to false fragment length determinations. Most common is an overload of the PCR product on the gel (Fig. 5.6). Because of a fairly limited range of sensitivity, the software algorithm partially inverts oversize peaks and "decides" that they should be mountain-chain-like formations, which usually results in a whole series of fragment length calculations, none of which truly represent the actual length of the PCR fragment (see Fig. 5.6). This can be overcome simply by taking into account the results of the agarose gel electrophoresis, which is used to determine the appropriate amount of PCR product that should be loaded into the high-performance electrophoresis unit.

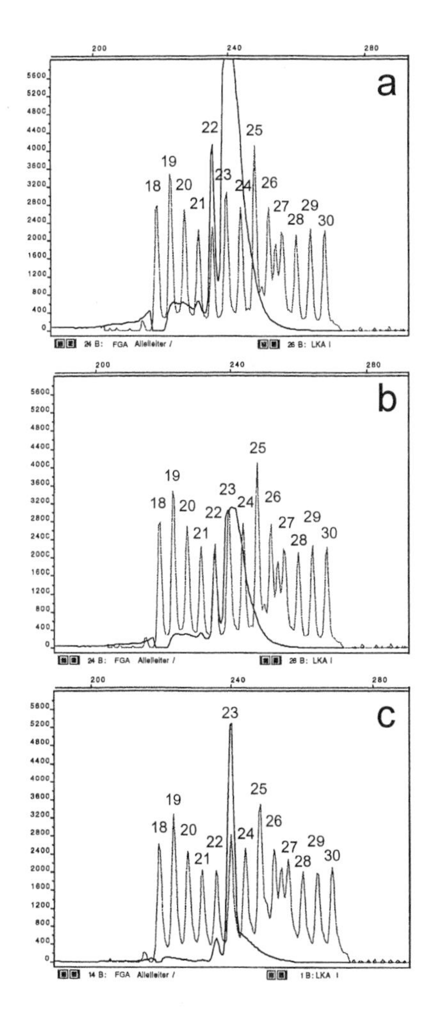

Fig. 5.6. The electropherograms show the same amplification product, but different amounts of the product were loaded to the PAA gel. On top (**a**), 1 µl of the autosomal STR amplification product was loaded, which is obviously a product overload since an allele determination is not possible. It is unclear whether the amplified allele matches allele 23 or 24 of the allelic ladder of the human STR FGA (= Fibra). In the middle (**b**), the height of the amplified allele product of (**a**) has had a software-based reduction to half of its original height. However, the peak only occurs compressed, but it still is not possible to assign the amplified allele. At the bottom (**c**), only 0.1 µl of the product was loaded to the PAA gel. This measurement now enables a sound allele determination, clearly showing the amplified product matching allele 23

Another reason that may significantly affect the size calculation and lead to ambiguous results is the miscalculation of a strong slope of the peak. Such slopes may be caused by poor separation. Examples of this are stutter products, which are a particular feature of dinucleotide repeats (cf. section 2.2.2), or other artifact products that are less numerous than the intended product but very similar in fragment length. In order to improve the electrophoretic separation, it often helps to reduce the amount of PCR product a little. Alternatively, the use of longer gel plates would increase the separation distance and resolve the problem.

One technical problem that is difficult to overcome is a result of older automated sequencing units, which work on a filter wheel basis. Those particular filter wheels have fixed wavelength optima, which do not exactly correspond to the modern fluorescent labels that are used in the current applications, e.g., the STR amplification kits. This results in an artifact that is referred to as "pull up" and may become a problem in multiplex analysis (Fig. 5.7).

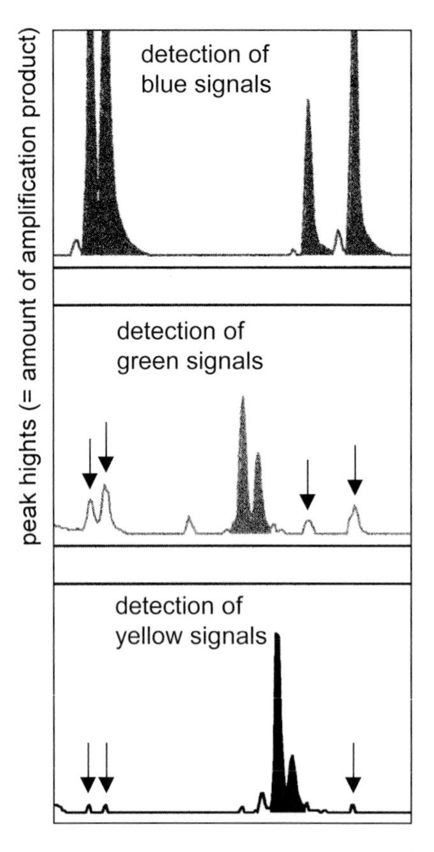

Fig. 5.7. Analysis of a multiplex amplification product that used "Fam"- (= blue), "Hex"- (= green) and "Ned"- (= yellow) dye-labeled primers on a 373 stretch automated sequencer. The pull-up peaks (indicated by *arrows*) that are a result of the strong blue signals are detectable in the wavelength analysis range for green signals and even the range of yellow signals. The pull-up signals reveal the same fragment length (horizontal scale = basepairs) as the blue peaks they originate from. Therefore, it is possible to discriminate these artifacts from true products that are Hex- and Ned-dye labeled (= green and yellow). Although these artifacts are usually stronger in machines using the filter wheel technology, even in capillary electrophoreses, they may occur in cases like this in which the very large amount of Fam-dye labeled products (= blue) represents a product overload compared to the lesser amounts of Hex- and Ned-dye labeled amplification products

Pull-up peaks occur because the wavelength optimum of the label does not fully match the optimum of the respective filter on the wheel, but is shifted. In such a case, the filter that has its optimum nearby will detect a strong signal, but a second filter will also detect the signal of the product label (Fig. 5.8). This is because the dye labels do not send monochromatic light except within a wavelength range; also, the filters detect a range of wavelengths. If, in addition to this, there is also an overloaded amount of product, then even the third window may still detect a signal.

Since all pull-up peaks have their origin in one amplification product, they are characterized by conveying almost exactly the same fragment length. In fact, the artifacts do reveal much lower peak heights, which clearly enables the identification of the peak representing the amplification product.

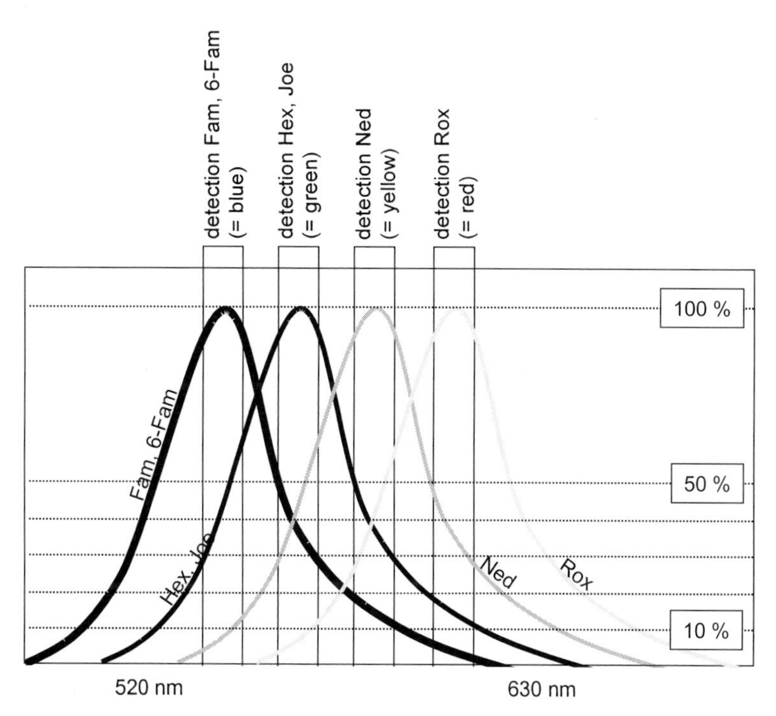

Fig. 5.8. Wavelengths sent out by the dye labels under laser excitation and their detection by filter wheels and slit technology. Each detected wavelength contains contributions from the adjacent dyes. A matrix standard is used to calibrate the system. It works perfectly as long as the intensities of the fluorescent emissions of the dye-labeled products are similar to the (homogenous) product amounts used in the matrix standard calibration. However, in aDNA analysis this may not be given because of the randomly occurring, considerable inhomogeneities in product amounts

The most effective way of reducing the occurrence of pull ups to a minimum is to calibrate the software using a "matrix standard," which should match the labels that have been used for the primers. If more than one filter registers a signal at the same time, the software is instructed by the algorithm created by the matrix standard to leave the most intense signal but to cut down the less intense signals by particular amounts. This solves the problem as long as the amount of product does not exceed the amount of matrix standard that was used for calibration. Once again, the occurrence of pull-up peaks is thought to be enhanced by too much product. However, since the amount of product usually strongly varies within an ancient DNA multiplex assay (ranging from 100–300 bp; cf. section 4.6.1, Fig. 4.17), compromises are necessary that require overloading short fragments in order to reach the detection range for the smaller amounts of the long fragments.

The pull-up peak problem is considerably reduced if the analysis is carried out on machines that allow a small wavelength range to be defined for each color that is to be detected by a software tool. This technique has been adapted in all capillary electrophoresis units. Again, matrix standards are necessary for calibration. Therefore, capillary electrophoresis is considered to be particularly suitable for ancient DNA analysis. However, in cases of complete or partial overload, the detection range for a second color may even be reached as well. In particular, partial overload may be unavoidable in aDNA analysis because of the extremely different rates of success in the amplification of fragments of differing lengths. In such cases, it is advisable to analyze two different aliquots of PCR product (e.g., 1 µl and 0.1 µl). The larger volume will enable the determination of the alleles of the long amplification products (>200 bp), while the small volume will suit perfectly the short but numerous products (<200 bp).

5.2 Sequence analysis

Sequence analysis, which is often referred to as sequencing, enables the analysis of the succession of bases within an amplified DNA fragment. It can discern all types of polymorphisms, such as nucleotide exchanges, insertions and deletions of bases. Depending on the precise nature of a polymorphism, alternative techniques may be used. For example, if an insertion or deletion of 2–6 bp is to be expected at a particular site of the sequence (cf. sections 2.2.1 and 2.2.7), and if this addition or lack of bases is the only information that is to be gained, a software-based fragment length analysis (cf. section 5.1) may be the method of choice. This situation is the case, for instance, in the analysis of ΔF508, where a single 3-bp deletion on chromosome 7 causes cystic fibrosis. Also, in cases of possible single-point mutations at known sites, a restriction fragment length polymorphism (RFLP) analysis (cf. sections 5.3 and 8.9) of the amplified product should be suitable. Other such examples might include the point mutations coding for the AB0 blood-group genes (cf. section 2.2.3) as well as those species identifications in which cytochrome-b analysis (cf. section 2.1.3) is used.

However, sequencing cannot be replaced in any type of analysis where multiple polymorphisms of any kind may occur at sites that are not necessarily known in advance, such as any site of an amplified product that is a potentially informative site. This applies to investigations of the mitochondrial hypervariable regions (cf. section 2.1.1) and those cases of species identification that do not start from a particular hypothesis (cf. section 2.1.3).

Sequencing can be carried out in different ways (Wu 1993). But due to the modern, automated sequencing technology that is in use, the so-called "Sanger sequencing" will be the only one discussed here. Sanger sequencing, which is also referred to as Taq-cycle sequencing, basically requires a second PCR following the initial one that served for the exponential amplification of the target sequence. In contrast to the first, the second PCR is asymmetric; that is, only one primer is used, resulting in a linear amplification of a single strand of the product. The most important point of this second PCR is to label all bases making up the sequence in a particular color depending on whether it is an adenine (A), cytosine (C), guanine (G) or thymine (T). This is achieved by not only including dinucleotides (dNTPs) but also fluorescent and multi-colored dye-terminating dideoxynucleotides (ddNTPs) as building material within the reaction. In contrast to dNTPs that enable a regular elongation, the addition of ddNTPs leads to an immediate termination of the elongation step, a so-called "chain interrupt." Provided that the dNTPs and ddNTPs are supplied in suitable proportions, the chain interruptions will occur randomly throughout all sites of the amplified PCR product. As a result, a population of single-stranded products is generated, representing all possible lengths starting from primer +1 to primer +*n*. The single-stranded, labeled sequences are finally electrophoresed and sorted by length, resulting in a succession of peaks of different colors representing the investigated sequence (Rosenblum et al. 1997; Brandis 1999; Li et al. 1999) (Fig. 5.9).

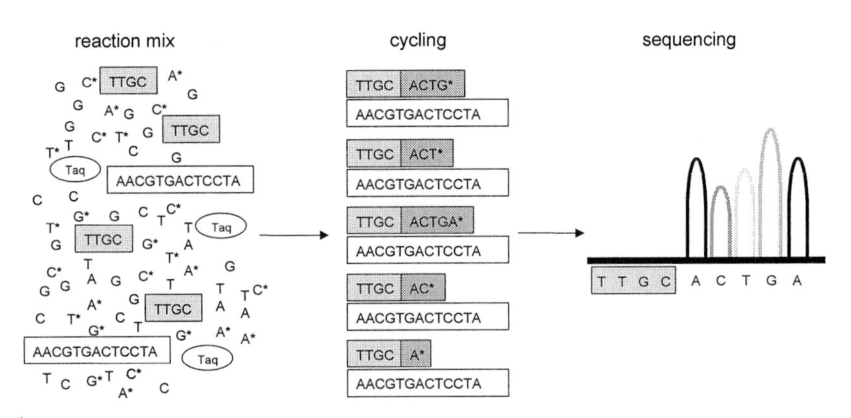

Fig. 5.9. Diagram showing the principle of Taq-cycle sequencing. Taq-cycle sequencing is a linear amplification reaction that uses only one primer and a mixture of dNTPs and dye-labeled ddNTPs (*). When the latter are built into the sequence, they cause a chain interrupt. Since the built-in occurs randomly, at the end of the Taq-cycle sequence reaction all possible lengths of product (primer +1 to primer +*n*) will be represented in the amplification product. If the product is separated by electrophoresis, the result of the analysis is the base succession

5.2.1 Direct sequencing vs. sequencing clones

The decision about whether a PCR product is directly sequenced or if a cloning procedure (Fig. 5.10) is necessary prior to sequencing depends on the scientific question and on the quality of the amplified product (Andersson and Gibbs 1994).

Basically, if the product is pure, i.e., the product originates from the target sequences of a single individual, direct sequencing is the most efficient way to generate correct sequence information (e.g., Bevan et al. 1992; Gill et al. 1994; Wilson et al. 1995; Weichhold et al. 1998; Fattorini et al. 1999; Bär et al. 2000; Anslinger et al. 2001). In fact, it is sufficient if the vast majority of products originate from a single organism, since the signals of the majority sequence superimpose possible minority sequences. Given that the ratio of majority:minortiy is at least 10:1, this generally results is a perfectly readable sequence. On one hand, this is encouraging if one takes into account the possible minority sequences originating from single-cell contaminations that are present in reaction tubes (cf. section 6.1.2) or due to Taq-polymerase-induced misincorporations of bases in the course of the PCR process (e.g., Smith and Modrich 1997) (cf. chapter 6). On the other hand, if a possible admixture is to be detected, such as in forensic evidence material or in archaeological samples of unknown composition, direct sequencing would not lead to a valid result.

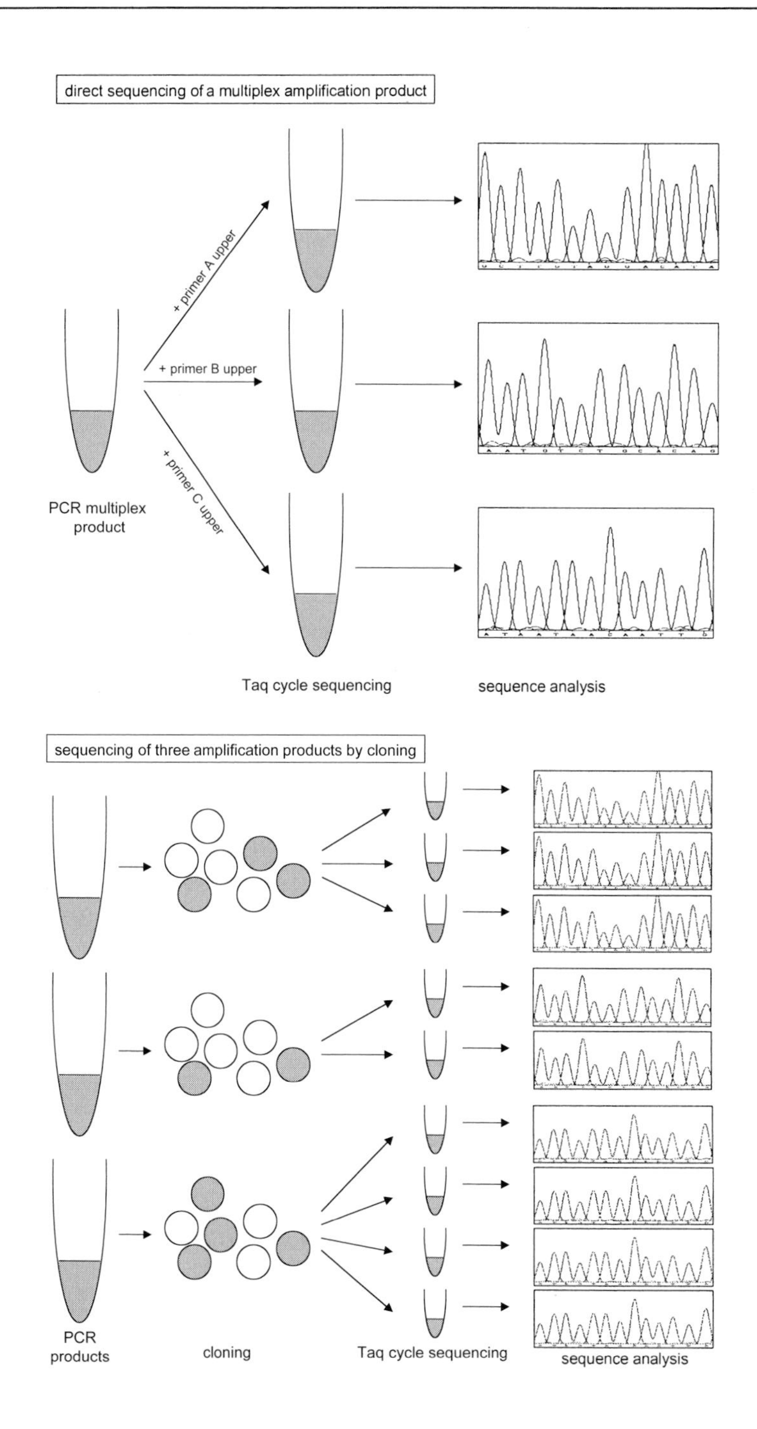

Fig. 5.10. Sequence analysis by direct sequencing and via a cloning step

Therefore, if the PCR product was generated from targets originating from more than one individual and the aim is to detect this, a cloning step prior to sequencing is the only option available. DNA mixtures may have occurred for several reasons, ranging from unintended contamination to mixed evidence sample materials or archaeological material such as feces (coprolithes), blood stains from mass disasters or historical glues and paints that could be expected to consist of DNA from more than one individual or species (e.g., Lukyanov et al. 1996). Also, in case of Taq-polymerase-induced base misincorporations represented by a major portion of the PCR product, sequencing clones may help to identify the problem. However, the cloning does not help to discriminate the artifact sequences from the authentic ones. Such artifacts may occur if the original matrix strands are severely damaged and if only very few targets are present at the beginning of the PCR. In such a case, a misincorporation in this very early stage of the cycle is not only more likely than in case of intact matrix strands, but also the product revealing the false sequence is a true competitor to the authentic sequence.

5.2.2 Overlapping sequences vs. a multiplex approach

Another challenge in ancient DNA analysis is the need to perform sequence analysis on long genomic regions. Since the information may be spread over hundreds of base pairs, the region must be targeted using multiple amplifications. Again, two different strategies are available that successfully combine the length requirements with those of authentication of the aDNA analysis. One strategy to adopt is overlapping amplification fragments; the other strategy would be to employ the multiplex approach (Fig. 5.11), which has proved to be a very powerful authentication tool in STR typing analysis (cf. section 6.2.2).

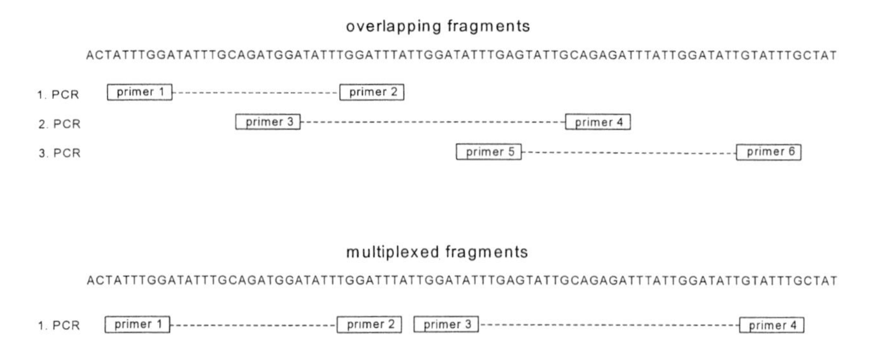

Fig. 5.11. Strategy of overlapping fragments and multiplex amplification

The first strategy, amplifying overlapping fragments (e.g., Singh et al. 1995; Krings et al. 1997; Nasidze and Stoneking 1999; Gabriel et al. 2001), has two clear advantages: the primer design can be carried out independently for each pair, i.e., all primer sets may run at their individual optimal reaction conditions instead of having the compromise in conditions that might be necessary when using a multiplex assay. Secondly, the overlap strategy is applicable to both direct sequencing and to sequencing clones. Although this strategy is theoretically unlimited in the amount of sequence information that can be generated (i.e., thousands or ten thousands of bases of information can be investigated this way), a clearly limiting factor is the amount of available sample or DNA extract. Practically, the sequence analysis has to observe all of the above-mentioned points, since all amplified products are independent. The decision about sequencing them directly or performing a cloning step in advance should be guided by the criteria suggested in section 5.2.1.

The advantages of the multiplex approach (Schultes 2000) is in the enhanced authentication potential, since all polymorphic sites of the region in question are amplified simultaneously from the same aliquot of target DNA (Fig. 5.12).

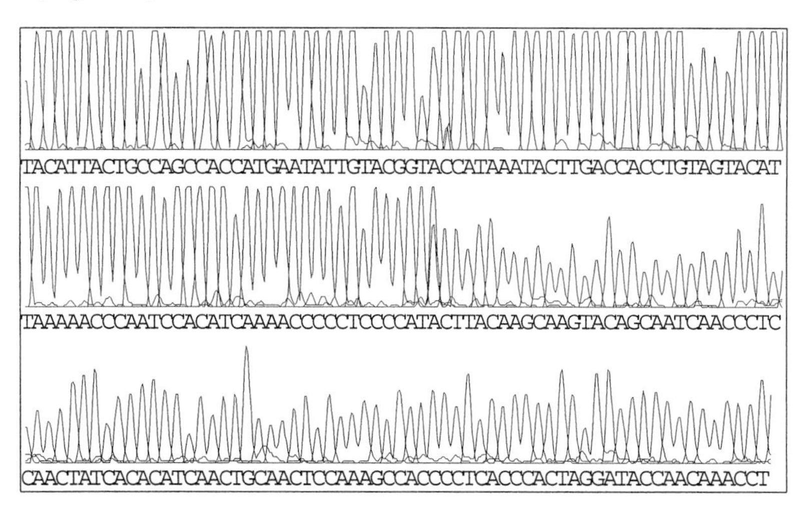

Fig. 5.12. Part of the direct sequencing from a multiplex-generated amplification product of the human hypervariable regions I and II (cf. section 2.1.1). The amplified products are the 168-bp product and the 312-bp product, which consists of the 168 bp, the 131 bp and the 13 bp intervening sequence. Here, the upper primer of the 168 bp region was used as the sequencing primer. Therefore, both the 168-bp and the 312-bp products contribute to the first part of the sequence (high peaks); later, only the 312-bp product contributes to the sequencing reaction (lower peaks)

Analogous to multiplex STR typing, the generation of multiple pieces of information simultaneously will strongly increase the validation power (cf. section 6.2) when compared to relying on a reproduction of results for just 1–2 polymorphic sites within an overlapping region. Another positive point is that a maximum of information can be gained from a minimum amount of sample material (cf. section 4.5). Since all the products are present within each of the cycle-sequencing reactions, a perfect primer design (cf. section 4.4) and stringent amplification conditions are absolutely mandatory in order to target just one of the multiplex products. If the aDNA reveals a good state of preservation, then additionally a combined product is generated that consists of both the single-targeted products and possibly intervening sequences (cf. section 2.1.1 and Fig. 5.12).

5.3 RFLP analysis

Restriction fragment length polymorphism (RFLP) analysis of a PCR product is an elegant way to investigate certain types of sequence polymorphisms, so-called point mutations (e.g., Meyer et al. 1995; Nishimukai et al. 1996). Single-point mutations are often observed in coding regions of the genome. For example, on chromosome 9, single-point mutations are found that determine the human ABO blood groups (cf. section 2.2.5). Further point mutations are found in the mitochondrial cytochrome-b gene, where different species and subspecies reveal variants at certain nucleotide positions (cf. section 2.1.3). Point mutations can be base exchanges, base deletions or insertions.

The basic mechanism of RFLP analysis relies on the ability of restriction enzymes, the "endonucleases," to cut double-stranded DNA as soon as a certain succession of bases is present. Usually this restriction reaction is carried out at 37°C, although this may vary depending on the restriction enzyme used. The process, which converts, for example, an original 103-bp amplification product into two restriction products of 66 bp and 37 bp in length, is called digestion (Fig. 5.13). In case the recognition site of the enzyme results in blunt-end cuts in the sequence, the restriction products need no further consideration before electrophoresis. However, if the products have sticky ends, they may re-anneal to one another if they are kept too long at room temperature. They should therefore be run on a gel as soon as possible or, alternatively, a denaturing agent (e.g., formamide) should be mixed into the digested products.

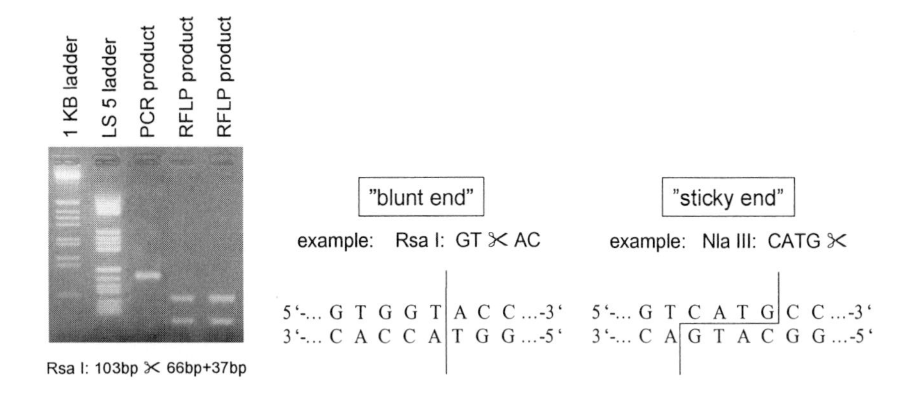

Fig. 5.13. RFLP analysis of an 103-bp amplification product (*left*, cf. also section 2.2.5). The restriction products that were received by a digestion of the PCR product through the endonuclease Rsa I reveal fragment lengths of 66 and 37 bp, respectively. Since the recognition site of Rsa I creates blunt-end fragments (*middle*), no further treatment of the restriction fragments is necessary. If as a result of the particular sequence polymorphism the RFLP analysis requires a restriction enzyme that creates sticky-end fragments (*right*), the addition of a denaturing agent such as formamide may be necessary to prevent renaturation of the fragments

One important prerequisite in performing successful RFLP analysis is an understanding of the region of interest. First, it must be known at which nucleotide position the possible mutation will occur and, second, what type of polymorphism may occur (e.g., C instead of A, deletion of G or insertion of T). The nucleotide position of the polymorphic site will determine the way in which a primer set is designed in order to end up with appropriate restriction products that can easily be distinguished from the original, undigested amplification product. Of course, the actual polymorphism resulting in different base successions is essential for choosing a suitable enzyme.

Furthermore, a successful RFLP analysis is dependent on an initial hypothesis: it must be ensured that the result is fully informative. A simple example will illustrate the situation. Given that dogs reveal a "G" at a certain nucleotide position, while cats, mice and elephants reveal a "C" at the same position, an RFLP result that indicates for "C" is only informative if either the result "dog" or "not dog" is sufficient or if the task had already been reduced to a binary problem, for example, where mice and elephants had already been excluded by anatomical features of the sample (Fig. 5.14).

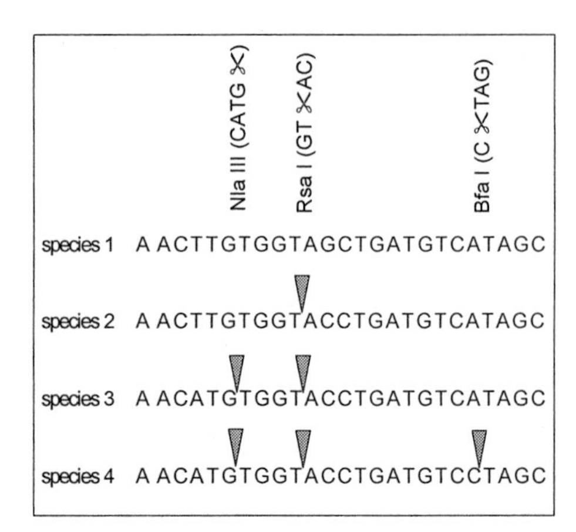

Fig. 5.14. Hypothetic sequence polymorphisms for different species. In case the sample reveals no morphological features that enable the exclusion of all except two possibilities, multiple RFLP analysis must be carried out (cf. also 2.2.5). In such cases, it is important that either further suitable polymorphisms are included by the amplified product or that otherwise further amplifications must be carried out

If the "dog, cat, mouse or elephant" problem is taken literally in such a way that more than two alternatives must be distinguished, then there are, of course, RFLP-based strategies to solve the problem. For example, a suitable number of polymorphic sites can be included in the amplification product, as we have demonstrated for different approaches to species identification (cf. sections 7.2.4 and 7.2.8).

A second method is to target a number of polymorphic sites by different PCRs and to work out the final result by successive exclusion. Basically, the strategy is similar to identification by SNP (single nucleotide polymorphism) typing (chapter 1). This strategy was done for ABO blood group typing (sections 2.2.5 and 7.4.3), where, at particular polymorphic sites, two PCR products and four endoculeases were involved.

If the "dog, cat, mouse or elephant" problem is seen in a more generalized way, it demonstrates the limitations of RFLP analysis. RFLP will only give an answer to the question that was originally raised. To be precise, RFLP does not even do this, since the true nature of an RFLP result is "as dog," "as cat," "as mouse" or "as elephant." If this is not a sufficient answer, the next option must be to do sequence analysis.

References

Andersson B, Gibbs RA (1994) PCR and DNA sequencing. In: Mullis KB, Ferré F, Gibbs RA (eds) PCR. Polymerase chain reaction. Birkhäuser, Boston

Anslinger K, Weichhold G, Keil W, Bayer B, Eisenmenger W (2001) Identification of the skeletal remains of Martin Bormann by mtDNA analysis. Int J Legal Med 114:194–196

Bär W, Brinkmann B, Budowle B, Carracedo A, Gill P, Holland M, Lincoln PJ, Mayr W, Morling N, Olaisen B, Schneider PM, Tully G, Wilson M (2000) DNA Commission of the International Society for Forensic Genetics: guidelines for mitochondrial DNA typing. Int J Legal Med 113:193–196

Bevan IS, Rapley R, Walker MR (1992) Sequencing of PCR-amplified DNA. PCR Meth Appl 1:222–228

Brandis JW (1999) Dye structure affects Taq DNA polymerase terminator selectivity. Nucleic Acids Res 27:1912–1918

Butler JM (2001) Forensic DNA typing. Biology and technology behind STR markers. Academic Press, San Diego

Fattorini P, Ciofuli R, Guilianini P, Edoni P, Furlanut M, Previdere C (1999) Fidelity of polymerase chain reaction direct sequencing analysis of damaged forensic samples. Electrophoresis 20:3349–3357

Gabriel MN, Huffine EF, Ryan JH, Holland MM, Parsons TJ (2001) Improved mtDNA sequence analysis of forensic remains using a "mini-primer set" amplification strategy. J Forensic Sci 46:247–253

Gill P, Ivanov PL, Kimpton C, Piercy R, Benson N, Tully G, Evett I, Hagelberg E, Sullivan K (1994) Identification of the remains of the Romanov family by DNA analysis. Nat Genet 6:130–135

Krings M, Stone A, Schmitz RW, Krainitzki H, Stoneking M, Pääbo S (1997) Neandertal DNA sequences and the origin of modern humans. Cell 90:19–30

Li Y, Mitaxov V, Waksman G (1999) Structure-based design of Taq DNA polymerases with improved properties of dideoxynucleotide incorporation. Proc Natl Acad Sci USA 96:9491–9496

Lukyanov KA, Matz MV, Bogdanova EA, Gurskaya NG, Lukyanov SA (1996) Molecule by molecule PCR amplification of complex DNA mixtures for direct sequencing: an approach to in vitro cloning. Nucleic Acids Res 11:2194–2195

Mertes G, Schäfer T, Schild TA, Schmidt G, Schuster D, vom Stein J (1997) Automatische genetische Analytik. Neue Methoden der DNA-Analyse basierend auf innovativer PCR-Technologie und multifluorophorer Detektion. Wiley-VCH, Weinheim

Meyer R, Höfelein C, Lüthy J, Candrian U (1995) Polymerase chain reaction-restriction fragment length polymorphism analysis: a simple method for species identification in food. J AOAC Int 78:1542–1551

Nasidze I, Stoneking M (1999) Construction of larger-size sequencing templates from degraded DNA. BioTechniques 27:480–484, 488

Nishimukai H, Fukumori Y, Okiura T, Yuasa I, Shinomiya T, Ohnoki S, Shibata H, Vogt U (1996) Genotyping of the ABO blood group system: analysis of nucleotide position 802 by PCR-RFLP and the distribution of ABO genotypes in a German population. Int J Legal Med 109:90–93

Rickwood D, Hames BD (eds) (1990) Gel electrophoresis of nucleic acids. Oxford University Press, Oxford

Rosenblum BB, Lee LG, Spurgeon SL, Khan SH, Menchen SM, Heiner CR, Chen SM (1997) New dye-labeled terminators for improved DNA sequencing patterns. Nucleic Acids Res 25:4500–4504

Sambrook J, Fritsch EF, Maniatis T (1989) Molecular cloning. A laboratory manual (2nd edn). Cold Spring Harbour Laboratory Press, New York

Schultes T (2000) Typisierung alter DNA zur Rekonstruktion von Verwandtschaft in einem bronzezeitlichen Skelettkollektiv. Dissertation, Georg August-Universität, Göttingen. Cuvellier, Göttingen

Singh PJ, Julien P, Mirault ME, Murthy MR (1995) Amplification of the entire mitochondrial DNA by polymerase chain reaction in two large overlapping segments. Anal Biochem 225:152–155

Smith J, Modrich P (1997) Removal of polymerase-produced mutant sequences from PCR products. Proc Natl Acad Sci USA 94:6847–6850

Weichhold GM, Bark JE, Korte W, Eisenmenger W, Sullivan KM (1998) DNA analysis in the case of Kasper Hauser. Int J Legal Med 111:287–291

Wilson MR, DiZinno JA, Polansky D, Replogle J, Budowle B (1995) Validation of mitochondrial DNA sequencing for forensic casework analysis. Int J Legal Med 108:68–74

Wu R (ed) (1993) Recombinant DNA. Part I. Methods in enzymology. Section I: Methods for sequencing DNA. Academic Press, San Diego

Ziegele JS, Su Y, Corcoran KP, Nie L, Mayrand PE, Hoff LB, McBride LJ, Kronick MN, Diehl SR (1992) Application of automated DNA sizing technology for genotyping microsatellite loci. Genomics 14:1026–1031

6 Authenticity of results

Contamination is the most serious hazard for any researcher working with the PCR technique. This is due to the sensitivity of PCR, a technique capable of generating large amounts of amplified product from as little as single cells (e.g., Higuchi et al. 1988; Li et al. 1988; Zhang et al. 1992; Hubert et al. 1992; Wages and Fowler 1993; Drury et al. 2000; Drury et al. 2001).

The fact that researchers working on contemporary DNA usually are not concerned about contamination is solely due to the fact that their sample targets simply outnumber the contaminating target sequences. This enables cycle numbers to be kept between 25 and 30, which is few enough not to be confronted with the problem of contaminants, given the average standards of modern DNA PCR laboratories. These average standards are not adequate for the analysis of degraded DNA, where cycle numbers up to 35, 40 or even 50 should be taken into account. When using PCR for modern DNA, these large numbers of cycles would not even be considered an option, because the results would be judged unreliable.

Unfortunately, ancient DNA researchers followed the same train of thought and for some time insisted that PCR should not be carried out for more than 30 cycles. Although this no longer appears to be required, the legacy remains, which is unfortunate for a number of reasons. Firstly, it reflects a common tendency to cure the symptoms instead of trying to understand the problem in the first place. Secondly, it means that all amplified DNA sequences should be evaluated in the same way, even though this is certainly not acceptable. DNA sequences that are not distinctive are undoubtedly much harder to verify than DNA sequences and DNA patterns that are unique. As a result, different types of sequences may require different measures.

Basically, any contamination source is avoidable (Kwok 1990), with the exception of the contaminants that are purchased with the PCR tubes (Hauswirth 1994; Schmidt et al. 1995). Although the latter contaminations are very inconsequential with respect to the number of contaminating targets, they have been found in the majority of commercially available reaction tubes, regardless of the particular brand (cf. section 6.1). This problem cannot be avoided, and here only the companies can set the standards[11]. Returning to the other types of contamination, although they are in principle avoidable, they may occur. How often they occur depends on the one hand on the standards set by the laboratory and on the other hand on the care that is taken by the respective researcher. Independent of this, the avoidable as well as the unavoidable contaminations must be detected. Whether this is successful or not depends on a sophisticated analysis strategy. This includes control samples but does not necessarily mean a standard set of negative controls consisting of no-template samples and extraction blanks. This is an important point, because in the case of any single product amplification, it is never possible to completely exclude extraneous sample contamination, independent of the number of negative controls that are processed. Regardless of whether all the controls turn out negative, the sample may have been contaminated or, if the negative controls are positive, the samples may not have been.

There are other situations where the use of controls is not necessary when markers such as STRs are used. The choice of control samples should therefore be guided by the particular nature and properties of the DNA sequences that are being analyzed (cf. section 6.2). If primer design includes ingenious controls that have been carefully considered, aDNA results and the publications on aDNA results do not necessarily have to end up like many of the earlier examples that were generated as a result of contamination.

However, reaching that improved standard also requires researchers who are involved in reviewing manuscripts to be prepared to evaluate research work not by the use of a standard protocol but rather with willingness to accept the existence of excellent control strategies different from their own. The principal reason for this is that different situations require different strategies.

6.1 Contamination sources

There are a number of different sources of possible contamination. All require different approaches to minimize both the risk of their occurrence and the adequate set of control samples and/or control amplifications. One of the most obvious kinds is contamination due to handling; the other is contamination within the actual PCR reaction tubes.

[11] Since 2001 so-called "PCR clean" reaction tubes (Eppendorf) are available. They are advertised as to not contain amplifiable human DNA sequences. The "PCR clean" reaction tubes have not yet been tested in our laboratory (cf. section 6.1.2)

6.1.1 Contamination through handling

Contamination that arises in PCR analysis through the handling of the samples is generally avoidable. However, in practice, it does occur. In order to recognize this source of contamination, it is important to be equipped with suitable detection strategies. This may be in the form of an adequate set of control samples and/or adequate control amplifications.

Contamination as a result of handling can be grouped into three types: cross-contamination between samples, contamination by persons handling the samples and/or chemicals and PCR product carry-over. The latter is probably the easiest to recognize, despite the fact that it has the potential to be the most disastrous.

Carry-over contaminations occur if amplification products are introduced to a pre-PCR analysis step. In the early days of PCR, there were attempts to control carry-over contamination by the use of dUTPs/uracil DNA glycosylase (UDG) digestion (e.g., Longo et al. 1990; Rys and Persing 1993; Niederhauser et al. 1994). However, this method was abandoned and a strict regime of separating pre- and post-PCR areas, including the assigned equipment, is widely favored today.

Regular controls without the addition of a template are suitable to monitor contaminations of the carry-over type. However, if this contamination type does occur, the worst case scenario is that all future amplifications using the particular primer pair or primer set will reveal exactly the genotype of the introduced amplification product. This could come about if for some reason major areas, equipment or chemicals in the extraction or PCR set-up laboratory are contaminated by an amplification product. If there is particular interest in the intended sequence or even parts of it, then the only solution in such a case would be to abandon the room, all its equipment and all chemicals and reagents and to try somewhere else. Therefore, all possible efforts should be undertaken to avoid contamination of the carry-over type. This can be done using a set of strategies, ranging from strictly separated laboratories to assigned and separate equipment, chemicals, reagents and machines. Additionally, adopting "one-way traffic" is essential, i.e., never going back to a laboratory that was assigned to an earlier stage in the analysis process in the same day. This would include anyone who may have a legitimate reason for entering a laboratory, such as cleaning and maintenance staff.

Cross-contaminations are a result of sample-to-sample contamination. They are not as devastating as PCR-product carry-over but are much more difficult to detect. In order to minimize the risk of cross-contamination, it is required that all pre-PCR analysis steps, starting with sample preparation, DNA extraction and finally the PCR set up, should be done with careful consideration. The requirements for avoiding cross-contamination are basically the same at all stages of the analysis: strict separation for the handling of the samples, effective cleaning management for all equipment that is in direct contact with the samples (e.g., mortars and mills), changing disposable gloves when handling different samples, under no circumstances touching the inner surface of the reaction tubes, precise pipetting and, finally, the use of positive displacement pipettes or aerosol-tight pipette tips. Either of the latter two possibilities should be used if non-individual-specific conservative sequences are investigated, since there is no possibility to monitor for cross-contamination in such cases. If individual-specific sequences are investigated, all of these measures of course should be followed with the same care. However, the monitoring situation for possible cross-contamination events is much improved if individual STR sequence investigations are used because of the high discrimination ability of these markers. If possible, non-individualizable markers (e.g., mtDNA markers) should be combined with individual-specific markers (e.g., STR markers) in multiplex PCRs (cf. sections 2.2.4, 2.2.6 and 2.2.7). Negative control samples such as no-template samples and extraction blanks do not monitor cross-contamination.

Contamination that may have occurred because of present or former lab personnel handling the samples and/or the chemicals can be approached in different ways. How the contamination occurred and, more importantly, how it can be detected and avoided depends on the stage at which it arose and the way it was introduced to the process in the first place.

If a chemical or reagent is contaminated during the course of the extraction phase, this will be revealed by amplification products in the blank controls. The no-template controls introduced to the analysis at the PCR set-up phase will remain blank. If individual-specific markers are used, it will be obvious, because all samples will reveal the same genotype or at least give an indication that a mixture of a second genotype is in all of the samples.

If a chemical or reagent used in the PCR set up is contaminated, both extraction blanks and no-template controls will show signals. Again, if individual-specific markers are used, all the samples should reveal the same genotype or respective mixtures, similar to negative control samples.

If a single or a few samples are contaminated directly at any stage of the analysis process, both types of negative controls should remain blank. This contamination type can only be detected by using highly variable, individual-specific markers and information on the genotypes of all persons who handled the samples. In cases where more general markers are analyzed (e.g., mtDNA haplotypes) and for some reason it is not possible to use individual-specific markers as well, the processing of the sample material in a second, independent laboratory may be useful, although the quality of the data validation will remain inferior.

If a single sample is severely contaminated due to handling by former unknown persons, this will neither be indicated through the use of negative control samples nor by processing the sample in a second laboratory. In such cases where serious contamination cannot be excluded (e.g., very valuable samples that passed through many hands), the only control that is possible is to process additional samples that are known to have the same history. Individual-specific markers can now indicate contamination events. If only semi-informative markers are analyzed (e.g., mtDNA HVR I and II), the additional control samples should preferably be of animal origin. In general, the analysis of single samples that may be a result of severe contamination because of their particular history is the most complicated to validate.

Another challenging situation is presented if the sample material is suspected to have been treated with a glue for conservatory reasons. Such glues used to be prepared from animal bones and have proved to contain DNA in amplifiable amounts (Burger et al. 2000; Burger 2000; Scholz and Pusch 2000). As described above, the only possible control to detect major contaminations of this type are individual-specific markers.

Minimizing risks of contamination by the introduction of cellular material is of concern in all steps, from sample preparation to PCR set up. Therefore, all sample surfaces that possibly may have been in contact with cellular contaminants must be removed. The chemicals and reagents used in all analysis steps should be stored as aliquots in order to minimize repeated handling. It is also necessary to wear disposable gloves, face masks, head-dress gowns and glasses during all chemical and reagent preparations, sample preparations, DNA extraction and PCR set-up stages (Kitchin et al. 1990). Whether lab coats or everyday clothing are worn will depend on the frequency with which the lab coats are washed. Laboratory coats look very professional; however, using suitable long-sleeved cotton T-shirts and trousers that have been freshly washed will transfer less cellular material than a lab coat that has been worn for weeks.

The benefit of irradiating work benches and equipment by ultra-short UV light (λ 254 nm) is unconvincing (Frothingham et al. 1992; Dwyer and Saksena 1992; Cone and Fairfax 1993; Niederhauser et al. 1994). In contrast to what is often assumed, UV irradiation of double-stranded DNA is only effective in aqueous solutions (e.g., Sarkar and Sommer 1990; Cimino et al. 1990; Sarkar and Sommer 1993) (cf. section 6.1.2). It destroys living organisms[12] or high molecular-weight DNA in solution by cross-linking the double strands, but it does not effectively affect cellular material that is already dried up or isolated, short DNA fragments. Furthermore, cross-linking of double-stranded DNA is only reached if the irradiation distance is very short (e.g., <5 cm with a 30 watt lamp), and the irradiation time is long (>2 h). On the other hand, 254-nm UV irradiation causes a high production of ozone (O_3) and can seriously affect all the common types of plastic (pipettes, tube racks, etc.). It therefore becomes a question of whether this procedure should be used, a prodedure that certainly appears respectable, although ineffective, but also generates unhealthy molecules and causes damage to laboratory equipment.

This is easily overcome, because it is effective to clean all surfaces of work benches and equipment that are used with aqueous solutions that are either alkaline (concentrated detergents) or acidic substances. Both alkaline and acidic aqueous solutions affect the DNA strands by denaturing or hydrolizing them, respectively. In contrast, alcohol-based methods do not affect DNA as they are part of the extraction procedure in all protocols and in fact help in preserving the DNA. They may only help to dry surfaces or equipment parts more quickly after they have been thoroughly cleaned using aqueous acidic or alkaline solutions and thoroughly rinsed with pure water.

6.1.2 Contamination in reaction tubes

Most of the publications that address the topic of contamination deal with strategies and measures to minimize or eliminate the risk of contamination that result directly from inadequate laboratory processing. In contrast, there are very few publications that have concentrated on the contamination sources that are beyond a laboratory's influence, i.e., contamination in reagents and disposable lab equipment purchased directly from the manufacturer. Examples are reports on false-positive reactions due to residues of *E.coli*-DNA sequences in recombinant DNA-Taq polymerase (Böttger 1990; Rand and Houck 1990; Schmidt et al. 1991; Koponen et al. 2002). In particular, PCR amplification tubes as possible sources of contamination are first mentioned by Hauswirth (1994). Our

[12] UV irradiation is a sterilization technique that, like autoclaving, has been transferred from microbiology to molecular genetics. The fact that autoclaving was recognized already in the early days of PCR as an unsuitable measure for reaching high analysis standards and that UV irradiation is still used (and even promoted by manufacturers) is because autoclaving certainly contradicts high-standard PCR, while UV light is only ineffective.

own initial experiments concerning contamination sources from disposable plastic equipment were carried out in 1994 (Schmidt et al. 1995) and were initiated in the first instance by employing primers for mitochondrial DNA amplification, a highly conserved D-loop sequence of the control region V (Wrischnik et al. 1987). Until this point, we had only worked on chromosomal aDNA, where contaminations that were monitored using negative controls occurred rarely and only in cases with very high cycle numbers (50–60). In these rare instances, we considered this to be due to mistakes in sample processing and repeated the experiments from the beginning rather than trace the problem. In the mitochondrial aDNA amplifications, the number of contamination events monitored by the negative control samples drastically increased, even when lower cycle numbers were carried out (35–45). On the other hand, amplifying chromosomal single-copy sequences such as amelogenin and VWA (Lassen et al. 1995; Zierdt et al. 1996) using exactly the same disposable material and reagent lots and aliquots (except primer pairs) revealed that contamination did not occur as frequently at all in the negative controls. At this time, there was no explanation for this significant disparity; therefore, we decided to set up experiments to trace back to the source of contamination.

Three basic reasons for contamination in the PCR analysis were considered to be air, reagents and disposable laboratory equipment.

The experiments were carried out with batches of 10 or 20 no-template controls (X1 – X10/X20), respectively. To monitor the suitability of the amplification parameters, two positive controls containing defined amounts of human target DNA were also incorporated into the experiment with each batch of negative controls. The amplified product was a 121-bp fragment of the D-loop of the human mitochondrial control region V. All PCRs were carried out for 45 cycles (Fig. 6.1).

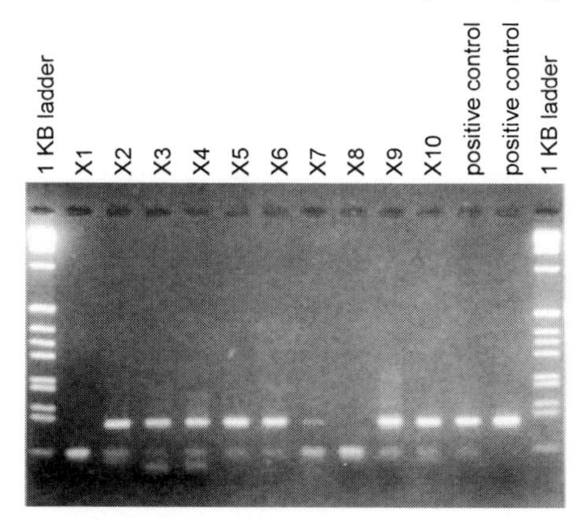

Fig. 6.1. Typical percentage (80%) of contamination in no-template controls after 45 amplification cycles for the 121 bp fragment of the human mitochondrial control region V (GeneAmp reaction tubes, Perkin Elmer)

To determine whether the contamination could be traced back to one or more of the amplification reagents, they had been successively replaced with reagents purchased from other suppliers. In the second part of the experiment, all reagents (except Taq polymerase) underwent a filtration treatment to avoid the possibility that all reagents from all suppliers might contain cellular DNA. Changing any of the reagents as well as the filtration treatments failed to alter the percentage of contaminated tubes (Fig. 6.2).

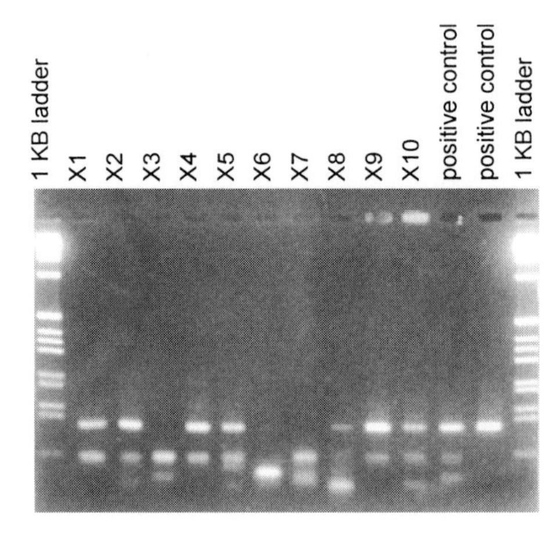

Fig. 6.2. Contamination in no-template controls after exchange of chemicals and reagents. Here, Taq-polymerase was substituted. Changes in the percentage of contaminated tubes compared to the initial situation (cf. Fig. 6.2) were not detectable. (GeneAmp reaction tubes, Perkin Elmer; human mitochondrial DNA, control region V, 121 bp)

Furthermore, it was tested whether DNA contamination in the amplification reactions was from airborne sources. Batches of 18 no-template samples not yet covered with mineral oil were exposed to laboratory air for 0, 5, 10, 15, 30 and 60 min in addition to the minimum time of about 1 min in total during which the reaction tubes are necessarily uncapped for pipetting. Three tubes were closed at each of the exposure times given above. This experiment was repeated three times. The results revealed no correlation between the percentage of contaminated tubes and the exposure time (shown in Fig. 6.3).

A final test was carried out in order to determine whether those laboratory disposables that are in direct contact with the PCR reaction mix were the sources of contamination. When the reaction tubes were replaced by other commercial brands, reproducible differences concerning the contamination rates were obtained (Table 6.1).

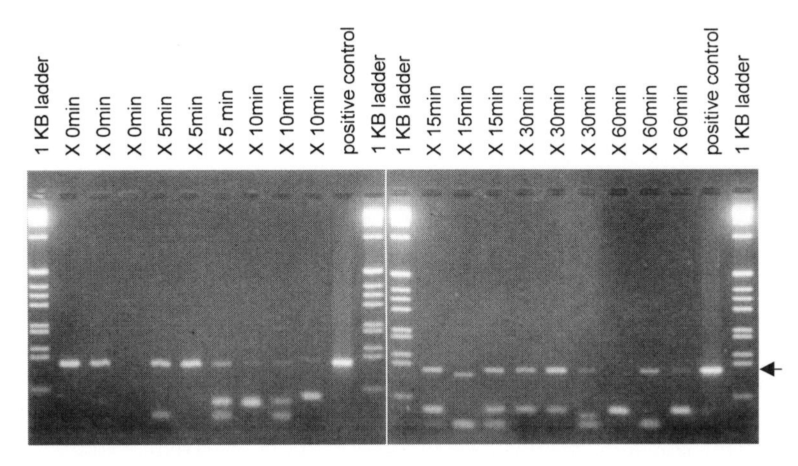

Fig. 6.3. Contamination in no-template controls that were exposed to laboratory air as indicated. An increase in the percentage of contaminated tubes with increasing time of exposure was not detectable (GeneAmp reaction tubes, Perkin Elmer; human mitochondrial DNA, control region V, 121 bp)

Table 6.1. Percentage of contamination in reaction tubes of different brands (45 amplification cycles, human mitochondrial DNA, control region V, 121 bp)

Brand	Manufacturer	Contamination rate[13]
Ultra-thin walled	Biozym	90%
GeneAmp	Perkin Elmer	80%
Thin walled	Perkin Elmer	80%
Safe-lock Biopur	Eppendorf	38%
Safe-lock regular[14]	Eppendorf	20%

[13] The data presented in the table are based on tests that were carried out in 1995. Later experimental testing (1998 and 2000) basically revealed the same results. However, it must be mentioned that a certain ranking in the quality of the tubes was not obtained with respect to the suppliers but with respect to treatments of the tubes, which are part of the manufacturing process. Tubes that do not undergo an autoclaving treatment generally reveal much lower contamination rates than those brands of tubes that are autoclaved (cf. footnote 14).

[14] The regular 0.5-ml reaction tubes from Eppendorf were the only tubes in this test that did not undergo autoclaving as part of the manufacturing process (cf. below).

Based on this result, which strongly suggested that the tubes were the origin of the contamination, a cross experiment was carried out (Fig. 6.4). Based on the results, it was most likely that the contaminations were of the cellular type. The sequencing of nine randomly chosen mitochondrial contamination products from the no-template samples pointed to a human origin for the vast majority of the contaminations.

Although some efforts to clean the tubes by rinsing with different fluids and by UV irradiation revealed decreasing rates of contamination, no method was successful enough to reach a zero or close to zero contamination rate.

Such tests have been repeated regularly in our laboratory, with the last one taking place in November 2000. At this time the tested brands were Eppendorf (Safe-lock regular) (also cf. Fig. 6.7), Sarstedt, Greiner and Lafontaine. The results were basically the same as 6 years ago, except that contamination rates as low as 20% were not achieved again. Another test series carried out in February 2000 also involved cytochrome b (Fig. 6.6). Sequencing those amplification products revealed that the reaction tubes are not only contaminated by human DNA, but also by the DNA of cattle (*Bos taurus*).[15]

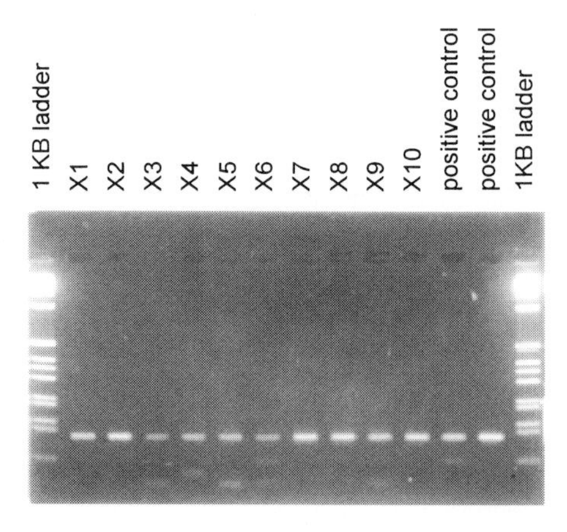

Fig. 6.4. Contamination in no-template controls after the cross experiment that was carried out by heating the aliquoted reaction mixes to 94°C, pooling and redistributing them. The heating caused a lysis, which led to all mitochondrial sequences then being uniformly distributed among all the tubes. This not only confirmed that the reaction tubes are a source of contamination but also pointed to the cellular nature of the contamination

In contrast, experiments testing the reaction tubes for human autosomal STRs revealed much lower contamination rates (20 tubes of each brand were tested). Those experiments were carried out in triplex amplifications (VWA, TH01, FES/FPS) using 60 amplification cycles. In those reaction tubes that revealed signals, usually only one of the targeted systems occurred (Fig. 6.6). In contrast to the experiments investigating the

[15] The experiments revealing cytochrome-b sequences in amplification reaction tubes were carried out by Dr. Joachim Burger.

mitochondrial contaminations, the tubes that underwent an autoclaving step by the manufacturer showed only slightly higher contamination rates (e.g., Sarstedt sterile, Perkin Elmer GeneAmp), while regular tubes showed lower rates (e.g., Greiner regular, Sarstedt non-sterile) (cf. Table 6.2.). As demonstrated by Schmidt et al. (1995), also these findings point to contamination on the single-cell level, where only very few nuclei but thousands of mitochondrial genomes are spread out by the autoclaving procedure.

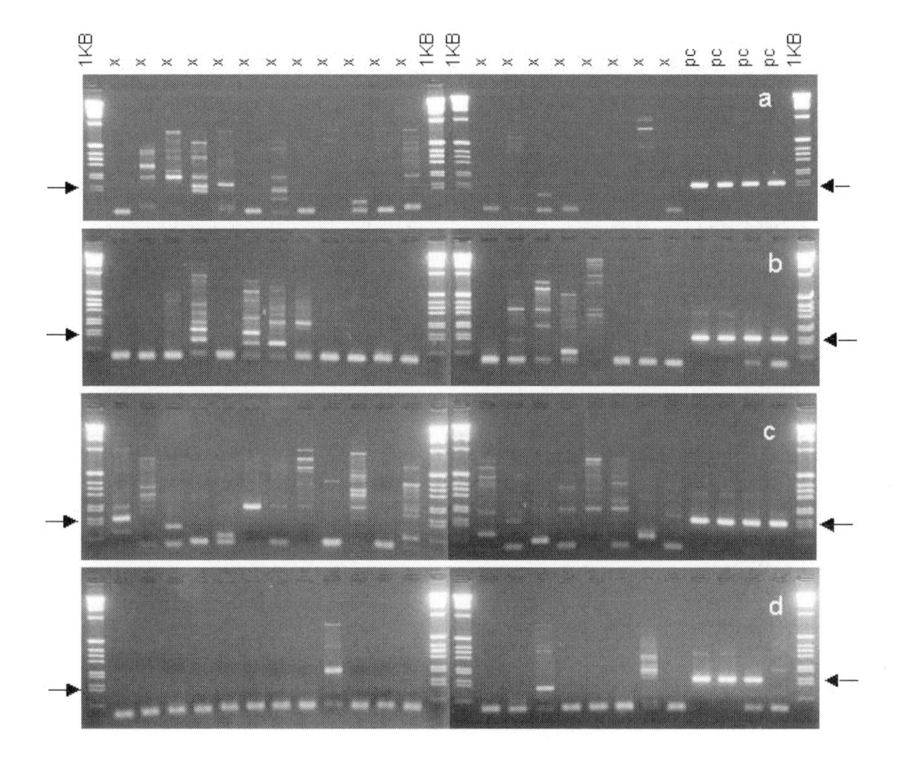

Fig. 6.5. Percentage of contaminations in amplification reaction tubes of different manufacturers and brands after amplification of a cytochrome-b sequence (134 bp, 52 cycles). **a** GeneAmp, Perkin Elmer, **b** Multiply pro, Sarstedt, **c** Safe-lock Biopur, Eppendorf, **d** Safe-lock 0.5 ml regular, Eppendorf. The direct sequencing of the amplification products showed that the majortiy of the tubes revealing amplification products was contaminated by *Bos taurus* sequences

Additionally, the triplex assays clearly indicated that also the contaminating cellular material in the reaction tubes is degraded: the shorter fragments of VWA and TH01, both covering lengths of up to about

170 base pairs, were present much more often than the longer fragments (>210 base pairs) amplified by FES/FPS. However, the patterns look quite different from those of authentic aDNA extracts. In authentic aDNA amplifications, usually the signals of all multiplex targeted sequences are represented with the characteristic decrease of peak heights above approximately 200 bp (cf. section 4.6.1, Fig. 4.17 and section 3.3.3, Fig. 3.15). Contrarily, the no-template samples generally only revealed single signals of a single STR, although the approaches were multiplex (cf. Fig. 6.6).

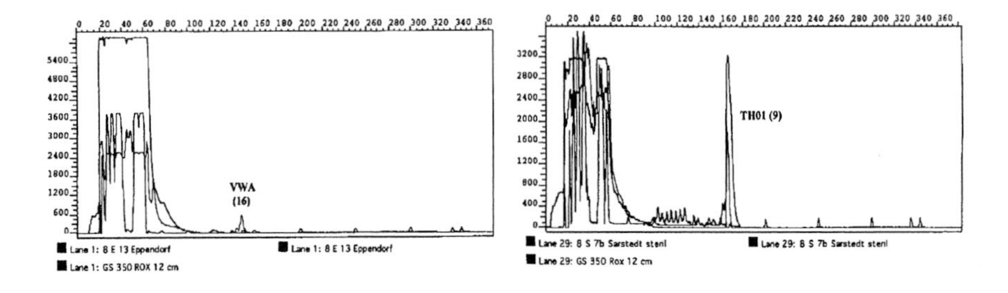

Fig. 6.6. Contamination in no-template controls after a 60-cycle multiplex amplification of three STR markers (VWA, TH01 and FES/FPS). Usually, only one of the STRs showed a signal, and in some cases, two of the STRs did. In none of the cases were all three STRs present, nor were heterozygote results obtained

Table 6.2. Contamination rates in reaction tubes of different brands (60-amplification cycles, human autosomal STRs, triplex PCRs)

		Triplex assay*		
Manufacturor	Brand	VWA	TH01	FES/FPS
Sarstedt	Regular (non-sterile)	10%	5%	5%
Sarstedt	Sterile	20%	10%	5%
Biozym	Regular	10%	15%	0%
Greiner	Regular	0%	10%	0%
Eppendorf	Safe-lock regular	20%	10%	0%
Perkin Elmer	GeneAmp	10%	30%	5%

* The triplex assay combined the most efficiently amplifying marker from the STR triplex consisting of VWA, FES/FPS, F13A1 and the STR duplex consisting of TH01 and CD4 (cf. section 8.5).

Talking to major manufacturers of reaction tubes (e.g., Eppendorf and Saarstedt) revealed that they are aware of the problem. Moreover, they know which steps in the manufacturing process are the critical ones where something could be done to avoid the problem (cf. footnote 11, chapter 6). In the case of the identification of cattle DNA in the tubes, it is obviously a consequence of the production factory being located in a rural area where there is extensive cattle breeding. On the other hand, their marketing analysis (Eppendorf) reveals that the portion of PCR users who are unhappy with this situation is too small to justify the major financial investment that would be necessary to change to a fully contamination-free manufacturing process, although in principle, this would be possible. The above experiments and the following paragraph name manufacturers, brands and product lines, in particular Eppendorf and their products. This is in no way intended to suggest that Eppendorf products or marketing politics are inferior to other brands. To the contrary, Eppendorf has been interested in talking to us candidly about the contamination problem, and the company appears to be generally very interested in selling products of the highest quality.

A major problem seems to be not only that supply guides demand, but also that demand directs supply. Therefore, it is not too surprising to learn that, for example, the Biopur product line of Eppendorf, which is sold as "DNA free," i.e., $< 5\times10^{-11}$ mg amount of DNA stated on the carton, actually undergoes an autoclaving process after the manufacturing process. This means that every possible cellular contamination, whether it is from something living or dead, is spread out through all tubes of the batch as opposed to being in a localized concentration in some tubes, as we had previously found. The certificate, which is in every carton of the particular type of tube sold, is based on microbiological testing, i.e., bacteria-culture growing tests. Autoclaved reaction tubes are, according to Eppendorf, what the majority of PCR users working with high standards of quality want. Therefore, this is what they get.

The situation is indeed improved if longer DNA fragments are amplified (>200–250 bp), since the DNA becomes fragmented during the autoclaving process. In fact, many of the modern intact DNA applications use fragment lengths of at least 250, 300 or 400 bp. In other words, it is to the advantage of these applications that the tubes are autoclaved, a treatment that appears to be totally incongruous to degraded DNA investigators or others who intend to amplify short fragments.

Thus, the small segment of PCR users who are not content suffers because of the wishes of the majority. Unfortunately, the majority of PCR users are not only satisfied, but they may not even be aware of the fact that a blank no-template control does not necessarily mean that there is no contaminated material present. This only means that nothing can be seen at this stage of the PCR process (the detection limit of DNA on ethidium-bromide stained agarose is 2 ng if transilluminated by 254-nm UV light).

In my opinion, a major problem lies in the fact that PCR is often judged to be "magic,"[16] meaning that irregular and unpredictable things happen. This is certainly not the case. To the contrary, PCR has been shown to be a predictable technique, particularly if sequences are amplified that do not come too much in conflict with the limiting parameters, such as the type of contaminations that are common in the reaction tubes, i.e., human mitochondrial and – at least from a German production plant – bovine mitochondrial DNA (cf. section 4.2). Whenever "magic" things happen (e.g., total failures from samples known to contain well-preserved DNA or unspecific bands from a primer pair that is usually absolutely specific), it regularly turns out that somebody made a mistake, ranging from using the wrong volume in pipetting to forgetting to include one of the reagents or the Taq polymerase to using an incorrect program in the cycler.

At present, it seems that using the PCR technique at its detection limits must continue to be restricted to those types of amplifications that enable a high power of discrimination for the results for data validation (cf. section 6.2). Alternatively, if highly conserved mitochondrial sequences from humans or common domestic animal species are amplified by many cycles, the experiments should be accompanied by a convincing control amplification, demonstrating that the reaction tubes enable high cycle numbers for the species in question without revealing unacceptably high rates of contamination. These findings indicate that such experiments cannot be replaced by repeated amplifications in conservative sequences, regardless of how often they are carried out, nor does the involvement of a second laboratory solve the problem. However, repeated analysis greatly improves the situatuion if polymorphic mitochondrial haplotypes such as the hypervariable region are amplified.

As long as the manufacturers do not offer truly DNA-free tubes, another solution may be improved decontamination treatments for the tubes.

6.1.3 Cleaning the PCR reaction tubes

The following protocol is intended to describe how to treat the tubes and reaction mix in order to improve the contamination situation in the tubes, a

[16] Even in a PCR manual co-edited by Kary Mullis, one of the inventors of the PCR, the amplification process is judged in this way: "Trial and error and black magic are often components of a PCR analysis in the lab..." (Garner, 1994).

prerequisite for the amplification of the low-target numbers of non-individual specific mitochondrial sequences (e.g., control region V, cytochrome b). Taking what was known from the experiments and cleaning efforts undertaken by Schmidt et al. (1995) and other publications on cross-linking double-stranded DNA in solution by 254-nm UV light (cf. section 6.1.2), we designed a new series of experiments. The following experiments have not been carried out as systematically as those of Schmidt. For example, we did not use tubes from different companies. In addition, we did not vary many of the treatment parameters. However, each experiment was reproducible, revealing some encouraging results, although we only achieved a clear reduction of the number of negative controls showing contamination but did not reach a truly satisfying situation. The parameters of the protocol that result in a reduction of the number of tubes containing amplifiable contaminations are given below.

The initial conditions are shown in Fig. 6.7, where an amplification was performed using the control region V primers for 50 cycles with no template added to the reaction tubes. In contrast to this, we reached a clearly improved situation (shown in Fig. 6.8) if the tubes and the master mix (without Taq polymerase) underwent the following treatment protocol prior to use in the amplification reaction:

- fill the tubes with Ampuwa water; heat up to 94°C for 15 min
- discard the water and remove small drops by pipetting
- prepare master mix without Taq polymerase
- aliquot the master mix to the treated tubes (mark the level)
- fill some further tubes with 50–100 μl Ampuwa (these will be needed for compensating for evaporation in the following step)
- place the tubes with master mix and tubes with water for 1.5 h under UV light (λ 254 nm, distance <5 cm); the caps must be open
- all tubes will generally lose approximately 10 μl to evaporation
- use the irradiated Ampuwa water tubes for setting up a diluted Taq-polymerase solution (this measure enables more precise pipetting of the Taq polymerase as a result of handling larger volumes)
- pipette the respective amount of diluted Taq polymerase (=2 U)
- fill up the remaining missing volumes using the irradiated Ampuwa (for no-template sample, i.e., a final volume of 50 μl)
- add DNA extract to the other tubes
- run PCR

Since the tubes show a substantial change in color (to yellow) as a result of the 1.5 h UV-light treatment, it must be assumed that they have been chemically altered as well. Therefore, it was not tested how much longer the tubes could withstand the UV irradiation, although this might further improve the contamination-rate situation. In the case of the brand we used for these experiments (Eppendorf standard 0.5-ml reaction tubes), the DNA amplification was apparently not inhibited by these chemical changes. This may be different if longer times for UV irradiation are used or if other brands are used, as different plastic components may be used for the manufacture. However, an elongation of the UV treatment seems to be unavoidable in order to achieve a better cross-linking rate.

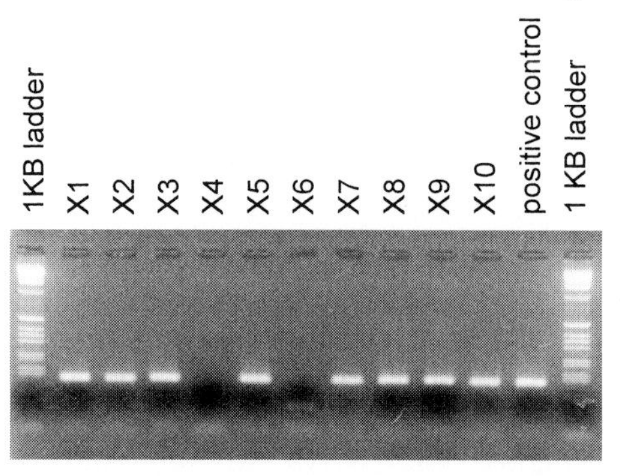

Fig. 6.7. Many of the non-treated tubes (70–80%) revealed contamination in the no-template samples that does not significantly differ from the positive controls after 50 cycles for the control-region V amplification product

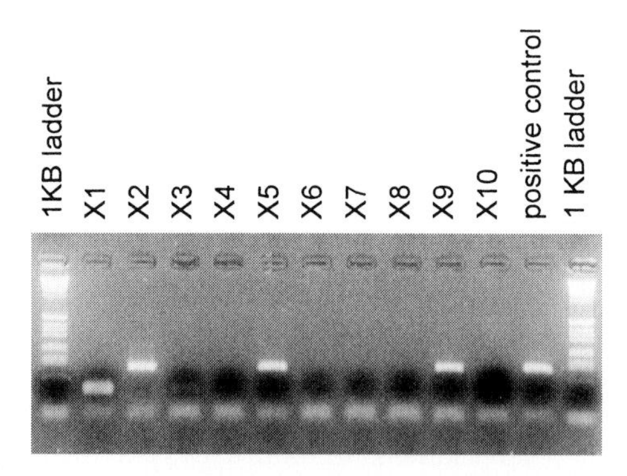

Fig. 6.8. Fewer of the UV-treated tubes (30%) show signals in the no-template controls. Although the tubes were severely affected by the UV treatment (change in color and decreased elasticity), they were still suitable for PCR without any inhibition of the Taq polymerase. The amplification consisted of 50 cycles targeting the control region V of the human mitochondria

6.1.4 PCR in glass tubes

Attempts to achieve contamination-free tubes by the use of glass reaction tubes have been completely successful. In general, DNA is known to adhere to glass at room temperature. However, the entire PCR process is carried out at a temperature level that does not allow DNA to stick to the walls of the glass reaction tubes. Therefore, there is no decrease in the reaction efficiency compared to the amplifications in plastic tubes (Figs. 6.9 and 6.10).

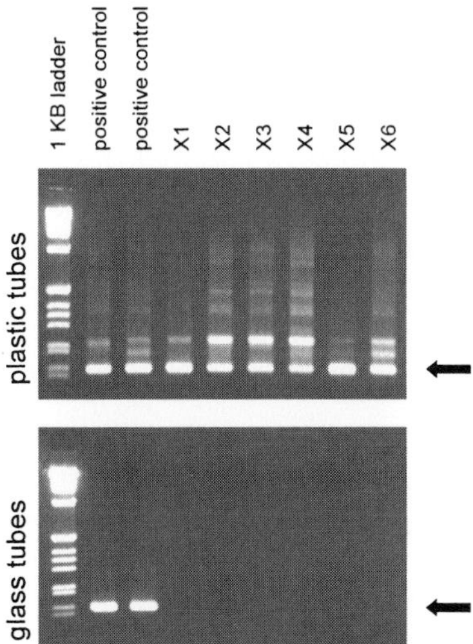

Fig. 6.9. Amplification of a 154-bp Y-chromosomal repeat sequence that is present 800–5,000 times in a single human genome. The results of positive and negative controls (X1-X6) in plastic tubes (Safe-lock Biopur, Eppendorf) and glass tubes that were individually manufactured by a glass blower are compared. The reaction mix had been prepared as a master mix before it was distributed to the different types of reaction tubes. The amplifications consisted of 50 cycles. The results clearly show that, in contrast to plastic tubes, the glass tubes enable contamination-free analysis

Unfortunately, glass tubes that fit the common thermocycling machines are not available from commercial suppliers. Therefore, it is necessary to have them manufactured by a glass blower. In order to enable the multiple use of the individually manufactured tubes, it is necessary first to quantitatively remove the amplification products of the reaction. This is done by a heat treatment procedure that is comparatively easy to carry out. The protocol of the procedure for the quantitative removal of amplification products from glass reaction tubes in order to enable their multiple use in PCR[17] is given below.

Most important concerning the use of glass tubes that are not commercially available but must be manufactured individually is the possibility to re-use them. The following protocol presents a cleaning treatment that quantitatively removes PCR amplification products. This procedure enables numerous series of amplification reactions within the same tubes to be carried out. Product carry-over and other kinds of contamination were never observed in the 60-cycle amplifications of the human autosomal STR system TH01 after the cleaning treatment of the glass tubes (Fig. 6.11). The following is the procedure for the quantitative removal of amplification products from glass reaction tubes:

- clean glass reaction tubes with a detergent (e.g., Alconox)
- rinse glass reaction tubes with H_2O_{bidest}
- heat glass reaction tubes to 600°C for 3 h
- the glass reaction tubes are now ready for use in PCR

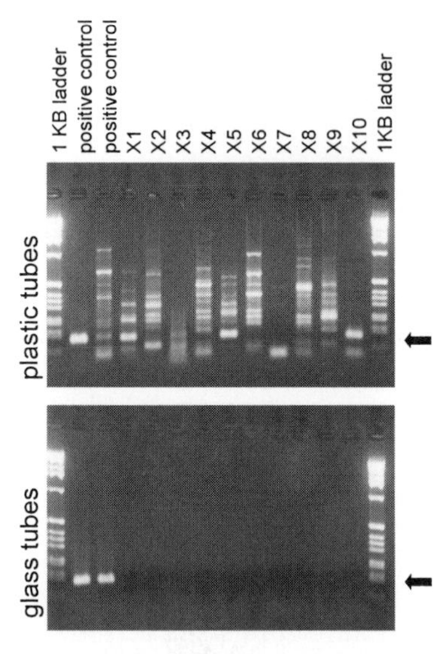

Fig. 6.10. Amplification of a 134-bp mitochondrial sequence of the cytochrome-b gene. Again, the results of positive and negative controls (X1-X10) in plastic tubes (Safe-lock Biopur, Eppendorf) and glass tubes (individually manufactured) are

[17] The procedure for the quantitative removal of amplification products from glass reaction tubes in order to enable their multiple use in PCR is the content of patent no. 198 15 668.5-42 ("Verfahren zur Amplifikation von Nukleinsäuresequenzen unter Verwendung geeigneter Reaktionsgefäße").

compared. The amplification consisted of 55 cycles. As in the previously presented experiment, the reaction mix had been prepared as a master mix before it was distributed to the different types of amplification tubes. Again, the glass tubes enable contamination-free analysis, while the plastic tubes reveal numerous contaminations, which result in specific and unspecific amplification products (also cf. Fig. 6.5)

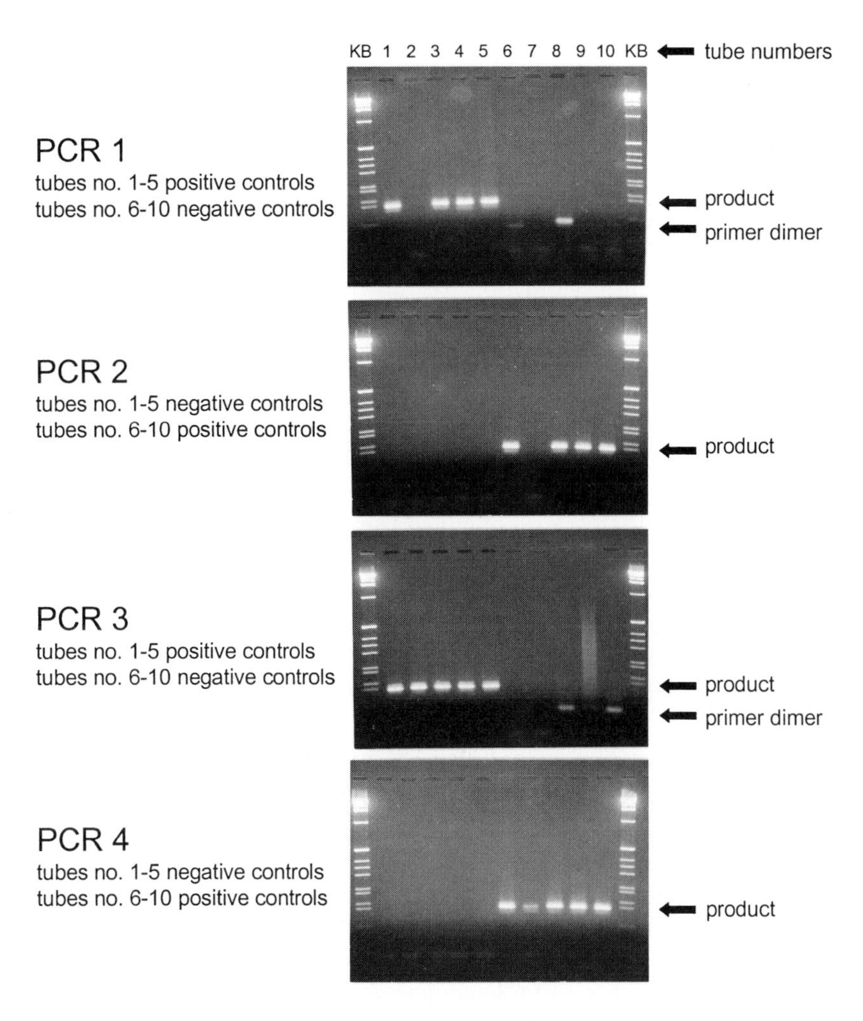

Fig. 6.11. In this experiment, the glass tubes were re-used in 60-cycles amplifications of the human autosomal STR TH01. Thus, those tubes, which contained the positive controls in one PCR run, served as tubes for the negative controls in the next PCR run. The heat treatment of the glass tubes in between the amplification runs ensures the quantitative destruction of any DNA molecule and thus the occurrence of any type of contamination, either carry-over or cellular

6.2 Validating aDNA results

In principle, data validation for aDNA amplification results should follow the same rules as any modern DNA amplification. It must be proven that the indigenous DNA was the only one amplified and not a contaminating source. This is regularly done using sets of control samples consisting of positive and negative controls. The positive control samples indicate that the reaction was set up properly and all parameters were adjusted in a suitable fashion. The negative controls monitor for systematic contamination. The extraction blanks allow all reagents used (throughout the entire extraction phase and PCR) to be monitored. The no-template controls will only confirm those reagents used in the amplification reaction mix. Nevertheless, the validation of an aDNA result is harder to accomplish due to the fact that the average aDNA amplification deals with comparatively few intact targets, i.e., even minor possible contaminations may become potent amplification competitors. However, the introduction of such minor contaminations may be a matter solely concerning the samples, not the chemicals and reagents. Therefore, negative control sample sets are not the suitable method at this point. Rather, it is necessary to control and evaluate the result obtained from a sample itself.

The only way to control the result of a sample is to repeat the entire experiment. This is called the "criterion of reproducibility," which is essential for all natural science experiments. If possible, the repeated experiments should also be carried out on a different sample from the same individual, preferably arranged as a blind test (e.g., Yang et al. 1997; Burger et al. 2001).

However, the validation power of repeatedly generated results derived from DNA amplifications very much depends on the particular information content of the sequences amplified (also cf. Hummel in press). If, for example, a highly conserved region such as control region V of the human mitochondria (cf. section 2.1) is repeatedly amplified, the validation power is comparatively low, because the information content (i.e., the discriminating power) of this sequence is low. To express this in terms of contamination, anybody could have contaminated the sample and the subsequent amplification and sequencing result would be the same regardless of whether the reproductions were carried out in a single or in two laboratories.

The situation already improves appreciably as soon as the mitochondrial hypervariable region (cf. section 2.1) is concerned. Many haplotypes exist in humans as well as in animals; therefore, the discrimination power of a repeatedly obtained result is much higher. However, the increased polymorphism of this sequence still reaches its limits in terms of data validation. This point is reached as single samples are studied. A well-known illustrative example is the investigation of Krings et al. (1997) on the Neanderthal specimen. If this specimen had revealed a haplotype matching to one of the persons involved in sample handling and processing or any other of the approximately 1,000 human haplotypes known worldwide, then, following the argumentation of the authors, this result would not have been accepted as an authentic one. The results were indeed evaluated this way by omitting the data of the second lab involved in the investigation. To play devil's advocate, what if it were the other way around, and the Leipzig result was just a matter of severe degradation and the Pennsylvania result the authentic one. Not to be misunderstood, I do not seriously consider that the Pennsylvania result was the correct one, since the evidence for contamination in this case was presented convincingly in the paper. This point is simply to demonstrate that only deviating or uncommon results can be verified in the investigations of single samples. It would have been almost impossible to validate a result derived from a single individual indicating that Neanderthals used to have the same haplotypes as modern homo sapiens do if the validation strategy as presented in the manuscript were used.

The situation is completely different if a series of samples are investigated. This is because they will show different results, except where they all derive from the same matrilineal line. In a way, the different results obtained for different individuals begin to develop a means for self-validation if they can be obtained reproducibly (e.g., Montiel et al. 2001).

This sort of self-validating capacity of a series of different results is more powerful as the number of samples increases and as the discriminating power of the genetic markers that are investigated increases. This is exactly the reason why STR typing (cf. section 2.2) is so compelling. Not because a single STR has an unquestionable power for discrimination, but because in their particular combination they are unique, which is the reason they are called genetic fingerprints.

This can be simply demonstrated. If a single dice is thrown, the chance of getting any particular number is 1:6. If a couple of dice are thrown, the probability of obtaining a certain combination of numbers is very much lower. In this case it does not matter which dice showed which number. If the same number of dice are thrown but with the stipulation that a certain combination of numbers must be reached by particular dice showing particular numbers, the probability of obtaining a result twice is incredibly low. This is exactly how multiplexing autosomal STRs operates.

Furthermore, STR markers are usually much more polymorphic than dice, i.e., more than six alleles are possible, and each STR actually consists of two alleles, analogous to a number of pairs of dice being thrown at the same time. For amplifying STRs, this is called a "multiplex amplification." Using this example, it becomes clear that it is incredibly unlikely that an autosomal-STR typing result obtained from repeated independent experiments would be obtained twice just by chance alone, i.e., due to sporadic contaminations. For the particular combination of the nine STRs in the AmpFlSTRProfilerPlus amplification kit, this dissimilarity can be numbered to average 10^{-11}. The results obtained from tri- or quadruplex STR typing have already reached an impressive discrimination power, which can be precisely calculated either for a particular allele combination or as an average value that represents a sort of prediction for any possible combination (cf. section 2.2.2).

Since multiplexing genetic markers increases their combined discriminating power, this strategy is also one solution for the systematic identification of non-individual-specific genetic markers (cf. sections 2.2.6 and 2.2.7). By linking such a procedure with a multiplex PCR, the results of the power of discrimination are transferred to the non-discriminating result. There is an important prerequisite for this, since the non-discriminating genetic marker must be of the identical basic type as the discriminating marker. When combining non-discriminating markers with STRs, the non-discriminating one must also be a single-copy sequence, as STRs are. If they are polymorphic and not of the binary type (present or absent), the polymorphism must be a length polymorphism in order to enable the analysis to be interpreted.

The following tables (Tables 6.3, 6.4 and 6.5) show the power of autosomal STRs in identifying contamination and thus their validation power (Hummel et al. 2000). The identifications of contaminations were not obtained from using an experimental approach, i.e., the results were neither expected nor did anybody search for them intentionally. On the contrary, the contaminations were identified in the regular course of the investigations on the Goslar series of skeletons (cf. section 7.4.2). The three samples presented here were the only results in this investigation that revealed ambiguous STR typing. However, the negative controls of all amplifications were blank.

It can be concluded that data validation for aDNA analysis cannot be done by control samples alone – as is possible in modern DNA analysis. It requires discriminating, individual-specific genetic markers to exclude contamination, which may be a matter of the samples only and is therefore not monitored by negative controls, as was shown by the previous tables.

The self-validating power of the data obtained from semi- and highly polymorphic markers increases in a sample series. The individual results that can be obtained from the different samples of the series indirectly prove that the source of the individual-specific results was not a person who had formerly worked with the samples (e.g., an excavator).

In general, multiplex experiments are superior to any singleplex amplification with respect to data validation, since they reduce the risk of getting particular combinations of a pattern by chance alone, i.e., because of a contamination event (also cf. Hummel in press).

Table 6.3. Identification of a sample-to-sample cross-contamination during the extraction step. Sample GS 88 was obviously contaminated by sample AK, which was processed along with GS 88

GS 88 (early modern)											
Extract no.	PCR no.	Amelogenin	D3S1358	D8S1179	D5S818	VWA	D21S11	D13S317	FGA	D18S58	D7S820
I	1a	X	15	13	12/13	18/19	29	10/13	21/24	13/14	8/12
I	1b	X	14/15	13	12/13	18/19	29	10/13	21/24	13/14	8/12
II	2a	XY	-	13	9/12	17/19	29	-	23	15/18	-
II	2b	XY	-	11/13	9/12	17/19	-	-	-	-	-
AK		X	15	13	12/13	18/19	29	10/13	21/24	13/14	8/12

Table 6.4. Identification of a direct contamination of a single amplification reaction (I-1b) of GS 64 by the person (BB) handling the samples

GS 64 (early modern)											
Extract no.	PCR no.	Amelogenin	D3S1358	D8S1179	D5S818	VWA	D21S11	D13S317	FGA	D18S51	D7S820
I	1a	X	-	14	-	16/17	28/29	9	-	12	8/12
I	1b	X	14/17	13/14	11/14	14/17	30/32.2	11/14	21/24	13/16	8
II	2a	X	16	12/14	10/14	16/17	28/29	9/13	20	12	8/12
II	2b	X	16	12/14	10/14	16/17	28/29	9	21	12	8/12
BB		X	14/17	13/14	11	14/17	30/32.2	11/14	21/24	13/16	8

Table 6.5. Identification of a sporadic contamination (II-2a and II-2b). Because of the non-reproducibility, the contaminating sequences most likely stem from the reaction tubes and further suggest that extraction II failed to isolate considerable amounts of DNA from GS 31

GS 31 (early modern)											
Extract no.	PCR no.	Amelogenin	D3S1358	D8S1179	D5S818	VWA	D21S11	D13S317	FGA	D18S51	D7S820
I	1a	X	-	11/13	10/13	16/18	29	9/11	19/22	14	10
I	1b	X	-	11/13	10/13	18	29	9/11	19/22	14	10
II	2a	-	14	10/13	-	15/17	-	-	-	17	-
II	2b	XY	17	13/15	-	-	29.2	-	24	-	-
III	3a	X	15/18	11	13	16/18	-	9/11	19/22	14	10
III	3b	X	15/18	11/13	13	16/18	29	-	19/22	14	10

Many non-discriminating markers can become part of a multiplex assay as a direct consequence of their amplification simultaneously with polymorphic markers. Reproducible results from independent experiments are always necessary. It is not necessary to carry out these experiments in different laboratories. This may only be desirable if single samples are investigated, since then validation through the different results of further samples of the same kind cannot be given. If different laboratories are involved in such case studies, the mere involvement is not sufficient, rather both laboratories must demonstrate the same results independently.

If mitochondrial DNA is amplified, the high risk of contamination through amplification tubes must be carefully considered. In such cases there should be sufficient polymorphism in combination with reproducible results carried out through independent experiments that improve data validation. However, if non-polymorphic sequences are amplified, there are a number of possibilities: (1) the contaminations from the tubes must quantitatively be inactivated, (2) a method must be found to efficiently clean the tubes, (3) the amplifications must be kept to a limited number of cycles, (4) alternative reaction tubes (glass) should be used that enable a thorough decontamination treatment by heat or (5) new brands of reaction tubes may be suitable (cf. footnote 11, chapter 6).

If the choice is a limited number of cycles, one should be aware that possible results that could have been gained will be lost. Which number of cycles appears to be "safe" is probably best determined by randomly sampling batches of no-template controls from their containers. This also applies to investigations on non-polymorphic markers of domestic animal species or their ancestors (cf. section 6.1.2). For animal studies, however, it may be much easier to find reaction tubes that are not contaminated or at least not by the species of interest.

6.3 Ring exercises

Taking part in certificated ring exercises is a recommended way to check the standards of a laboratory. This includes sample processing and the technical standards as well as the scoring abilities of the personnel. For example, the samples that are given to the participants in the German DNA Profiling ring exercise (GEDNAP) on human STR typing (autosomal and Y-chromosomal) usually consist of a set of seven samples. Some of the samples represent individual genotypes of single persons only. These are dried blood, saliva or sperm stains, usually spread out on plain cotton cloth. Furthermore, there are samples that consist of mixed stains as they may occur in a crime context. Those samples come on pieces of paper, cigarette butts, stamps, etc. The alleles of all samples must be determined and assigned to each other, and relative amounts of DNA in case of mixtures must be determined. Two ring exercises are carried out each year. The German ring exercise on STR determination is organized by the Institute of Legal Medicine at the University of Münster[18] (also cf. Rand et al. 2002).

A similar ring exercise has been set up for mitochondrial haplotyping of the human mitochondrial HVR I and II. This is organized by the Institute of Legal Medicine at the University of Magdeburg[19].

References

Böttger EC (1990) Frequent contamination of Taq polymerase with DNA. Clin Chem 36:1258

Burger J (2000) Sequenzierung, RFLP-Analyse und STR-Genotypisierungen alter DNA aus archäologischen Funden und historischen Werkstoffen. Dissertation, Georg August-Universität, Göttingen

---

[18] http://medweb.uni-muenster.de/institute/remed/

[19] http://www.med.uni-magdeburg.de/image/e32.htm
or dieter.krause@medizin.uni-magdeburg.de

Burger J, Hummel S, Herrmann B (2000) Palaeogenetics and cultural heritage. Species determination and STR genotyping from ancient DNA in art and artefacts. Thermochimica Acta 365:141–146

Cimino GD, Metchette K, Isaacs ST, Zhu YS (1990) More false-positive problems. Nature 345:773–774

Cone RW, Fairfax MR (1993) Protocol for ultraviolet irradiation of surfaces to reduce PCR contamination. PCR Meth Appl 3:S15–S17

Drury KC, Liu MC, Zheng W, Kipersztok S, Williams RS (2000) Simultaneous single-cell detection of two mutations for cystic fibrosis. J Assist Reprod Genet 17:534–539

Drury KC, Liu MC, Lilleberg S, Kipersztok S, Williams RS (2001) Results on single cell PCR for Huntington's gene and WAVE product analysis for preimplantation genetic diagnosis. Mol Cell Endocrinol 183 [Suppl 1]:S1–4

Dwyer DE, Saksena N (1992) Failure of ultra-violet irridation and autoclaving to eliminate PCR contamination. Mol Cell Probes 6:87–88

Frothingham R, Blichington RB, Lee DH, Greene RC, Wilson KH (1992) UV absorption complicates PCR decontamination. BioTechniques 13:208–210

Garner HR (1994) Automating the PCR process. In: Mullis KB, Ferré F, Gibbs RA (eds) PCR. Polymerase chain reaction. Birkhäuser, Boston

Hauswirth WW (1994) Ancient DNA. Experientia 50:521–523

Higuchi R, von Beroldingen CH, Sensabaugh GF, Erlich HA (1988) DNA typing from single hairs. Nature 332:543–546

Hubert R, Weber JL, Schmitt K, Zhang L, Arnheim N (1992) A new source of polymorphic DNA markers for sperm typing: analysis of microsatellite repeats in single cells. Am J Hum Genet 51:985–991

Hummel S, Bramanti B, Schultes T, Kahle M, Haffner S, Herrmann B (2000) Megaplex DNA typing can provide a strong indication of the authenticity of ancient DNA amplifications by clearly recognizing any possible type of modern contamination. Anthrop Anz 58:15–21

Hummel S (in press) Ancient DNA: recovery and analysis. (Ref. no. 342) Encyclopedia of the Human Genome. Nature Publ Group (spring 2003)

Kitchin PA, Szotyori Z, Fromholc C, Almond N (1990) Avoidance of false positives. Nature 344:201–201

Koponen JK, Turunen AM, Yla-Herttuala S (2002) *Escherichia coli* DNA contamination in AmpliTaq gold polymerase interferes with TaqMan Analysis of lacZ. Mol Ther 5:220–222

Krings M, Stone A, Schmitz RW, Krainitzki H, Stoneking M, Pääbo S (1997) Neanderthal DNA sequences and the origin of modern humans. Cell 90:19–30

Kwok S (1990) Procedures to minimize PCR-product carry over. In: Innis MA, Gelfand DH, Sninsky JJ, White TJ (eds) PCR protocols. Academic Press, San Diego

Lassen C, Hummel S, Herrmann B (1995) Comparison of DNA extraction and amplification from ancient human bone and mummified soft tissue. Int J Legal Med 107:152–155

Li H, Gyllensten UB, Cui X, Saiki RK, Erlich HA, Arnheim N (1988) Amplification and analysis of DNA sequences in single human sperm and diploid cells. Nature 335:414–417

Longo MC, Berninger MS, Hartley JL (1990) Use of uracil DNA glycosylase to control carry-over contamination in polymerase chain reaction. Gene 93:125–128

Montiel R, Malgosa A, Francalacci P (2001) Authenticating ancient human mitochondrial DNA. Hum Biol 73:689–713

Niederhauser C, Höfelein C, Wegmüller B, Lüthy J, Candrian U (1994) Reliability of PCR decontamination systems. PCR Meth Appl 4:117–123

Rand KH, Houck H (1990) Taq polymerase contains bacterial DNA of unknown origin. Mol Cell Probes 4:445–450

Rand S, Schürenkamp, Brinkmann B (2002) The GEDNAP (German DNA profiling group) blind trial concept. Int J Legal Med (Online First: http://link.springer.de/link/service/journals/00414/contents/02/00285/)

Rys PN, Persing DH (1993) Preventing false positives: quantitative evaluation of three protocols for inactivation of polymerase chain reaction amplification products. J Clin Microbiol 31:2356–2360

Sarkar G, Sommer SS (1990) Shedding light on PCR contamination. Nature 343:27

Sarkar G, Sommer SS (1993) Removal of DNA contamination in polymerase chain reaction reagents by ultraviolet irradiation. Meth Enzymol 218:381–389

Schmidt TM, Pace B, Pace NR (1991) Detection of DNA contamination in Taq polymerase. BioTechniques 11:176–177

Schmidt T, Hummel S, Herrmann B (1995) Evidence of contamination in PCR laboratory disposables. Naturwissenschaften 82:423–431

Scholz M, Pusch CM (2000) Contamination through preparation: risk of molecular genetic studies by using biological preservatives for museum collections. Anthrop Anz 58:225–235

Wages JM, Fowler AK (1993) Amplification of low copy number sequences. Amplifications 11:1-3

Wrischnik LA, Higuchi RG, Stoneking M, Erlich HA, Arnheim N, Wilson AC (1987) Length mutations in human mitochondrial DNA: direct sequencing of enzymatically amplified DNA. Nucleic Acids Res 15:529–541

Yang H, Golenberg EM, Shoshani J (1997) A blind testing design for authenticating ancient DNA sequences. Mol Phylogenet Evol 7:261–265

Zhang L, Xiangfeng C, Schmitt K, Hubert R, Navidi W, Arnheim N (1992) Whole genome amplification from a single cell: implications for genetic analysis. Proc Natl Acad Sci USA 89:5847–5851

Zierdt H, Hummel S, Herrmann B (1996) Amplification of human short tandem repeats from medieval teeth and bone samples. Hum Biol 68:185–199

7 Applications

The following sections give examples from our own work on applications of aDNA analysis to the fields of historical anthropology, archaeozoology, conservation biology, primatology and food technolgy. Of course, the presented examples only represent a small part of the wide variety of possible applications that can help to answer old questions with this comparatively new molecular approach.

For historical anthropologists, skeletons are the major source of material to examine the past. For decades, knowledge about the lives of our ancestors was restricted to the biological data that could be obtained through metric data and morphological inspection. These techniques, of course, allow anthropologists to draw significant conclusions about changes in living conditions by following, for example, the changes in body heights throughout the ages. Through age and sex determination by morphological methods, it has also been possible to establish the average risk for females not to survive their young adulthoods because of pregnancies and births. However, using these more traditional approaches, some information has always remained beyond reach. For example, how were people related to each other? Did they marry within their own social class, and did they invest equally in all of their children, or did they differentiate in this depending on the child's gender? Was professional expert knowledge a matter of heritage or was it imported? Did our ancestors systematically breed and trade in livestock, and since when has this taken place and with whom? These questions reflect the way in which a human society is organized, and the answers are part of our cultural history. Molecular methodology will enable us to address some of these questions. In the following sections of this book, I would like to give several examples of what some of these answers could look like and how they could be achieved and interpreted. Detailed protocols, including all primer sequences, are given in chapter 8.

7.1 Sex determination

Of interest in any anthropological diagnosis of a skeleton, age and sex determination is a basic feature of biological information (Hummel and Herrmann 1991; Hummel 1994; Lassen et al. 1996; Stone et al. 1996; Faerman et al. 1997). These data are vital for paleodemography, which enables anthropologists to recognize trends in relation to life expectancy for both genders. This in turn enables conclusions to be drawn about changes in historic living environments and medical knowledge, for example.

An experienced anthropologist would face no major problems in determining the sex of an adult skeleton using morphological techniques. This was shown by a study that examined sex determination by morphological traits and compared these results using independent molecular data (Hummel et el. 2000). However, sex determination on skeletal material from infants remained a problem although there have been improvements in this method (Schutkowski 1990; Schutkowski 1993) compared to the morphological methods, which are unable to do this. Therefore, the approach to sex determination at the molecular level remains a valuable tool.

7.1.1 The children of Aegerten

Burials of children can tell us much about the history of family life and its sociocultural changes. However, this has not always been a concern for archaeologists or anthropologists. There is even an indication that such remains were often not excavated. This situation was entirely different in the excavation of the churchyard of Aegerten in Switzerland (Bacher et al. 1990), where the skeletons of about 120 infants, newborns and even stillborns from the second trimester of pregnancy were found. The preservation of this exceptionally well excavated and diagnosed skeletal material (Ullrich-Bochsler 1997) was certainly enhanced by the specific situation of this site. All infant burials were found in several layers near the church walls (Fig. 7.1), a tradition that goes back to the time of the Roman Catholic faith, although the parish of Aegerten was already reformed in the days of these particular burials, which date from the 16th to 19th centuries. Burying the very young infants close to the church walls was thought to ensure that they would be baptized as a result of the "blessed water" running down the roof of the church. For this reason, they were called "Traufenkinder." This was considered to be a particularly good solution for guaranteeing eternal blessing if the child had died before it could be baptized, which certainly happened often in the remote living areas in the mountains.

Further indication that this was the motivation for choosing this particular place for the burial of children was the fact that 69% of the "Traufenkinder" were newborns and premature births (Fig. 7.2), although this age group made up only 15% throughout the regular burial site. This indicates that the people of Aegerten cared a great deal about their children.

Therefore, it was surprising that the morphometric sex determinations revealed a large excess of female children (≈65%) (Ullrich-Bochsler 1997). This result deviates greatly from the natural mortality rates of both genders, which could only be explained by an intended neglect of female babies, or even infanticide. One possible reason could lie in the law of succession. An example of a particular law of succession causing major differences in the mortality rates of male and female infants comes from the Krummhörn, a region in northwestern Germany, where historic demographers investigated the topic by analyzing numerous parish registers (e.g., Voland 1984).

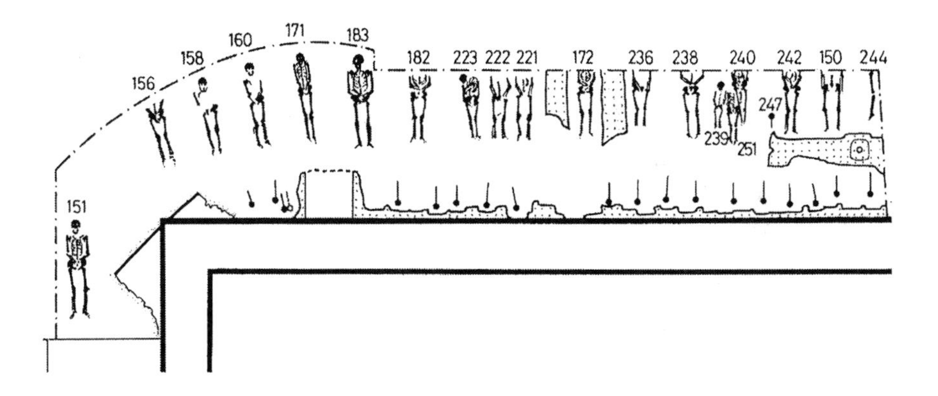

Fig. 7.1. Sketch of a part of the historical church and graveyard at Aegerten[20]. The skeletons of most of the very young infants and stillborns are situated immediately along the walls of the church. Because of their location and the blessed water from the church roof dripping down on them, they were called the "Traufenkinder"

[20] The figures 7.1 and 7.2 are from Bacher et al. (1990)

Fig. 7.2. In situ view of a burial of two young infants at the Aegerten site. The tiny bones clearly reveal a very good state of preservation

Therefore, the skeletons from all child burials in Aegerten were sampled in order to carry out a molecular sex determination by examining the amelogenin gene alleles.[21] This was successful in about 90% of the individuals. Only those results that gave at least three positive and reproducible analysis results, meaning an average of six amplification attempts, are detailed (Fig. 7.3). From those results, it became evident that the morphometric sex determination had failed, because there were more male stillborns and newborn infants than female (Table 7.1) (Lassen 1998; Lassen et al. 2000).

Overall, the male/female mortality ratio was calculated to be 1.75. This seems high at the first glance, but the great number of premature births should be considered. These data correspond closely with studies on the sex ratio in spontaneous abortions and the WHO data on perinatal male mortality excess (male:female 1.5) when intensive medical care is not

[21] At the time these experiments were carried out (1996), multiplex amplifications were not known to work on ancient DNA; therefore, the amplifications were carried out as singleplex reactions consisting of 40–60 cycles. At a later stage of the investigations the multiplex approach became known through STR typing. Eighteen of the remaining extracts were chosen randomly and typed by duplex PCRs consisting of VWA/FES-FPS and CD4/TH01. Amplification products were obtained from 12 of the samples that revealed reproducible results.

available. From these results it can be concluded that the people of Aegerten cared a great deal about their children, irrespective of the gender. This is consistent with the care that is also shown in the choice of preferential burial sites next to the church.

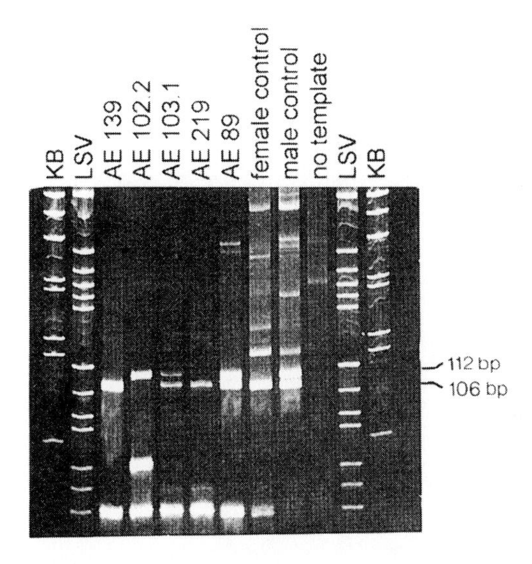

Fig. 7.3. Gel electrophoresis of the amelogenin amplification products. For the sample AE 102.2, an allelic dropout of the X-chromosomal fragment (106 bp) is seen. In principle, if the Y-chromosomal marker (112 bp) is shown to be reproducible, the sex determination can be carried out. In contrast, the sex determination should be treated carefully if the Y-chromosomal fragment is suspected to have failed to amplify

Table 7.1. Sex determinations by morphometric and molecular methods (*lm* lunar month)

		Morphometric method by discrimination analysis		Molecular method by amelogenin	
	Age	Males	Females	Males	Females
Premature Newborns/stillbirths	5.5–6.5 lm	2	3	2	3
	7.0–7.5 lm	2	3	3	2
	8.0–8.5 lm	4	9	11	2
	9.0–9.5 lm	8	25	18	15
Mature newborns	10 lm	11	12	16	7
Infant	0–6 month	7	7	8	6

7.1.2 Sex determination by multiplex amplification

The example of the burial site at Aegerten has shown that sex determination may be an important consideration in an anthropological investigation. However, this type of investigation was remarkably improved by more modern advances in genotyping. For such investigations, as well as those involving the sexing of single individuals, an improved sexing method compared to any singleplex analysis of the amelogenin locus was desirable. This was provided by a 45–55 cycle multiplex analysis of the amelogenin locus, two X-chromosomal STRs and two Y-chromosomal STRs[22]. By choosing X-chromosomal markers with high heterozygosity rates, the chance of obtaining a positive proof for female individuals receiving at least one heterozygote result is very high. This is shown by the application of the multiplex PCR to specimens from various burial sites (Fig. 7.4).

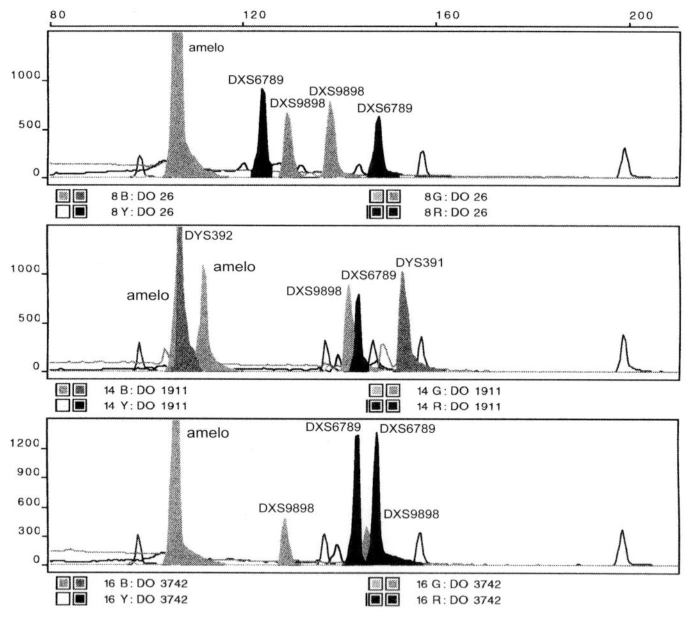

Fig. 7.4. Electropherograms showing the results of three Bronze Age samples. DO 26 (*top*) and DO 3742 (*bottom*) are female individuals, with each showing two heterozygotic X-chromosomal STR markers. As expected, the Y-chromosomal STRs (*bottom line*) show no signals. DO 1911 (*middle*) is a male individual showing 106 bp and 112 bp for amelogenin and one peak each for all STR markers

---

[22] The design of the multiplex PCR was carried out by Diane Schmidt and the author. More typing results will be published elsewhere.

By choosing the X-STR markers from closely neighboring regions of the X chromosome where linkage must be assumed (Hering and Szibor 2000; Hering et al. 2001; cf. section 2.2.4), we decided to lower the increased discrimination power of independent STR markers. This enabled us to recognize a sort of "haplotype," which is passed from parents to their children. By using this method, the identification of the father-daughter kin relationship, which is the most difficult one to identify, has been much improved. This quasi-haplotype can be used like most mitochondrial and Y-chromosomal haplotypes in order to trace family lineages (cf. section 7.3.3).

7.2 Identification

Identification by means of molecular genetics can be sought on various levels depending on the intention, i.e., the question that is asked. Questions may range from individual identification of either human or animal remains to identification at the species level. The following applications give examples that span the entire spectrum of questioning and are entitled: "A disturbed burial site," "Duke Christian II," "Pieces of parchment," "Sheep, goats and cattle," "Species, sex and individual identification," "Dog or fox" and "Rock wallabys." In the latter, two examples of species identification were carried out by means of sequencing, i.e., the analysis was not restricted to the identification. The sequence data could also serve population genetics. Nevertheless, they were included in this section, because the actual applications were singular identification cases.

7.2.1 A disturbed burial site

The Lichtenstein cave is a collective burial site located in the southwestern Harz mountains in Lower Saxony, Germany. Due to numerous finds of jewellery, tools and ceramics, the site has been dated to the Late Bronze Age (1000–700 B.C.) (Fig. 7.5) (Flindt 1996; Flindt 2001).

Fig. 7.5. Pottery found in the Lichtenstein cave. In particular, the vessels on the far *left* and *right* are characteristic for the Urnfield period in central Europe

The skeletal material of both human and animal origin that has been found in the cave is exceptionally well preserved due to a constant low temperature of 6–8°C and the fact that many skeletal elements were covered by a layer of gypsum sinter (Fig. 7.6). Both of these conditions are highly suited for DNA preservation (section 3.1). All skeletal elements found in the cave during the excavation campaigns (1995-2001) have been stored at -20°C within 24 h after removal from the site at the latest.[23]

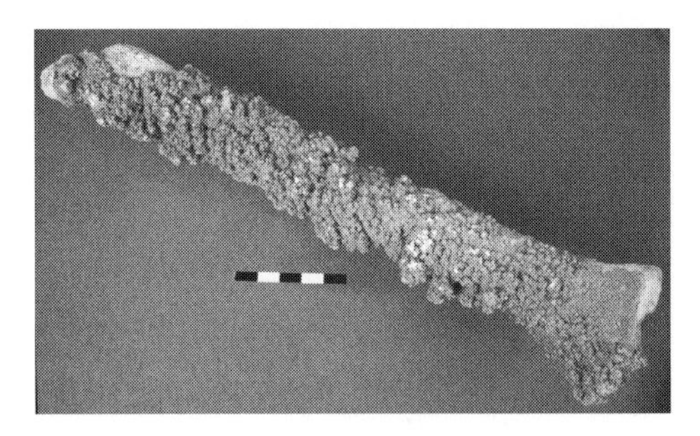

Fig. 7.6. Diaphysis covered by a thick layer of gypsum sinter

The skeletal elements in the cave were more or less dislocated from their original anatomical positions due to burial habits (Fig. 7.7). In order to test whether it is possible to assign skeletal elements to each other by STR typing, samples were taken from two femurs, two tibias and two hip

[23] We are most grateful to Dr. Stefan Flindt, archaeologist from Osterrode/Harz, for providing us with the skeletal material from the Lichtenstein cave, which he and his colleagues excavated in a number of expeditions. Thanks to the great and objective interest of Dr. Flindt in modern natural science methodology, it was possible to develop many methodological improvements for molecular analysis on archaeological finds using the human and animal skeletal material from the Lichtenstein cave. This is given as evidence in many applications within chapter 7.

bones. While the bones of the lower limbs showed highly similar morphological traits, it was unclear whether one of the hip bones would match.

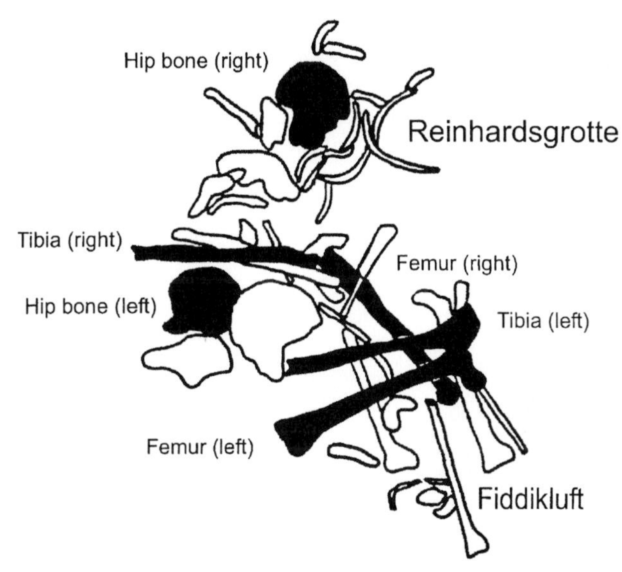

Fig. 7.7. Sketch of an in situ situation in two chambers (Reinhardsgrotte and Fiddikluft) of the Lichtenstein cave. To date, more than 10,000 skeletal elements from humans and animals have been excavated at this site

The genotyping was carried out using 55 amplification cycles of multiplex and singleplex assays employing STR loci (VWA, FES/FPS, CD4, TH01 and D3S366). For each PCR assay, two independent amplifications were carried out. From the STR typing it became clear that all of the long bones and one of the hip bones (Os coxa 1) belonged to a single individual (Table 7.2) (Schultes et al. 1997a; Schultes et al. 1997b).

Table 7.2. Assignment of skeletal elements by autosomal STR typing

	VWA	FES/FPS	TH01	CD4	D3S366
Femur (r)	16/16	10/12	7/9.3	5/10	12/16
Femur (l)	16/16	10/12	7/9.3	5/10	12/16
Tibia (r)	16/16	10/12	7/9.3	5/10	12/16
Tibia (l)	16/16	10/12	7/9.3	5/10	12/16
Os coxa 1	16/16	10/-	7/9.3	5/10	12/16
Os coxa 2	14/16	10/11	7/9.3	5/10	10/12

7.2.2 Duke Christian II

Duke Christian II of Braunschweig-Wolfenbüttel was a Protestant known for his brutal raids on Catholic churches and monasteries (Mayer 1996). During the Thirty Years War (1618-1648), his left arm was severely wounded during the Battle of Fleurus in 1622 and, according to the historical records, had to be amputated right on the battlefield "four fingers above the elbow" (Königl. Akad. Wiss. Hist. Komm., 1876) (Fig. 7.8). After this, the fate of his limb remained unknown.

Fig. 7.8. Duke Christian II of Braunschweig-Wolfenbüttel (1599-1626)

In 1995, when the sarcophagus was opened in the Beatea Mariae Virginis Church in Wolfenbüttel, Lower Saxony, the partially decomposed skeleton of Christian was found with the left humerus showing healed amputation marks. Additionally, the skeletal remains of an isolated left forearm were recovered that were mounted in an anatomically correct fashion with copper wires (Fig. 7.9). The forearm bones exhibit a glossy surface, indicating the wear of handling.

The objective of the following study was to determine if the isolated limb could be assigned to Christain, an assumption that was seemingly plausible considering the configuration of the fractures, the stage of maturity and the overall impression of the bone elements.

For aDNA analysis, bone samples were taken from the proximal diaphysis of the intact right humerus and from the ulna of the mounted left forearm. Several 60-cycle triplex amplifications employing autosomal STRs (VWA, FES/FPS and F13A1) and several 55-cycle singleplex STR amplifications for the autosomal STR D1S1656 were carried out on both samples. For sample preparation, DNA extraction and amplification conditions, see the respective protocols described in chapter 8.

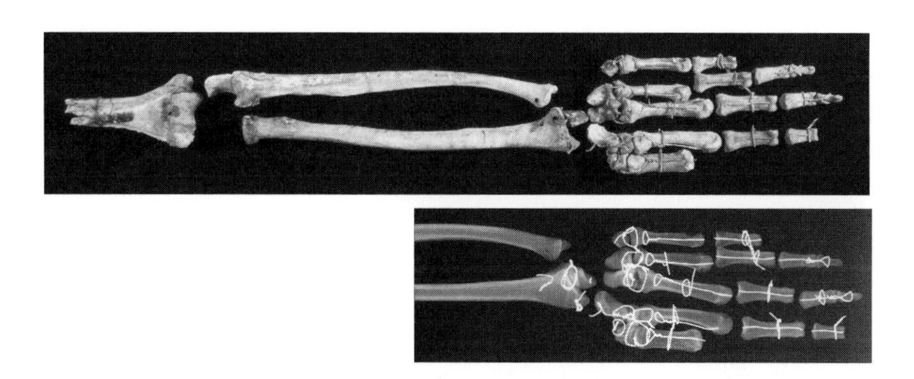

Fig. 7.9. Amputated forearm found in the sarcophagus of Duke Christian II. The lower picture shows an X-ray exposure of the specimen

The results of the allele determinations unequivocally showed that the isolated left forearm was from the skeleton of Christian II (Table 7.3) (Gerstenberger et al. 1998). This leaves room for speculation that the Duke had his arm deliberately prepared to be carried in a sling (cf. Fig. 7.8) and shown off for reasons of bravado or perhaps a bizarre sense of humor.

Table 7.3. STR allele determinations from humerus and ulna samples

	VWA	FES/FPS	F13A1	D1S1656
Humerus	14/16	10/11	5/7	13/18.3
	14/16	10	5/7	13/18.3
	13/14/16	-	5	13/18.3
	14/15/16	-	5/7	13
	14/16	10	5/7	
	14/16	10	7	
Ulna	14/16	10/11	5/7	13/18.3
	14/16	10/11	5/7	-

The matching probabiltity (P_m) for the particular genotype is P_m=4.4×10^{-7}, i.e., this genotype is generated from 1 out of 40,000,000 persons (cf. section 2.2.2). Due to the lack of historical allelic frequencies, the calculation is based on modern population data

7.2.3 Pieces of parchment

Historical records describe a Chinese servant in the court of Tsai Li to be the first person who ever manufactured paper. This occurred in the year 105 AD. It was more than 1,000 years later, in the 14th century, that the art of manufacturing paper came to Europe via the Islamic world. Until then, parchment was the only writing material and continued to be used for another 400 years, when the Frenchman Robert invented the first semi-automated machine for paper production.

Since parchment was prepared from animal skin, it was a valuable commodity and was predominantly used in monasteries for hand-written books and by civil servants for documentation of royal wills, treaties and civil contracts. Because of the degradation of this writing material, historical and paleographical investigations often need to assign fragments of parchment to each other before the interpretation of the written messages can begin. Usually, the meaning of the words themselves leads to an unambiguous assignment. However, sometimes the meaning of a text is not enough to decide to which page or scroll the fragments belong. In this latter case, a molecular analysis using autosomal STR typing may prove helpful (Fig. 7.10).

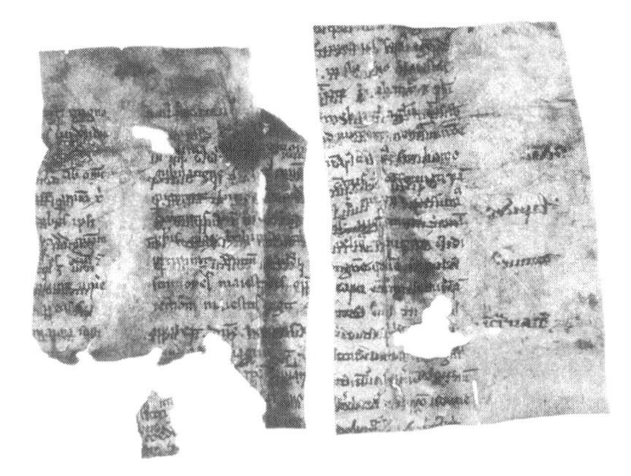

Fig. 7.10. Historical parchment revealing fragmented areas

The following experiment using autosomal STR typing was carried on two historic, early-modern parchment samples (Burger et al. 2000a; Burger et al. 2000b; Burger et al. 2001) that previously had been identified as being manufactured from calf skin (*Bos taurus*). This was done by an RFLP analysis on a 195-bp cytochrome-b amplification product (cf. section 7.2.4). The STR typing experiment was carried out as a blind test in order to find out about the identification power of genetic typing on minimum amounts of sample material. Therefore, the parchment samples were each cut into six small pieces of approximately 2 mm². The pieces were assigned random sample numbers and were extracted using Chelex100. The typing was carried out by amplifying nine dinucleotide STR markers that are commonly used in veterinary medicine for paternity testing (Stock marks kit, Applied Biosystems) (also cf. Heyen et al. 1997). The person who carried out the allele determinations was not aware that those 12 samples originated from only two different parchments.

The results (Table 7.4) of the STR typing demonstrated that allele determinations could be carried out for all STRs except one, which is supposed to amplify fragments longer than 250 bp. However, due to the many STRs amplifying shorter fragments, it was possible to unambiguously assign the 12 samples of the experiment to the original two parchments. Furthermore, a comparison of the allele frequencies also revealed that the alleles that were found in the historic samples are very rare and almost absent in modern cattle breeds (0.005–0.1).[24]

Table 7.4. STR typing of two historic parchments in a blind test experiment

	No..	TGLA 27	BM 2113	ETH 225	MTG 4B	TGLA 53	TGLA 122	BM 1824	INRA 23
Parchment 1	2	93/97	129/137	(144)/148	134	-	143/(153)	182/190	-
	3	93/97	129/137	(144)/148	134	156/158	143/(153)	182/190	-
	6	93/97	129/137	144/148	134	156/158	143	182/190	-
	7	93/97	129/137	144/148	134	156/158	143	182/190	196
	11	93/97	129/137	144/148	134	156/158	143	182/190	-
	12	93/97	129/137	144/148	134	156/158	141	182/190	-
Parchment 2	4	89/91	139/141	142/150	134/144	164/180	139/151	178/(188)	-
	5	89/91	139/141	142/150	134/144	164/180	139/151	178/(188)	-
	9	89/91	139/141	142/150	134/(144)	164/(180)	139/151	178/(188)	-
	10	89/91	139/141	142/150	134/144	164/(180)	139/151	178/190	-
	15	89/91	139/141	-	-	-	139/151	178/(188)	-
	16	89/91	139/141	-	-	-	139/151	178/(188)	-

[24] The experiments were part of the Ph.D. work of Dr. Joachim Burger, who carried them out in cooperation with Dr. Ina Pfeiffer of the Institute of Veterinary Medicine of the University of Göttingen.

7.2.4 Sheep, goats and cattle

Finds of animal bone can usually be identified easily by an experienced archaeozoologist. However, species identification may become difficult if only diaphysis fragments are left. In particular, the long-bone remains of two common animal species that have been domesticated from a long time, sheep (*Ovis aries*) and goats (*Capra hircus*), may be difficult or even impossible to distinguish from another.

Therefore, a RFLP-based analysis of a cytchrome-b amplification product was developed and carried out on animal bone samples, some of which were of uncertain species origin (Burger 2000). The diaphyses were excavated at the Lichtenstein cave, a Bronze Age site showing exceptionally good preservation of the skeletal material (cf. section 7.2.1).

The results from the ancient samples (DoT) show that the technique is perfectly suitable to enable species determination from archaeozoological skeletal material on the basis of a hypothesis (here: sheep vs. goat) (Fig. 7.11). Additionally, the same assay also enables investigators to distinguish cattle (*Bos taurus*).

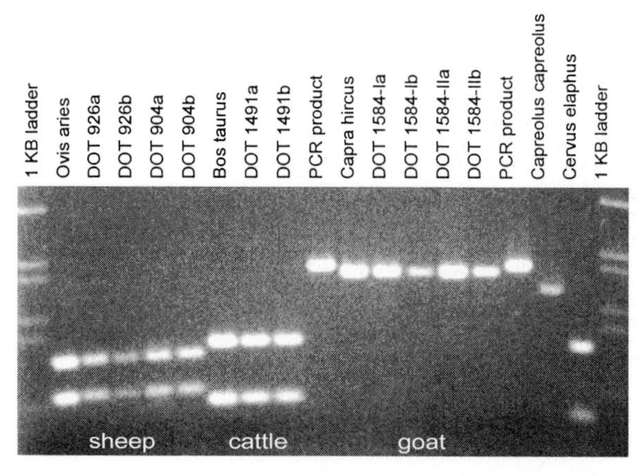

Fig. 7.11. RFLP analysis of 195-bp PCR amplification products of the mitochondrial cytochrome b. Clearly distinguishable are the results of sheep, cattle and goats. The Bronze Age samples presented here were also assigned to one of the respective species by morphological traits[25]

[25] We are most grateful to Dr. Reinhold Schoon for his cooperation and for providing the morphological species determinations.

7.2.5 Species, sex and individual identification

Maximizing the information from a minimum of sample material while improving data validation are the main advantages of multiplex PCR assays. The following experiments demonstrate these in two multiplex assays. The first multiplex PCR combines species identification with sex determination, in other words, for the first time a sequence and a fragment length analysis are combined. In the second multiplex PCR, three dinucleotide autosomal STR loci (cf. section 7.2.3) were amplified.[26] The materials for testing the multiplex PCRs were samples of parchments from various ages that came from the University of Göttingen library.

A prerequisite for combining sequence and fragment length analyses is that there must be sufficient differences in the fragment lengths to be analyzed on agarose gels. This is necessary in order to avoid labeling primers with the fluorescent dye, which would interfere with Taq-cycle sequencing of PCR products with dye-labeled ddNTP (section 8.7). These conditions were successfully met by combining a cytochrome-b sequence (195 bp, cf. section 7.2.3) with ZFX/Y (115 bp) and SRY (149 bp).

The results of testing this 40-cycle assay on the parchment samples are shown in Fig. 7.12. The results when including those of the STR multiplex amplifications are presented in Table 7.5.

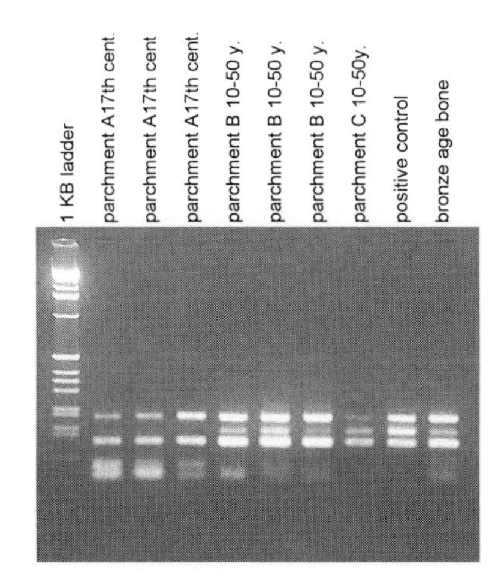

Fig. 7.12. Agarose gel electrophoresis of the multiplex amplified cytochrome-b and the ZFX/Y and SRY sequences. The samples can already be sex-determined from the agarose gel because of the bands of 115 bp (both sexes) and 149 bp (male individuals only). For species determination, the 195-bp amplification product has been directly sequenced from the multiplex assay

[26] The multiplex strategy for species and gender information by combining sequence and fragment length analysis was developed by Dr. Odile Loreille, who also carried out the experiments for the STR multiplex assay. The results were presented at the 5th International Ancient DNA Conference in Manchester, UK, July 2000.

Table 7.5. Species, sex and individual identification by multiplex PCRs

	Multiplex 1 Species and sex		Multiplex 2 STR typing		
Parchment	Species	Sex	BM2113	BM1824	ILSTS00
A (17th cent.)	*Bos taurus*	Female	133/135	178	185
B (10–50yrs)	*Bos taurus*	Male	139/141	178	183
C (10–50yrs)	*Capra hircus*	Male	131/133	170	181
Bronze Age bone	*Capra hircus*	Male	-	170	181

7.2.6 Dog or fox

Archaeozoological finds are not restricted to domestic animals, but may be co-mingled with wild animals. In most cases this situation is not a problem for archaeozoologists. However, in the case of determining the species of origin from fragmented and incomplete material, the members of the canidae family may be difficult to distinguish and are often recorded simply as *canidae spec.*

In the situation presented here, the objects in question were a Bronze Age mandibula fragment (DO 3379) and a tibia fragment (DO 87) (Figs. 7.13 and 7.14). It was difficult to determine if they originated from a dog (*Canis familiaris*) or a fox (*Vulpes vulpes*). Additionally, several other bone specimens were analyzed from the medieval Plesse Castle in Lower Saxony and the Bronze Age site in the Lichtenstein cave (cf. section 7.2.1). All of these were unambiguously assigned to either species using morphological traits (foxes: PL 1, PL 2 and DO 1766), (dogs: DO 4838 and DO 1 705).[27] The DNA of all bone and teeth samples was extracted employing the phenol-based extraction method (section 8.3).

Thirteen contemporary foxes served as control samples[28] and were found to represent four haplotypes; the saliva samples from four contemporary dogs represented two haplotypes. DNA was extracted from milligram amounts of soft-tissue samples from the ears using the

[27] We are indebted to Dr. Reinhold Schoon for providing us with the sample material and the species determinations.

[28] The analyses of the archaeological samples were carried out in the course of the diploma thesis of Andrea Bartels on reconstructing the relatedness of individuals in a fox population from the Black Forest by STR and mitochondrial markers in cooperation with Dr. T. Kaphegyi, the Insitute for Forest Zoology, University of Freiburg. The results on those aspects will be published elsewhere.

QIAampTissue kit (Qiagen) and the saliva DNA by Chelex100. A primer set (*V.vulpes*-upper and *V.vulpes*-lower) that had been designed to amplify the hypervariable region of *V. vulpes* was used on the ancient specimens (45 cycles) and modern controls (35 cycles) (section 8.5). All amplification products were directly sequenced.

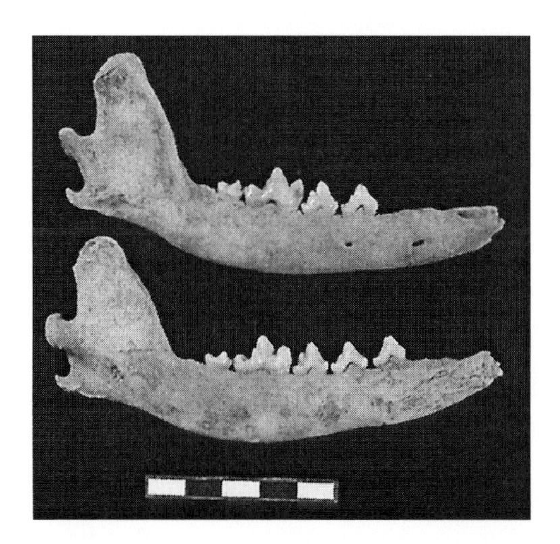

Fig. 7.13. Right and left mandibula of Do 3379. It was questioned whether the animal was a dog or a fox. A tooth root was taken for DNA extraction

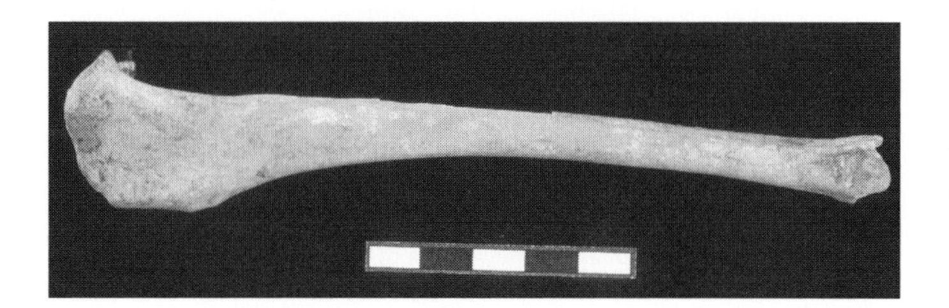

Fig. 7.14. Tibia of an animal that could not be determined unambiguously by morphological traits. It was questioned whether it came from a dog or a fox

The sequences obtained from the aDNA extracts clearly resembled either the mutation hot-spot clusters (as has been found in the fox population from the Black Forest) or were comparable to modern day domesticated dogs (Fig. 7.15). In the cases of the questionable mandibula and tibia, the determination clearly showed that the animals were from *Vulpes vulpes* (Bartels 2002).

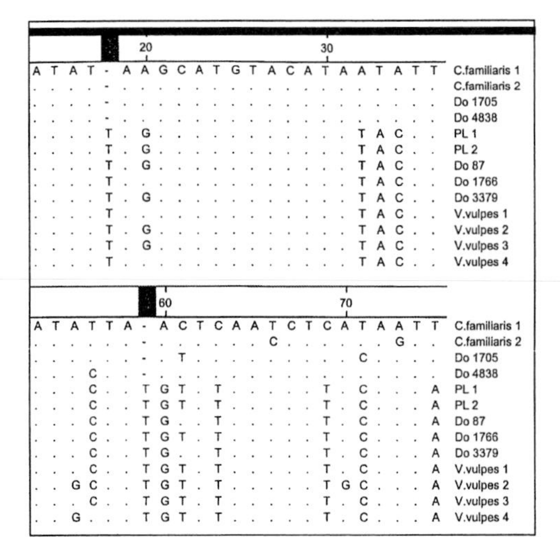

Fig. 7.15. Sequence alignments of parts of the HVR I of the modern control samples (*C. familiaris* 1, 2 and *V. vulpes* 1-4) of the medieval foxes (PL 1 and PL 2) and of the Bronze Age dogs (DO 1705 and DO 4838) and fox (DO 1766). From these results, it is evident that both DO 3379 and DO 87 are foxes

7.2.7 Elephants and rhinoceroses

African elephants (*Loxondonta africana*) and rhinoceroses (*Ceratotherium simum, Rhinoceros unicornis and Diceros bicornis*) are species that have been protected for many years by the International Convention for the Trade and Exchange of Endangered Species (CITES) in Washington. However, the trade in ivory objects that have been manufactured from the tusks of African elephants (Fig. 7.16) has continued. The same holds true for the sale of pulverized rhinoceros horn with its supposed "medicinal properties" (Fig. 7.17). This substance, which is mainly produced and traded throughout East Asia, supposedly acts as an aphrodisiac.

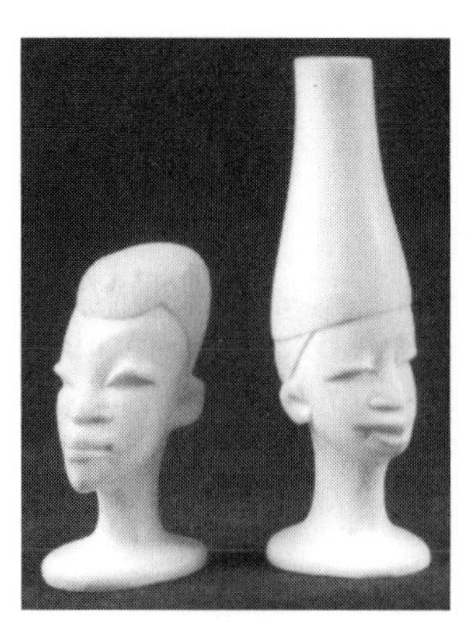

Fig. 7.16. Statuettes handcrafted from ivory. For the statue on the right, the sampling was done at the top.

Fig. 7.17. Horn of a rhinoceros. The sampling was carried out at the bottom part

One of the main problems in the enforcement of conservation biology treaties and trading bans is the fact that the products often can no longer be identified because of processing, as in the case of rhinoceros-horn powder. On the other hand, as in the case of elephants, the products cannot be assigned to either of the subspecies (*L. africana* or *Elephas maximus*) (e.g., Barriel et al. 1999; Thomas et al. 2000; Whitehouse and Harley 2001) by morphologically detectable means. Therefore, molecular methods can contribute to the identification of the confiscated products that fill the customs store rooms worldwide. In many cases, a reliable molecular identification would enable more effective law enforcement.

Two sets of primers were designed to amplify different regions of the cytochrome-b gene (Bollongino 2000; Bollongino et al., submitted). One region spanning 152 bp enabled the discrimination between African and Indian elephants (Fig. 7.18); the other region was able to discriminate the three subspecies of rhinoceroses by spanning 192 bp (Fig. 7.19). With respect to such objects that have been handled extensively by humans during manufacturing processes and trading, both primer sets were designed not to amplify human DNA when the amplification was carried out under stringent conditions. This was done by building in mismatches at the 3'-ends of the primers (cf. section 4.4.4).

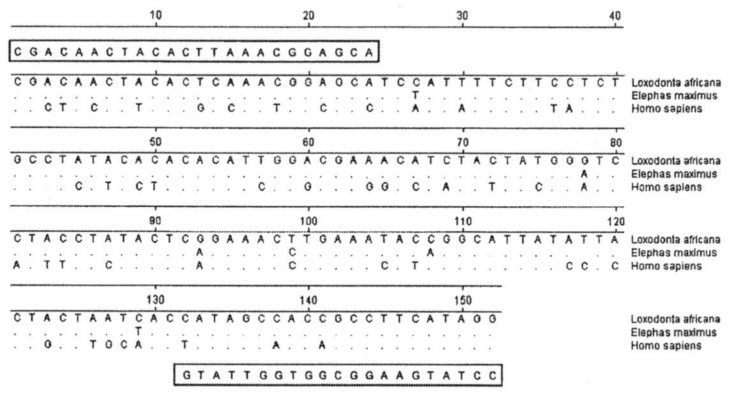

Fig. 7.18. Alignments of multiple GenBank sequences revealing the polymorphisms of African and Indian elephants (*Loxodonta africana* and *Elephas maximus*). Indicated are the mismatch sites at the 3'-ends against *homo sapiens* at the primers CBelephant-upper and CBelephant-lower

Samples were taken from the objects shown in Figs. 7.16 and 7.17. The samples cut off from the ivory statuettes were extracted following the phenol-based extraction method described in section 8.3. The horn samples were also extracted using the phenol-based extraction method, except that the initial steps using EDTA incubation and proteinase-K digestion were replaced with an incubation of the powdered horn in ATL buffer (QIAamp tissue kit, Qiagen) and proteinase K for 3 h. The 45-cycles amplifications revealed distinct amplification products that were directly sequenced. The alignment results allowed unambiguous identification of *Loxodonta africana* as the source of both statuettes and of *Diceros bicornis* as the source of the horn (Figs. 7.20 and 7.21).

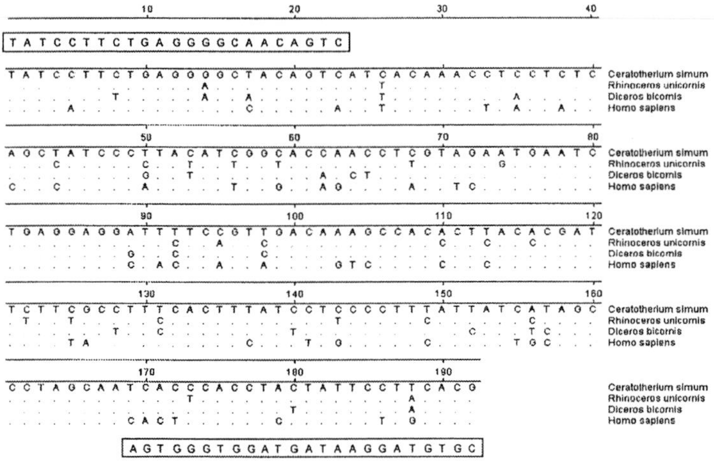

Fig. 7.19. Alignments of GenBank sequences revealing the polymorphisms of the three rhinoceros subspecies. Indicated are the mismatch sites at the 3'-ends against *homo sapiens* at the primers CBrhino-upper and CBrhino-lower

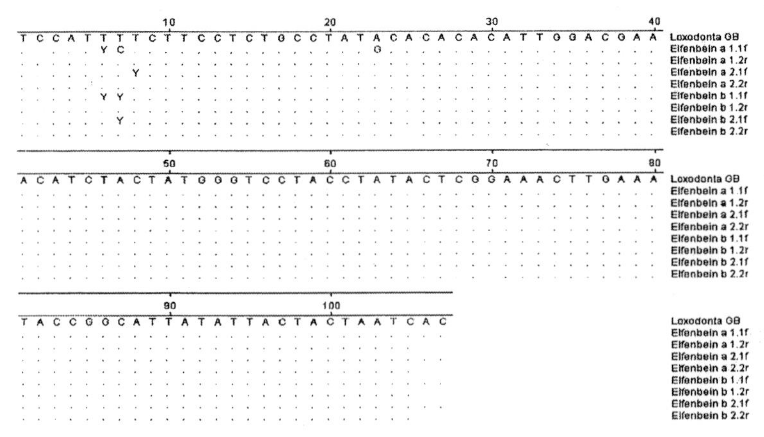

Fig 7.20. Alignment of the direct sequencing results derived from the extracts of the ivory statue ("Elfenbein") on the right: they show that the statues were manufactured from the tusks of the African elephant (*Loxodonta africana*)

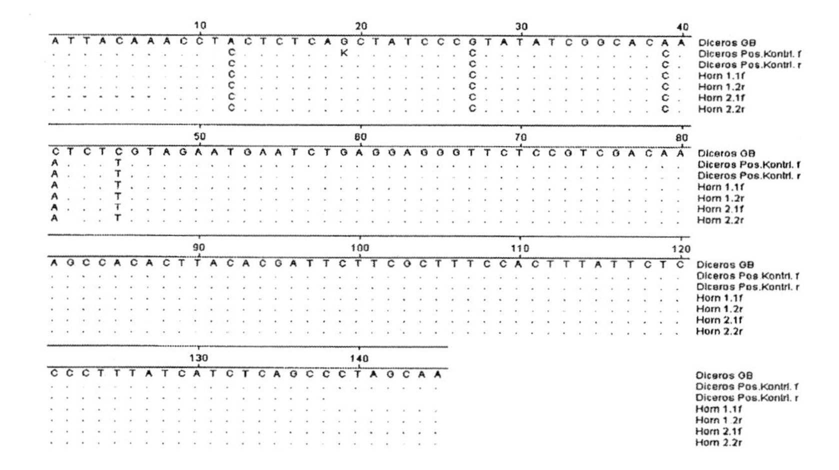

Fig 7.21. Alignment of the direct sequencing results derived from the rhinoceros horn sample. It shows that the horn is from *Diceros bicornis*. "Diceros GB" is a GenBank sequence, (Acc. no. X56283), which is either rare or incorrect. "Diceros Pos.Kontrl" are samples of *Diceros bicornis* from the zoological gardens of Berlin and Cologne

7.2.8 Beef in the sausage

High-quality control of food is of major interest to consumers for many reasons, such as the possible transmission infectious diseases, religious regulations, cultural agreements on taste or conservation biology (e.g., Allmann et al. 1995). Due to intensive processing, the DNA extracted from food may be severely degraded (e.g., Branciari et al. 2000; Calvo et al. 2001; Sebastio et al. 2001). In order to enable scanning tests on food, a primer set was designed that amplified all mammal and two bird species that are commonly consumed in Europe; this included cattle (*Bos taurus*), pigs (*Sus scrofa*), sheep (*Ovis aries*), goats (*Capra hircus*), chickens (*Gallus gallus*) and turkeys (*Meleagris gallopavo*). Primers that amplify a 248-bp fragment of cytochrome b (Hcytb526 and Lcytb777) were designed in a way that enabled the PCR products to be analyzed using a single RFLP digestion (Bfa I). This allowed us to distinguish all the named species by length-specific patterns (Fig. 7.22 and Fig. 7.23)[29]

.

[29] The experiments were part of the qualification thesis work of Diane Schmidt, which mainly focused on testing extraction methods for highly processed foods, e.g., canned or freeze dried. Those results will be published elsewhere.

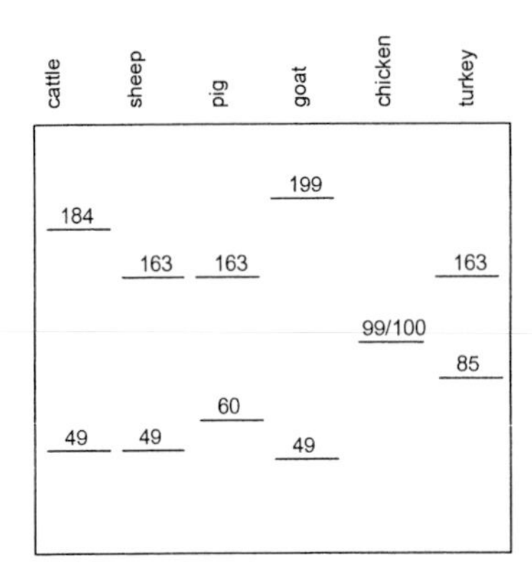

Fig. 7.22. Diagram showing the different restriction sites for the Bfa I endonuclease for the six domestic animal species

The PCR-based RFLP analysis (e.g., Meyer et al. 1995) was applied to different sausages that were declared not to contain beef. In addition to this, we analyzed milk and finally cheeses that were declared to be processed from pure cow milk only. Milligram amounts of the respective foods were extracted using the QIAamp Tissue kit (Qiagen) and amplified by a 35-cycle PCR (Schmidt 2001). All declarations by the manufacturers were found to be correct (Fig. 7.24).

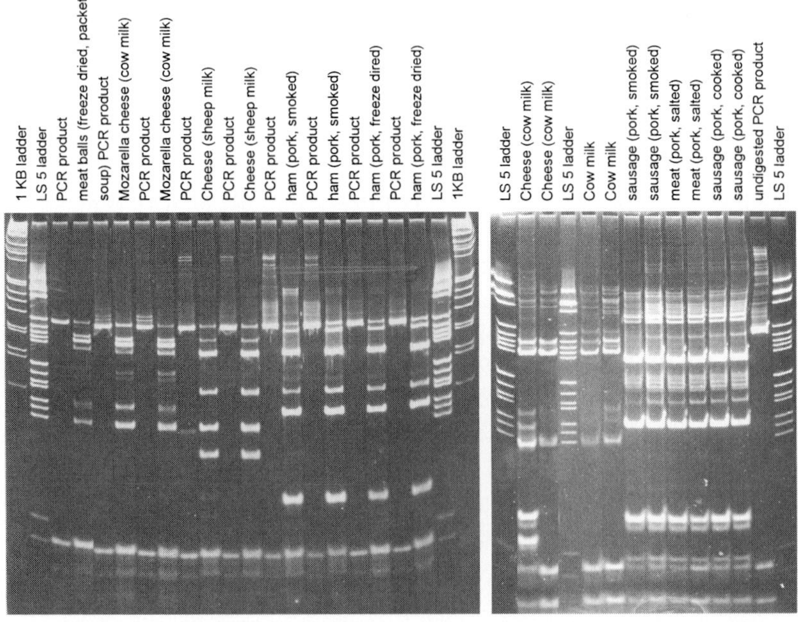

Figs. 7.23 and 7.24. RFLP digestion patterns from the 248-bp amplification products derived from freeze-dried products, sausages and several milk products. On the PAA gel on the *left (Fig. 7.23)*, the RFLP analysis of processed food

originating from different species is presented. The species had been declared for those food items. On the PAA gel on the *right (Fig. 7.24)*, two different sausages (smoked and cooked) are seen in double samplings. For these sausages, it was unclear if they derived from pork only or if they may have consisted of beef as well. The results show that they are obviously from pork only, since they show only the restriction pattern of the pork control sample (salted), but none of the restriction fragments that are typical for cow (*bos taurus*), which are shown by "cow milk" and "cheese (cow milk)"

7.2.9 Rock-wallabies on Depuch Island

Australia has had an appalling number of mammal extinctions since 1600, accounting for about one-third of all extinctions worldwide. The most severely affected Australian mammals have been the marsupials (Spencer et al. 1995; Houlden et al. 1996; Sinclair 1997; Firestone et al. 1999; Spencer and Bryant 2000). One recent example of this was a species of rock-wallaby (*Petrogale lateralis*), a small kangaroo species from Depuch Island in western Australia (Spencer et al. 1995). This taxa only recently vanished, probably as a result of predation by the introduced foxes. One of the problems was that its taxonomic status remained unknown. It could be one of two possible subspecies of *P. lateralis*, each possessing a different mitochondrial pattern. There are a number of different molecular markers, including the hypervariable or D-loop region, which contain well-defined regions for discriminating nucleotide polymorphisms among marsupial taxa. Therefore, in order to determine the subspecies status of this rock-wallaby, DNA was extracted from a tooth root and a compact bone sample of the femur using the automated phenol:chloroform technique and amplified using a 40-cycle PCR. The jaw and femur were then reconstructed by a taxidermist to their original appearance (Fig. 7.25).[30] We also designed a multiplex set of primers to amplify short sections of the hypervariable control region (cf. sections 4.4 and 8.5).

The amplified fragments could easily be resolved on an agarose gel, revealing three discretely sized fragments (103, 156, and 209 bp in size). Sequencing these fragments allowed us to distinguish the species as *Petrogale lateralis lateralis* (Fig. 7.26), which is still found on mainland Australia.

[30] The experiments were part of a collaborative study by Dr. Peter Spencer, Zoological Garden Perth, Australia, and our department. The detailed results will be published elsewhere.

One of the aspects that must be very appealing for any museum curator is the demonstration that reconstruction of the bones was achievable, so from a morphological perspective, the process can be carried out without the need for obvious sample destruction. Furthermore, this type of approach to conservation biology and systematics will become more widespread, because the subfossil material may represent a large proportion of or, in many cases, the only material that is available to work within Australia.

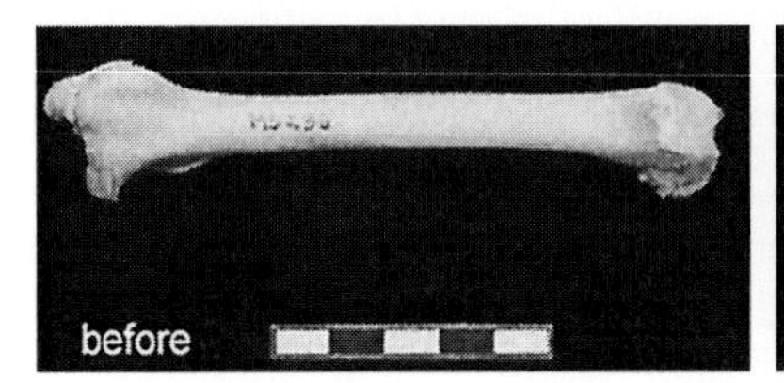

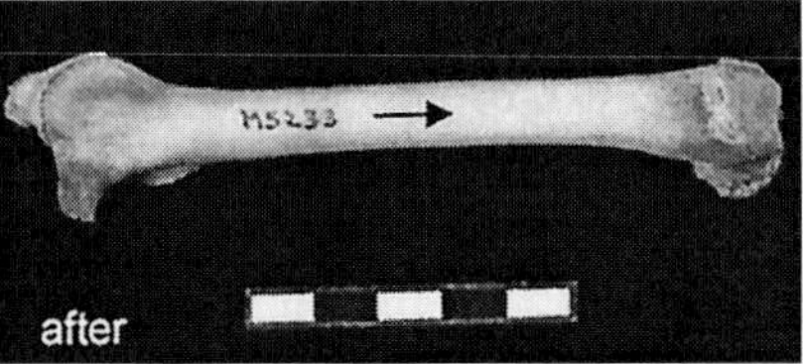

Fig. 7.25. A comparison of before and after the sampling of the femur bone from a museum specimen of the black-flanked rock-wallaby, a small kangaroo that inhabited Depuch Island. It can be clearly seen that degraded DNA analysis does not necessarily result in the destruction of museum specimens.

```
P. sp. unknown*   AAATTCTTACTCATGTCAGTATCAAACCACCTACCGTACCTAG
P. l lateralis1   ...........................................
P. l lateralis2   ..........................................A
P. l. hacketti3   ...........................................
P. l. pearsoni4   ...........................................
P. penicillata5   ...........T.....................T........A
P. herberti6      ...........T...................C.TG.......A


P. sp. unknown    TTTATTTTGTTTTTTT-ATGCACGTACATATATATGTATAACT
P. l lateralis    ................T..........................
P. l lateralis    ...........................................
P. l. hacketti    ...G..........................C..........TC
P. l. pearsoni    ...G............T.............C..........T.
P. penicillata    ........-.GC....T-.....T..T.C............T.
P. herberti       .........-CT....-.....C..T.C............T..
```

Fig. 7.26. Alignment of a partial sequence from a 156 bp-fragment of the mitochondrial DNA control region (D-loop) in seven species of rock-wallaby. The sequence shows that the museum specimen of the Depuch Island rock-wallaby (denoted with asterix) has greatest similarity to *Petrogale lateralis lateralis* (99%). Sequence identity of the other subspecies (GenBank accession numbers: *1*=AF348679; *2*=AF348675; *3*= AF348683; *4*=AF348681; *5*=AY040887; *6*=AF357284) with the museum specimen sequence (*P. sp. unknown*; cf. also Fig. 7.25) is indicated by dots.

7.3 Kinship

Kinship analysis in the sense of reconstructing genealogies from historical burial sites has always been one of the desiderates of anthropology. There have been many efforts to use morphological traits that reflect genetics for discovering the relatedness of individuals and populations (e.g., Ullrich 1972; Rösing 1986; Alt and Vach 1994; Alt and Vach 1998). However, it is the nature of phenotypic traits to display similarities. Despite the progress of the Human Genome Project, the genetic backgrounds of most phenotypic traits are still unknown. Therefore, the discovery of non-coding, highly polymorphic DNA sequences that enable the reconstruction of kinship through individual identification represents an enormous breakthrough for historical anthropology. The following sections will show examples of kinship analysis through the application of autosomal STR typing, as well as Y-chromosomal and mitochondrial haplotyping of skeletal human remains from various places and ages.

Kinship analysis is also the aim of many behavioral sciences. For example, in primatology it is in particular the male reproductive success that is the focus of interest, since it provides insight into an important part of social strategies and structures of the animal society. The last subchapter presents an example of how degraded DNA is extracted from feces samples and allows the reconstruction of the genealogical kinship of orangutans.

7.3.1 A Merovingian family

Construction work in Kleve-Rindern, North-Rhine Westphalia, led to the discovery of the skeletal remains of five individuals that were dated to the Merovingian age. Morphological traits enabled the identification of an adult female and a male, both in their early 40s, and three children aged 18 months, 3 to 4 years and about 9 years (Fig. 7.27).

The in situ situation strongly suggests that the burial of all five individuals happened at the same time, because the forearm skeletal elements of two of the individuals (NI 29 and NI 28) are laid tightly across one another without showing any sign of displacement, as would be expected if there had been separate burials. Because of this and the ages of the individuals, it was assumed that they were possibly from a small family consisting of the parents and three children.

The DNA analysis was carried out on tooth and bone samples. Unfortunately, only very few skeletal elements of both of the younger children were found as a result of the construction work. In particular for NI 31, the 18-month-old baby, only some tiny skull fragments were found.

The molecular typing used four STR markers (VWA, TH01, FES/FPS and CD4) [31] and the sex-determining X- and Y-chromosomal marker, amelogenin (Hummel et al. 1995; Herrmann and Hummel 1997). The results (Table 7.6) enabled sex determination for four of the individuals and proved with a high level of probability that NI 30 and NI 28 were the parents of NI 29, the 9-year-old girl. In relation to the the parenthood of NI 25, the STRs did not lead to an exclusion, because some of the alleles were missing. Therefore, a reliable probability of parenthood could not be calculated. Unfortunately, it was not possible to obtain reproducible amelogenin and STR data from the skull fragments of NI 31.

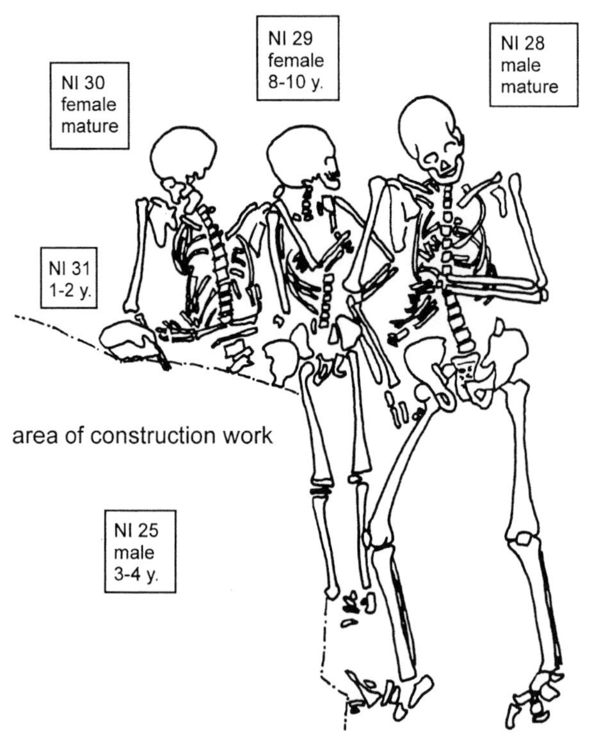

Fig. 7.27. Sketch of the in situ situation of the discovery of five Merovingian individuals in Kleve-Rindern, North-Rhine Westphalia

[31] The STR data have been accumulated from singleplex and duplex amplifications. However, they were the same primer sequences as those that are listed for the respective triplex and duplex PCRs in section 8.5. Deviating from the amplification protocols, slightly higher primer concentrations were used in 40–50 cycles amplification reactions

Table 7.6. Results of the sex determination and STR typing

	Age	Morphol. sex	Molecular sex	VWA	FES/FPS	TH01	CD4*
NI 25	Infans 1	-	XY	16/-	11/-	10/-	2/6
NI 28	Maturitas	Male	XY	16/18	10/11	7/10	2/6
NI 29	Infans 2	-	XX	17/18	11/11	7/10	1/6
NI 30	Maturitas	Female	XX	17/19	9/11	7/9	1/6
NI 31	Infans 1	-	X -	-	-	7/-	-/-

* the nomenclature of CD4 has meanwhile been renewed (cf. section 2.2.2)

7.3.2 The earls of Königsfeld

In 1993, excavations in the St. Margaretha church in Reichersdorf, Bavaria, led to the discovery of eight skeletons assumed to represent the earls of Königsfeld. The earls of Königsfeld were from a noble family who used the small church as their traditional family burial place over a period of seven generations from 1546–1749.

According to the inscriptions on seven memorial stones in the chancel, eight male members of the House of Königsfeld were laid to rest in St. Margaretha (Fig. 7.28).

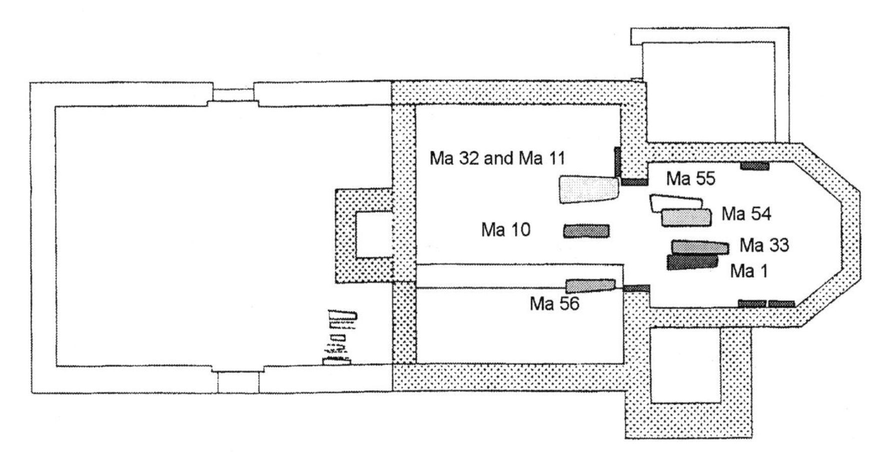

Fig. 7.28. Excavation plan of St. Margaretha, Bavaria

Five of the excavated skeletons could be identified as male individuals solely from the morphological traits of the individuals. In the cases of two other individuals, a sex determination was not possible using morphological criteria. However, it was assumed that those also were the individuals whose names were inscribed on the tombstones. One grave had been destroyed by grave robbers, so no skeletal elements were left.

Concerning the eighth skeleton that had been excavated, morphological traits and the fact that remains of female garments were recovered gave reason to believe that the individual from grave MA 1 was an approximately 30-year-old female. Since the historical records did not include the burials of female family members in the church, it could only be speculated whether this individual was affiliated with the House of Königsfeld.

The genealogical positions of the male family members were known from historical sources (Hobmaier 1889), as indicated by the family tree (Fig. 7.29).

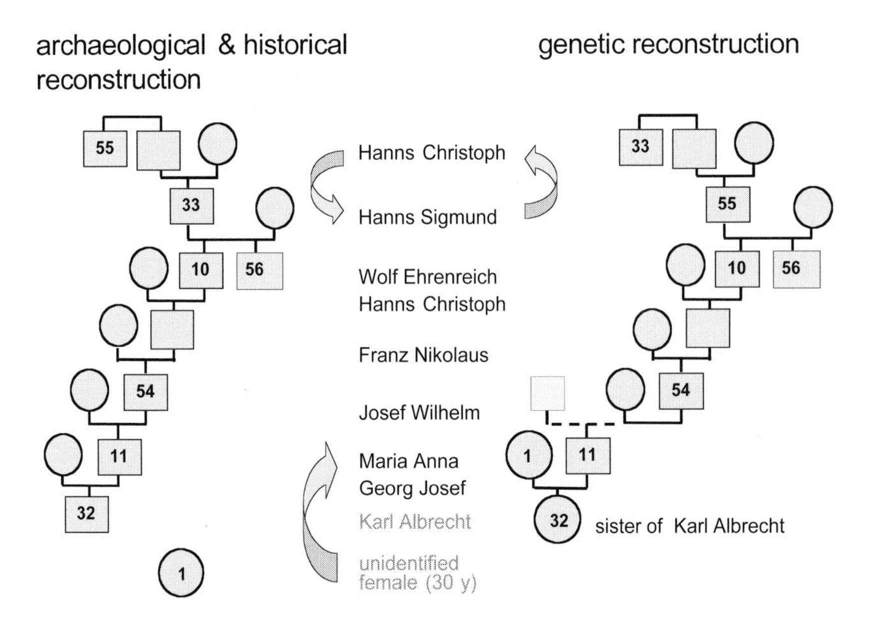

Fig. 7.29. The comparison of the historical, archaeological and genetic reconstructions reveals four deviations: (1) the tombstones of Hanns Christoph and Hanns Sigmund have obviously been exchanged, most probably during an earlier restoration of St. Margaretha; (2) Joseph Wilhelm is not the biological father of Georg Joseph; (3) instead of Karl Albrecht, the last male descendant of the family who died at the age of 13, one of his sisters, who also died at this age, was buried at St. Margaretha; (4) the unidentified female skeleton was revealed to be the corpse of Maria Anna von Königsfeld (1686-1722). For the trio case (MA 1, MA 11 and MA 32), a probability of paternity (W=85% for MA 11) and a probability of maternity (W=98% for MA 1) can be obtained based on modern allele frequencies

Therefore, a dual strategy, in the attempt to determine relationships among the individuals was followed. Y-chromosomal STRs were examined for confirmation that all male individuals were representatives of the same paternal lineage. Additionally, autosomal STRs were analyzed to verify the presumed kinship structures. The analysis was performed on DNA extracted from bone fragments and isolated teeth. Singleplex amplifications were carried out for the amelogenin gene locus to enable sex determination (60 cycles) and on the autosomal STR markers TH01, CD4 and D1S1656 (each 55 cycles). Further, triplex PCRs were performed for VWA, FES/FPS and F13A1 (60 cycles). The Y-chromosomal markers DYS19, DYS 389I+II and DYS 390 were amplified in a quadruplex reaction (Schultes et al. 1999). All amplifications were carried out repeatedly. The surprising results are shown in Fig. 7.29 and Table 7.7 (Gerstenberger et al. 1999).

Table 7.7. STR typing and sex determination for the earls of Königsfeld

	Autosomal STRs						Y-STRs				
MA no.	VWA	F13A1	FES/FPS	CD4	TH01	D1S1656	DYS 19	DYS 390	DYS 389I	DYS 389II	Amel
1	14/17	6/ 7	10/11	10/11	6/8	15.3/19.3					X
10	16/18	5/7	10/11/12	5/11	6/8	12/18.3/19.3	14	23	12	28	XY
11	14/17	6/7	10/11	5/12	-/-	11/18.3/19.3	14	24	13	33	XY
32	14/17	6/7	10/11	5/11	9/-	17/18.3/19.3					X
33	15/17	5/5	10/11	6/10	9/9	-/-	14	23	12	28	X(Y)
54	14/17	5/7	10/11	6/11	-/-	18.3/19.3	14	23	12	28	XY
55	16/17	5/5	10/10	5/6/8	6/9.3	-/-	14	23	12	28	XY
56*	14/-	-/-	-/-	-/-	-/-	-/-					

* MA 56 did not reveal sufficient extract to carry out Y-haplotyping

7.3.3 The Lichtenstein cave – a place of sacrifice or a burial site?

The Lichtenstein cave is 140 m long, extended, narrow and gap-like, and it is formed, particularly towards the back region, into a succession of small chambers. It runs through a rock formation mainly consisting of gypsum sinter and has temperatures of between 6–8°C. In 1972, the present entrance to the cave was found, but because of the very narrow and hidden passages, it took nearly a decade for the first rear chambers to be discovered. They contained human skeletal material. Only very few skeletal elements that were covered by strong layers of gypsum sinter (cf. Fig. 7.5) were removed from the cave at that time (cf. section 3.3.2). Systematic archaeological excavations began in 1995, revealing that the finds are nearly 3,000 years old, dating back to the early Bronze Age (Flindt 2001). In total, skeletal elements from 36 individuals were found, almost all of them dislocated from their anatomical positions (Fig. 7.30). Due to archaeological findings that pointed to a ritual use of the cave, the site was assumed to be a place of sacrifice. This assumption is supported by the fact that during the early Bronze Age, regular burials were very rare. This age belongs to the Urnfield culture, i.e., cremation was the common burial rite.

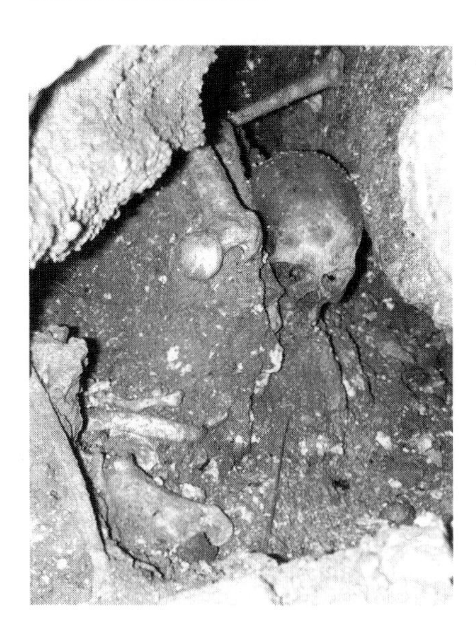

Fig. 7.30. Skeletal elements of 36 individuals were found in the Lichtenstein cave. The site was assumed to be a place of sacrifice. Almost all skeletal elements were dislocated from their anatomical positions

To investigate whether the site was either a common burial place or served as a sacrificial site, the remains of all individuals were genetically typed by autosomal STRs, Y-chromosomal STRs and mitochondrial HVR markers (Schultes 2000). This was carried out on compact bone samples of all 36 left femoras that were found throughout the cave system. Additionally, a classical anthropological investigation was carried out on all skeletal elements found in the cave. This included histological and morphological inspection to enable age determinations (Fig. 7.31) and to detect possible pathogenic deviations or possible lesions due to injuries.[32]

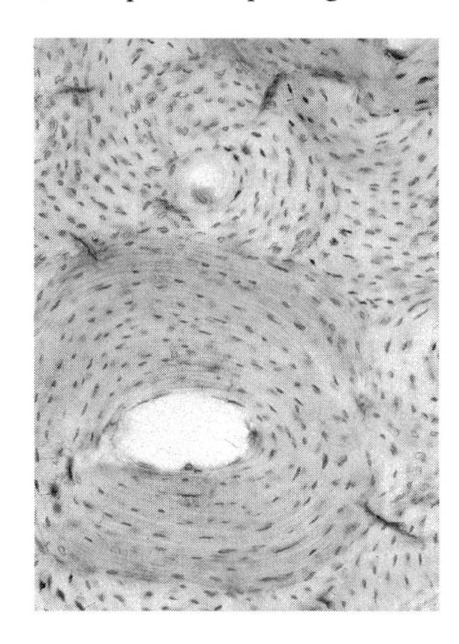

Fig. 7.31. Thin cross sections (80–100 μm) of femoral bone enable age determinations because of systematic changes in the bone microstructure (e.g., Uytterschaut 1993; Herrmann et al. 1990). The thin sections of the femora samples from the Lichtenstein cave (here DO 1911) also revealed perfectly intact microstructures not showing any signs of major microorganism activity (cf. section 3.2)

The morphological findings concerning age and sex distribution revealed no unusual results. Individuals of both sexes and all ages were represented, ranging from young infants to two old females, most probably in their 70s (Table 7.8).[33]

Table 7.8. Age and sex distribution of human finds

	Infans 1	Infans 2	Juvenis	Adultas	Maturitas	Senilis
Male	-	1.5	2.5	3	3	-
Female	-	2.5	1.5	5	-	2
nd	3	5	4	1	2	-
Total		9	8	9	5	2

[32] The morphological and histological investigations were carried out by Diane Schmidt, Dr. Cadja Lassen and the author.

[33] Most of the sex determinations given in Table 7.8 are based on the molecular genetic results, since unambiguous morphological sex determinations on isolated femoras were only possible in very few cases. These cases were confirmed by the genetic results.

There were no lesions caused by injuries that were found to be the direct cause of death. In one case, a surgical opening of the skull, a trepanation, was found in the left os parietale of a child about 8 years old. The trepanation margins reveal signs of bone remodeling; thus, the child had initially survived the treatment. However, it may have died as a result of indirect consequences; a severe infection had left its traces on the skull bone (Fig. 7.32). The only further indications of traumatic events were common and well-healed fractures on a radius and a clavicula of two adults (Fig. 7.33).

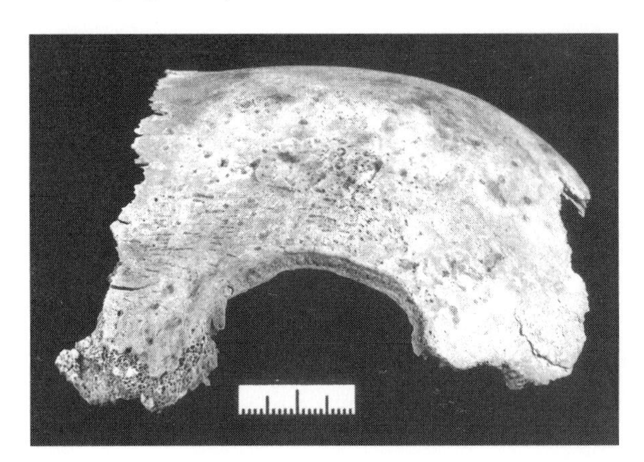

Fig. 7.32. Trepanation at the parietal bone of an approximately 8-years-old infant (DO 78), who obviously survived this operation for quite some time. This conclusion can be derived from the skull margins revealing signs of healing. Trepanations are thought to have been carried out mainly as a curative measure

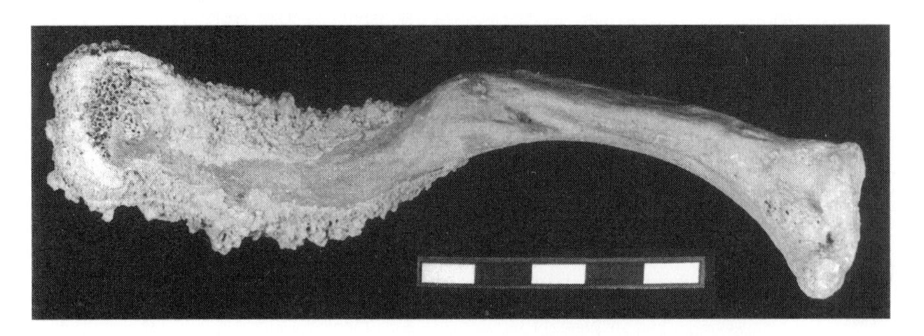

Fig. 7.33. Well-healed fracture of a clavicula bone of an adult man (DO 1478)

The molecular investigations were carried out using 35–40 cycle amplifications employing the AmFlSTRProfilerPlus kit (Applied Biosystems) (Fig. 7.34). Fifty-cycle amplifications were used for the quadruplex analysis of Y-chromosomal STRs (Schultes et al. 1999) (Fig. 7.35) and 35-cycle amplifications for multiplex analysis, including direct sequencing of the mitochondrial HVR I and II (Schultes 2000) (Figs. 7.36 and 7.37).

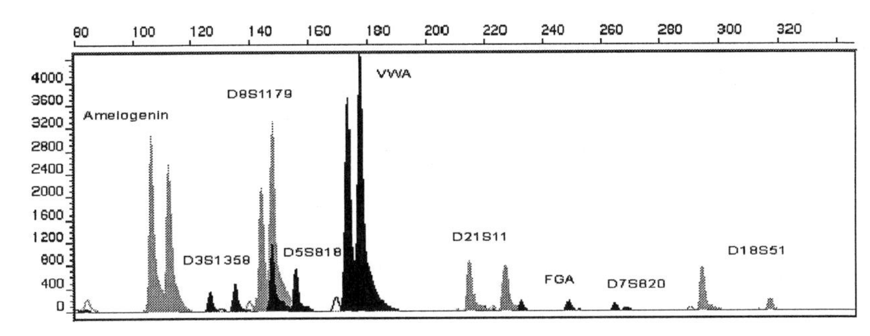

Fig. 7.34. Autosomal STR fingerprint of DO 1076, an adult male individual

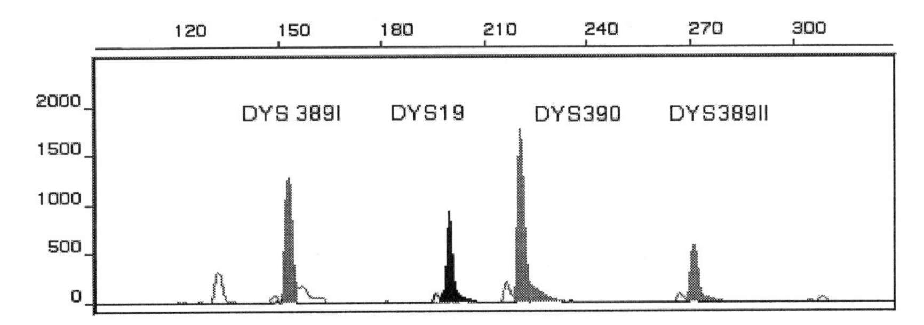

Fig. 7.35. Y-chromosomal STR haplotype of DO 1076

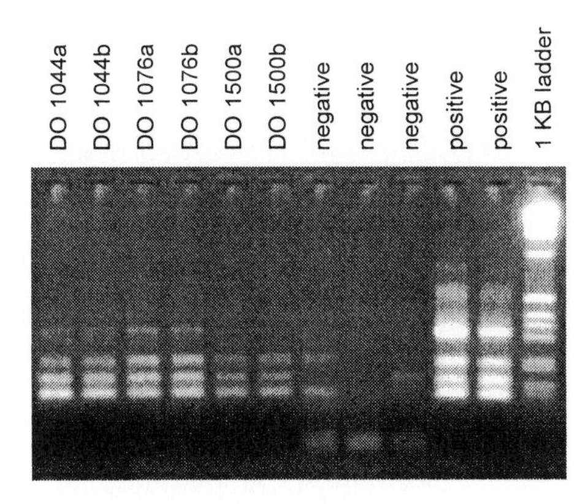

Fig. 7.36. Agarose gel electrophoresis of the multiplex-generated amplification products of the HVR I and II. It can be clearly seen that in those samples that reveal excellent DNA preservation, even the fourth 312-bp amplification product that spans both targeted sequences on HVR I (131 bp and 168 bp) is additionally present. For the amplification products seen in the negative controls cf. section 6.1.2

The typing for all markers was carried out at least four times, two amplifications each from two independent DNA extracts. On this basis, paternal and maternal lineages were identified, and autosomal STR typing was used for the reconstruction of genealogical relatedness within those groups (Fig. 7.38). A full genetic typing consisting of two marker systems (females) or three markers (males), respectively, was possible for 21 individuals. The genealogical reconstruction enabled the immediate linking of 14 of the individuals, as shown in Fig. 7.39.

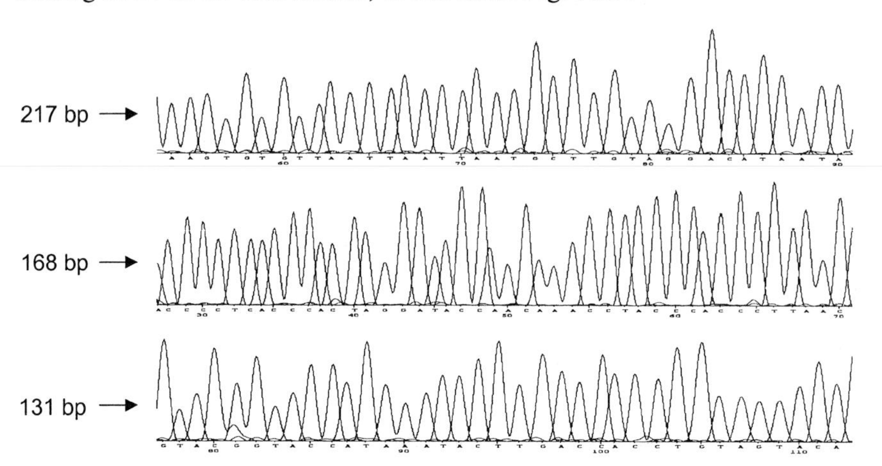

Fig. 7.37. Direct sequencing results for DO 1076. The Taq cycle sequencing reaction was carried out in three assays, each consisting of the respective upper primers for the three multiplex products (217 bp, 168 bp and 131 bp) (cf. section 5.2.1).

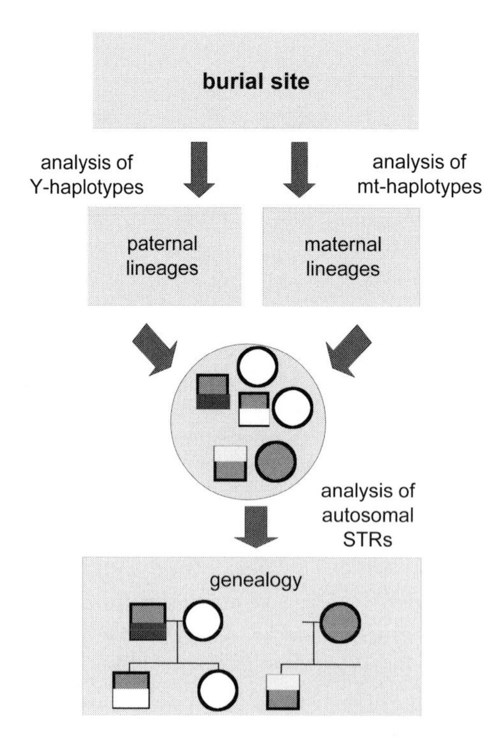

Fig. 7.38. Strategy for reconstruction of genealogical kinship. The individuals are typed by Y-chromosomal STRs and the mitochondrial HVR I and II. Now, they can be assigned to family groups. Within these groups the genealogical reconstruction is carried out on the basis of autosomal STR typing data

The findings suggest that the Lichtenstein cave was a regular burial site revealing the patterns that would be expected from a small Bronze Age population. This interpretation is supported by the morphological investigation that revealed no signs of traumatic violence such as had been found in skeletal collections from the nearby Kyffhäuser caves (e.g., Geschwinde 1988; Flindt 1996), where skull impression fractures were found as well as cut marks on the skeletal elements of many individuals.

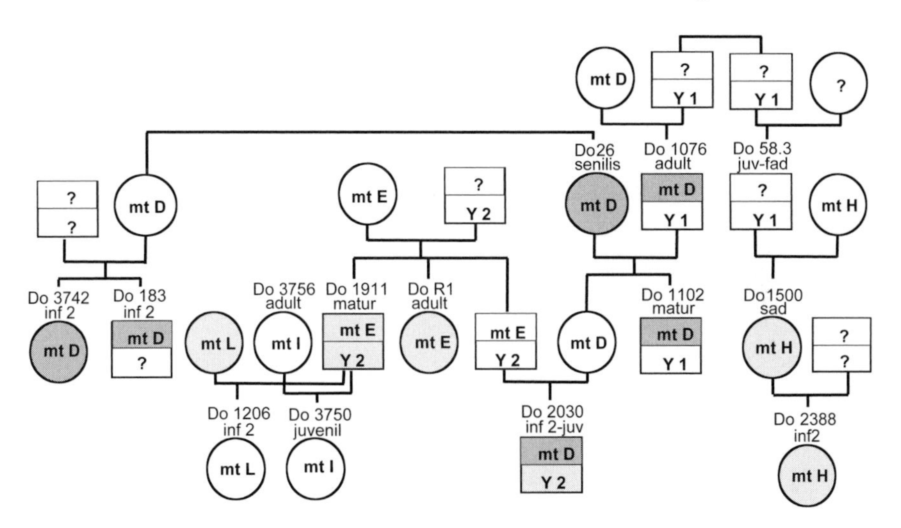

Fig. 7.39. Genealogical reconstruction of kinship for 14 individuals (*stained symbols*) from the Lichtenstein cave. From the results, the genotypes of the missing individuals (*blank symbols*) can be deduced

7.3.4 Widukind

Widukind lived about 1,200 years ago during the 8th century. He was the military leader of the Saxons fighting the Frankish troops of Charlemagne, who occupied the Saxon countries. After 7 years of fighting, Widukind was subdued by Charlemagne and agreed to be baptized in 785. It has always remained unclear if he truly converted to Christianity or if the baptism was just a political act.

After his death, which is believed to date to the 1st decade of the 9th century, his body was buried at the chancel in the collegiate church of Enger in Westphalia. Historical record tells us that Widukind initiated a competition among three families who were founding churches at that time; he wanted to be laid to rest in the church that was finished first. This was accomplished by the founding family of the small town of Enger.

Although Widukind underwent baptism, he became an idealized hero for his resistance against the Frankish Empire. Many later German kings claimed to be descendants of Widukind (Althoff 1983; Schmid 1964). This lasting glorification was also the reason that as many as 3 centuries later, a sarcophagus with a lid showing an image of Widukind (a so-called gisant) was donated to the church of Enger (Fig. 7.40).

Fig. 7.40. The gisant located in the church of Enger, North-Rhine Westphalia, showing the image of Widukind. At the end of the 8th century, Widukind was the military leader of the Saxons in their battles against the Frankish emporer, Charlemagne

Beside the skeletal remains of Widukind, two other skeletons of male individuals were found in the chancel. From the historical record, it is not clear whether these men were related to Widukind, who was said to have had close relatives at his side during his battles. Otherwise, the other buried persons could have been members of the family that founded the church.

By morphological inspection, one of the individuals was discovered to have died in his youth, at about 16 years of age. Because of the histological characteristics of the compact bones, the second skeleton showed the features of an individual about 60 years old, the same age as determined for the skeletal remains thought to be those of Widukind (Fig. 7.41).

The morphological inspection also provided some other highly interesting facts. The vertebrae columns of both of the elderly individuals show severe degeneration, which would be consistent with a life on horseback, just as is described for Widukind and his comrades. But most interestingly, the 60-year-old EN 462 shows a rare detail: one of his right fingers was once fractured and healed in a dislocated position (Fig. 7.42). If this is compared to the image on the sarcophagus lid, one might think that Widukind is perhaps EN 462 instead of EN 463.

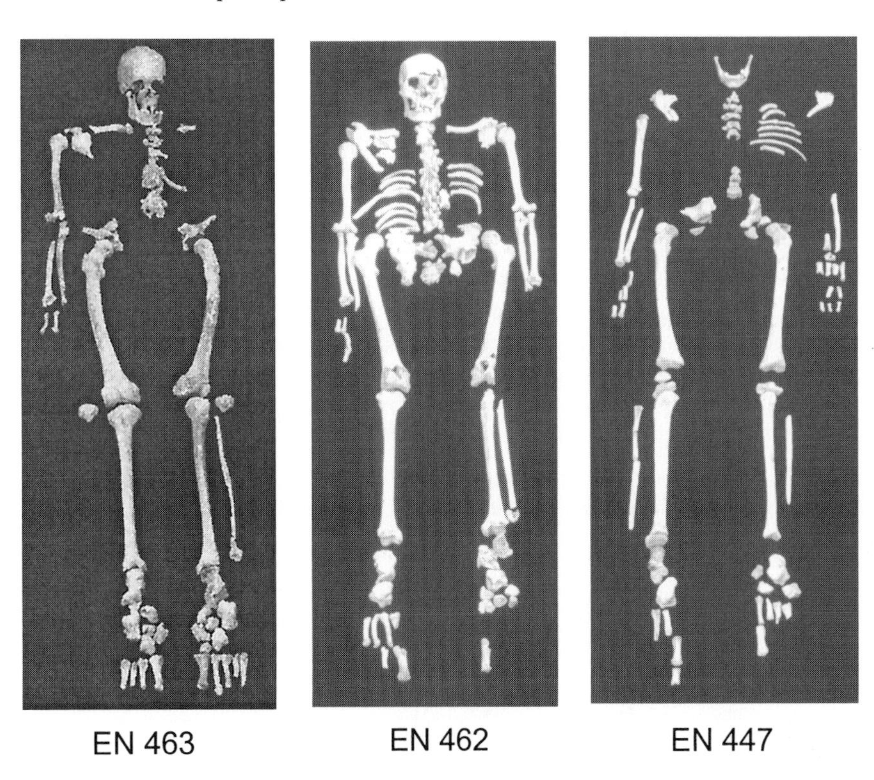

Fig. 7.41. Skeletal remains of three individuals buried at the collegial church of Enger, North-Rhine Westphalia. EN 463 is thought to be the remains of Widukind; EN 462 and EN 447 were either male relatives of the Saxon or members of the founding family that supported the construction of the church

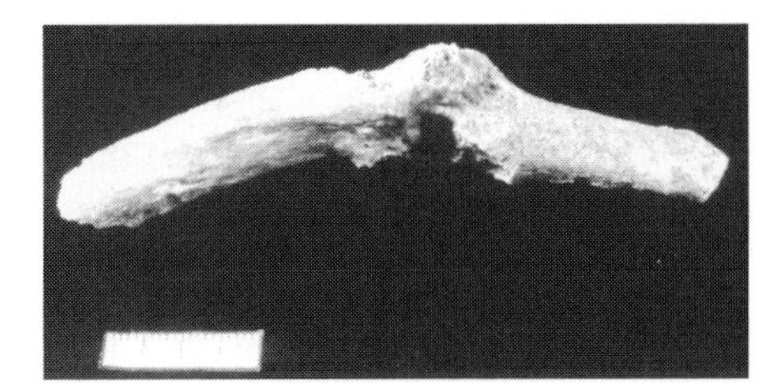

Fig. 7.42. Digit bone of EN 462. The metacarpus and the phalanx are healed in a dislocated position, most probably after a fracture of the bones close to the joint. This anatomical characteristic is strikingly close to the way Widukind is depicted on the lid

of the sarcophagus, for example. This gives reason for a reconsideration of whether Widukind is EN 462 instead of EN 463. Possibly, the molecular data of mitochondrial and Y-STR haplotype analysis that allow the drawing of conclusions about the geographical origin of the individuals will give further hints

The molecular investigations, although still in progress at this time, already have revealed some most interesting results. By autosomal STR typing, which was carried out with the AmpFlSTRProfiler Plus amplification kit, it could be proven that none of the individuals are related to one of the others in a father-son relation. This has already been confirmed by quadruplex Y-chromosomal STR typing; all three individuals revealed different alleles, although the haplotypes are not complete yet. However, the sequence analysis of the multiplex amplifications of the hypervariable regions I and II clearly shows that EN 463 and EN 447 have the mitochondrial haplotype, and EN 462 reveals a single deviation only (Table 7.9). This indicates that at least EN 463 and EN 447 are related through the maternal lineage; the closest possible relation would be as an uncle and nephew (Fig. 7.43)[34].

Table 7.9. Mitochondrial haplotyping (cf. section 2.1.1). [*nps* nucleotid position, *REF* reference sequence (Anderson et al. 1987), *Y* = C or T, *ns* not sequenced]

HVR	HVR I fragment 1 [131 bp] nps 16234-16316	HVR I fragment 4 (1+2) [312bp (168bp+131bp)] nps 16049-16316		HVR II fragment 3 [217 bp] nps 150-322	
np	16311	16196	16311	263	310
REF	**T**	**G**	**T**	**A**	-
EN 463	C	A	C	G	Insert C
EN 462	*	A	C		Insert Y
EN 447	C	A	ns	G	Insert C

Another promising aspect can be explored with the Y-chromosomal haplotypes, since they may give information about the geographical regions of the paternal origin of the individuals. At present, although only a few Y-STR alleles could be determined so far, the particular partial

[34] The morphological and the molecular investigations were carried out by Diane Schmidt and the author. Detailed results will be published elsewhere.

haplotypes of EN 463 and EN 447 seem to be very rare in Europe. A contemporary database shows that these allele combinations are known in only very few European individuals (http://ystr.charite.de), most of them living in the towns of Leipzig and Magdeburg. If this can be confirmed by the full data sets through Y-chromosomal multiplex haplotyping (cf. section 7.4.6), the paternal origin of these two individuals would not be in the geographical region where they most probably lived and were buried, but in this more eastern region.

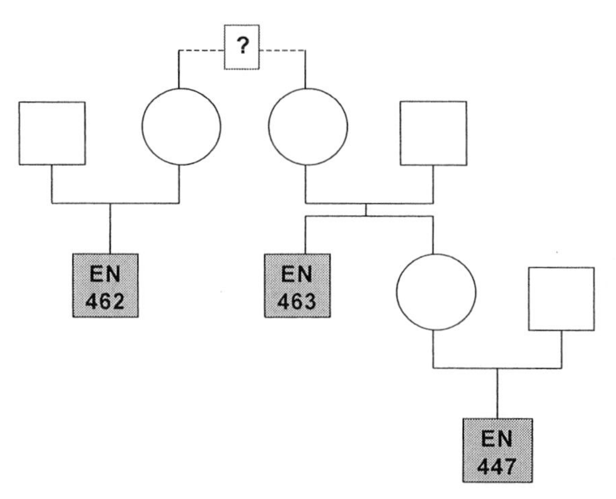

Fig. 7.43. Closest possible relation of the individuals EN 463, EN 447 and EN 462 derived from the mitochondrial haplotyping of the hypervariable regions I and II. Given that the individuals were buried within a time span of no more than a generation, EN 463 and EN 447 may be uncle and nephew. The haplotype of EN 462 is at least very similar, revealing a single base substitution only (cf. Table 7.9). This indicates that there is at least most likely a close relation through the maternal lines, which were divided by a single mutation event only

7.3.5 Orangutans

Genetic analysis from sample materials that can be collected non-invasively is preferred in many scientific contexts that focus on wildlife animals. Typical tissues that may serve as sample material would be shed hair and feces (e.g., Kohn et al. 1995; Hopwood et al. 1996; Taberlet et al. 1997; Wasser et al. 1997). Unlike shed hair, fecal samples will not only allow mitochondrial DNA analysis but also enable access to chromosomal genome information. This is interesting in particular if the reproductive success of male animals is investigated, which is one of the focuses in ethological studies of promiscuous, non-human primate societies. Besides access to the chromosomal genome, fecal samples also provide information about nutritional habits and infections, such as those from endoparasites (Poinar et al. 1998; Loreille et al. 2001). Such an approach to the genetic information about animals may be just as interesting for conservation biology studies.

In this study, fecal samples from 16 orangutans living in two German zoological gardens were investigated in order to test whether fecal samples enable a sound genetic typing by human STRs, which serve identification purposes and enable the reconstruction of kinship (Immel et al. 1999; Immel et al. 2000). The genealogical kinship of the animals was known from the studbooks (Fig. 7.44).

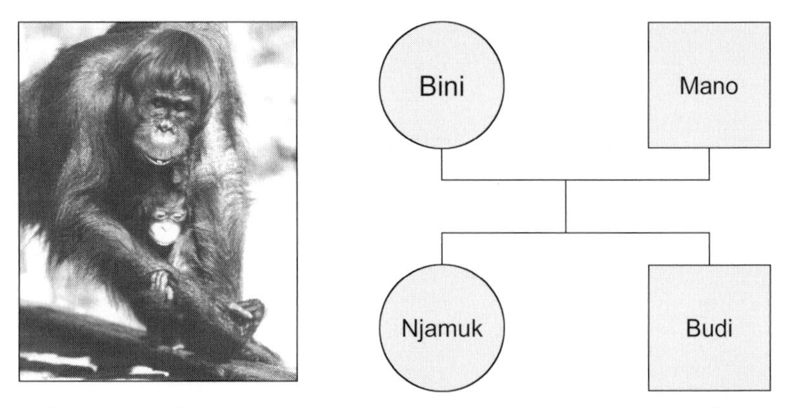

Fig. 7.44. Genealogical relation of an orangutan family quartet living at the zoological garden of Berlin. Fecal samples of the animals were collected in order to test whether the sample material is suitable for genetic typing by human-specific STRs, which would enable the reconstruction of kinship on the DNA level

The fecal samples were stored in 99% ethanol at room temperature prior to extraction; this technique had already proved to be successful in earlier genetic studies of fecal samples collected from langures (Launhardt et al. 1998). The extraction of the fecal samples derived from the orangutans was basically carried out as described in section 8.3, except that initially the homogenized sample material (70 mg) was suspended in CTAB buffer (1.4 M NaCl, 20 mM EDTA; 100 mM Tris, pH 8.0) instead of 0.5 M EDTA.

The autosomal STR typing was carried out as a multiplex assay with the AmpFlSTRProfiler Plus kit (Applied Biosystems) and revealed perfectly distinguishable alleles for five of the nine STR systems (Fig. 7.45). The total failure of the VWA and the D21S11 STR systems, for example, was cross-checked by singleplex amplifications with primer pairs that hybridize at sites other than the respective primer pairs in the kit. The amplifications with the different primer pairs now also revealed reproducible results. This was interpreted as a clear indication that the amplification failures in the multiplex approach were just a matter of sequence polymorphisms within the primer hybridization site. The complete amplification failures with the primers in the kit point to polymorphic sites most likely close to the 3'-ends of the primers (cf. section 4.4.4).

The allele determinations of the reproducibly amplified STRs were fully consistent with the known kinship of the animals. Therefore, multiplex STR typing from feces samples proved to be a reliable method that can be applied in primatology as well as in conservation biology contexts (Kohn et al. 1995).

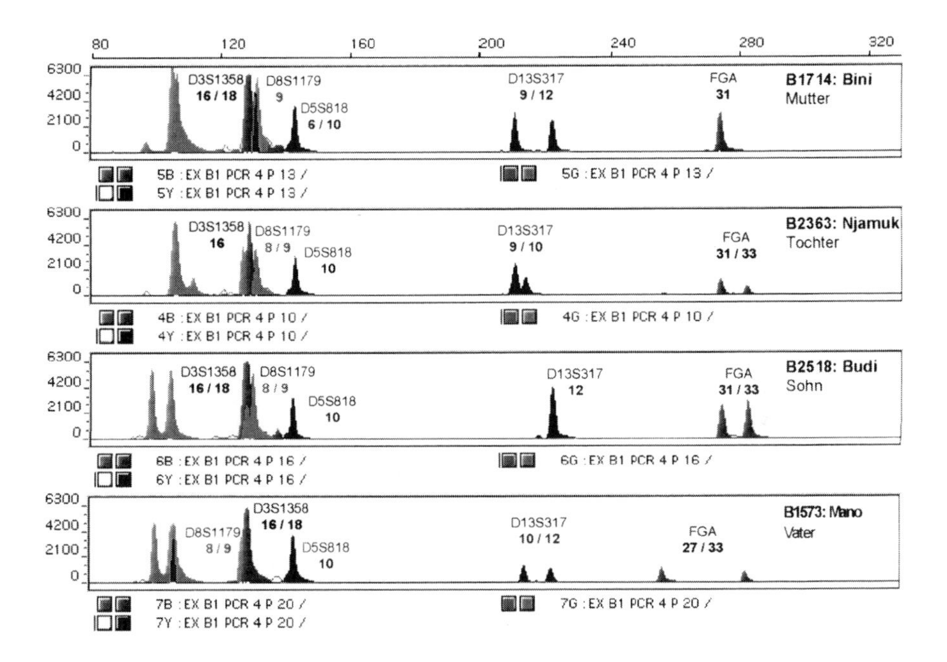

Fig. 7.45. Electropherograms of STR amplifications that were carried out with DNA extracted from feces of orangutans. Since the great apes are phylogenetically close to humans, it was possible to use the Amp*Fl*STRProfiler Plus amplification kit, which is designed to type human DNA. The genealogic relation of the animals was known from the studbook (Bini and Mano are the parents of Njamuk and Budi; also cf. Fig 7.44). The allele determinations also given in the figure (numbers below the STR names) clearly show that the amplification results are fully consistent with the kinship of the animals

7.4 Population genetics

7.4.1 Marriage patterns in Weingarten

The archaeological site of Weingarten, a small town close to Lake Constance, dates back to the 5–8th centuries. In total, more than 800 individuals were buried on this Alemannic site. Due to the particular period of time, almost all burials were well equipped with artifacts, ranging from small items such as rings and needles to large weapons such as saxes, which are long swords.

The archaeological investigations at the site found a reliable intra-serial chronology that suggests a social sub-stratification of the population (e.g., Roth and Theune 1988). This can be derived mainly from the different articles that were also present in the graves, ranging from the belongings of the rich (Fig. 7.46) to those of the poor (Fig. 7.47).

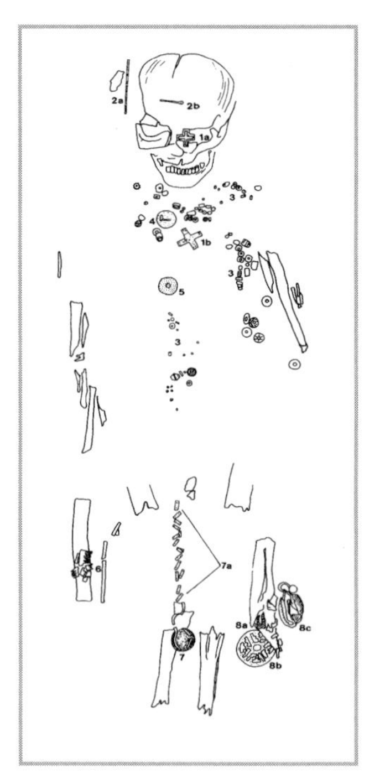

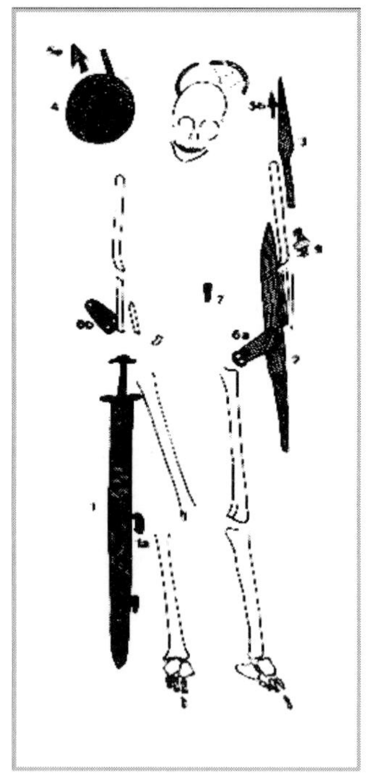

Fig. 7.46. Richly equipped graves at the medieval site of Weingarten. On the *left*, the burial of a female individual is seen, revealing several pieces of golden jewellery, a silver belt and rock crystal. On the *right*, the burial of a male individual reveals several weapons, arrow points and a belt set

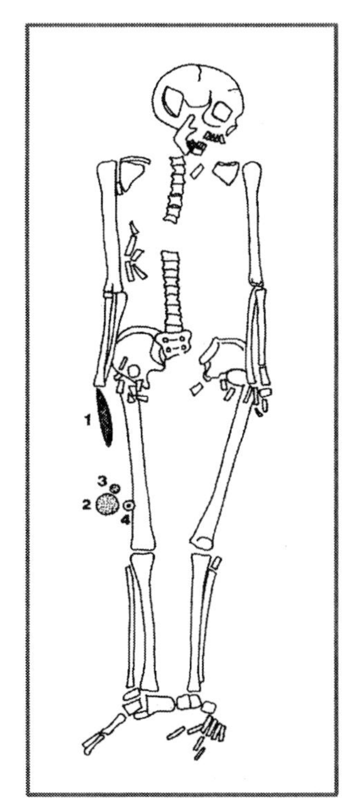

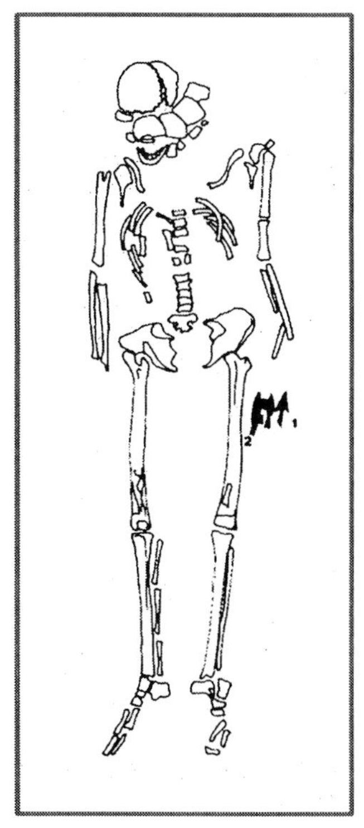

Fig.7.47. Poorly equipped graves at the medieval site of Weingarten. On the *left*, the burial of a female individual only reveals a few coins and a comb. On the *right*, the burial of a male individual reveals some arrowheads

Although STR analysis had been carried out on degraded skeletal remains in the early 1990s (Hagelberg et al. 1991; Kurosaki et al. 1993; Gill et al. 1994), large scale STR-based studies on historic material were not available. Therefore, the aim of an initial study on almost 200 randomly chosen individuals was to test the general suitability of STR investigations on ancient skeletal material with respect to population genetics (Zierdt et al. 1996). For this, extracts from compact bone and teeth were amplified with the VWA locus (Fig. 7.48). Alleles derived from 76 individuals were determined with the help of allelic ladders using ethidum-bromide-stained PAA gels.[35]

In this initial study, the amplifications on the Weingarten samples were carried out only once and, as we know today, under sub-optimal

[35] At this time the laboratory possessed no automated sequencers equipped with fragment length detection software.

amplification conditions.[36] Therefore, the comparatively low rate of heterozygosity (40.8%) that was found does not come as a surprise. It was interpreted as an artifact, most probably due to allelic dropout events. This was concluded from the fact that "homozygotic" individuals were found throughout all alleles. This would be different in an inbred population, where homozygotes should concentrate on two to three predominantly represented alleles (cf. section 2.2.2).

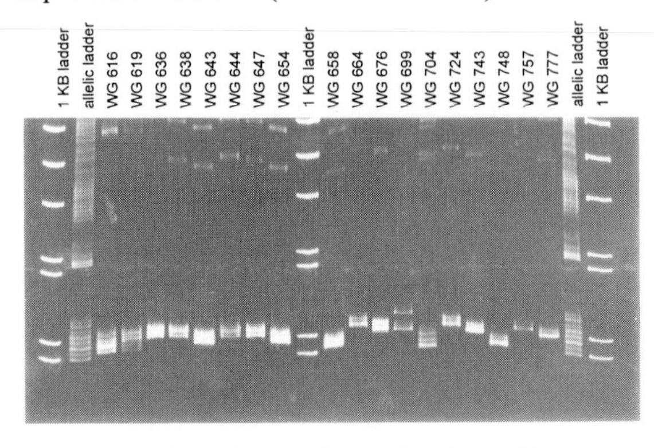

Fig. 7.48. PAA gel electrophoresis of the VWA amplifications from the medieval samples of Weingarten

A second study on the series from Weingarten was recently carried out (Gerstenberger 2002).[37] The aim of this study was to gain insight into the marriage customs of the historical, socially stratified population. Marriage can be regarded as an important social institution that is found in almost every human population. A significant part of social affiliations among groups of individuals is sustained by marriage relationships between individuals. Marriages between kinship groups can lead to the establishment of social alliances (Levi-Strauss 1981).

One of the objectives was to find out if marriages took place between individuals of the two archaeologically determined groups at all or if isogamous marriage rules existed. This would mean that only individuals of the same social status were allowed to marry each other. A further aim of the study was to obtain information about residence patterns in the

[36] The upper VWA primer (Kimpton et al. 1992) consisted of 24-mer oligonucleotide (cf. section 8.5). Although this is a common length for primers, the particular neighboring region to the VWA region requires a longer primer because it is extremely rich in A and T. Therefore, this primer reveals a much lower annealing temperature optimum than the lower VWA primer for this set. Later studies, such as the ones on Duke Christian II and the earls of Königsfeld, were already carried out with a 30-mer oligonucleotide that better served the needs of well- balanced primer sets.

[37] The study is part of the Ph.D. work of Julia Gerstenberger. The detailed results will be published elsewhere.

Alemannic population. By examining the variability of mtDNA- and Y-haplotypes in a population, it is possible to find out about where a newly wed couple lived after marriage. In the case of patrilocal residence, the couple chose the locality of the husband as their postnuptial domicile. The consistent practice of patrilocal residence creates groupings of kin related through the paternal line (Fig. 7.49), each residing at the same locality. Matrilocal residence, on the other hand, is residence in the wife's home area. Just as patrilocal residence keeps men of a patrilineal group together and disperses the women, so matrilocal residence keeps women related by the maternal line together (Fig. 7.50) and disperses the men.

The question about whether marriages between individuals of different social statuses were common and also if status was determined by birth are still matters of lively discussions in the archaeological scientific community (Siegmund 1998; Steuer 1982).

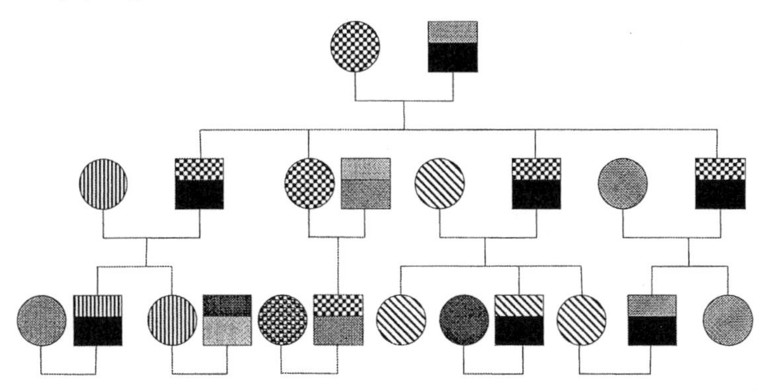

Fig. 7.49. In a society revealing patrilocal marriage patterns, comparatively high variability for mitochondrial haplotypes and low variability for Y-chromosomal haplotypes are expected

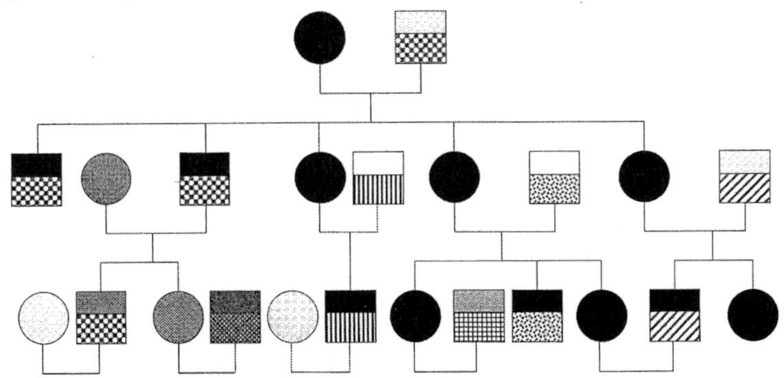

Fig. 7.50. In a society revealing matrilocal marriage patterns, comparatively low variability for mitochondrial haplotypes and high variability for Y-chromosomal haplotypes are expected

The results of the study allow the following statements to be made: based on a statistical evaluation of the autosomal STR data (AmpFlSTRProfilerPlus, Applied Biosystems), it seems very unlikely that the population of Weingarten was made up of two groups of individuals with restricted marriage rules. The analyzed population appears to be genetically homogenous, which speaks for extended marriage relationships among the Weingarten individuals regardless of their social standing. Although the material evidence, i.e., the grave articles, suggests the existence of two divergent groups according to their social ranks, no genetic subgroups were detectable in the population sample.

As a second result, it can be deduced that patrilocality was practised in the medieval society of Weingarten. Y-STR data (cf. also Schultes et al. 1999) were obtainable for 28 male individuals. From these data, it seems justifiable to assume that the men of Weingarten stayed in their birthplace, while the women changed residence and moved to their husbands' families after marriage. The analyzed men can be assigned to 16 paternal lineages, but belong to 21 different maternal lineaeges, with the higher variability of the mt-types speaking for patrilocality. Furthermore, 18 of the men belong to mt-types that are represented by more than one individual in the population. This could be evidence for the fact that these individuals were not "newcomers" into the society but were affiliated to a kinship group already rooted in the population.

The issue concerning whether social status in the Alemannic society was determined by birth and was thereby fixed for life or not is still an open question among archaeologists. The presented study points in the direction that social rank was not a predetermined fixture, but was indeed subject to change during the lifetime of an individual, probably due to social achievements. Various putative father-son pairs were determined by aDNA analysis, and several of those pairs are made up of individuals who were assigned to different social ranks according to archaeological classification. It can therefore be concluded that members of the same family did not belong to the same social group, thus revealing that social rank was not predetermined by birth.

7.4.2 Expert knowledge in Goslar

During the 11th century, the originally small village of Goslar in the northern Harz mountains, Lower Saxony, grew to be a flourishing city, the largest outside the former Roman Empire. This was due to the enormous silver reserves that had been discovered in the nearby Harz mountains.

During this time, the town acquired characteristics that can still be seen there today in the different areas, such as market districts, craftsmens' quarters and the lower town (Griep 1998).

In 1993, a burial site called "Hinter den Brüdern," consisting of 93 skeletons, was discovered. This churchyard area was used as a burial site for about 50 years during the early 18th century. Church documents testify that only metallurgists and their family members were allowed to be buried at this place. This group of highly specialized workers (cf. Fig. 7.51) lived in the Frankenberg quarter and were the only ones who had permission to work on the valuable metals that were mined in the nearby Rammelsberg mountain.

Fig. 7.51. Metal workers at the ore mine and at the furnace. Had this highly specialized work become a family tradition within the small group living in Goslar, or did experts continue to move into the town from all over, as was common in the Middle Ages

A genetic study was carried out on these samples of skeletal remains to determine whether these experts represented in the graveyard population had continued to come from different regions and had moved to the town of Goslar, as was common in the Middle Ages, or if they had developed a family tradition in a socially and genetically close society.

This was investigated by evaluating allele frequencies and heterozygosity rates with five of the STR markers of the AmpFlSTRProfilerPlus kit (Applied Biosystems), which was used for amplification of the ancient DNA (Bramanti et al. 2000a; Bramanti et al. 2000b). All 26 of the Goslar samples were reproduced using two independent extracts and produced genetic fingerprints suitable for the intended statistical tests. As a reference sample, a present-day German population composed of 28 unrelated individuals was used. The tests on the Hardy-Weinberg equilibrium showed no significant deviations in the historical sample compared to the modern reference sample (Table 7.10).

Table 7.10. Hardy-Weinberg equilibrium (HWE) in Goslar

STR loci	HWE (H_0=Het.deficit)	HWE (H_0=Het.excess)
D3S1358	0.035±0.002	0.965±0.002
D8S1179	0.293±0.006	0.720±0.006
D5S818	0.148±0.006	0.876±0.006
VWA	0.765±0.007	0.300±0.008
D21S11	0.346±0.009	0.718±0.007

However, Wright's F statistics revealed statistically significant differences in the expected and observed heterozygosity rates when the historical samples' F-values were compared to those of the modern-day reference sample. In the historical Goslar sample, F statistics generally predicted a rate of homozygosity higher than that from random mating alone (Table 7.11). The result was confirmed by a "hierarchical" analysis measured by Wright's F statistics. Drift phenomena were excluded by F_{st} values.

Table 7.11. Expected and observed heterozygosity and Wright's F in Goslar

	Expected heterozygosity		Observed heterozygosity		Wright's F	
	Modern population	Historic population	Modern population	Historic population	Modern population	Historic population
D3S1358	0.786	0.784	0.786	0.682	-0.000	+0.133
D8S1179	0.807	0.748	0.893	0.684	-0.108	+0.088
D5S818	0.697	0.723	0.607	0.632	+0.132	+0.129
VWA	0.809	0.844	0.857	0.889	-0.061	-0.054
D21S11	0.810	0.831	0.679	0.786	+0.164	+0.056

These results suggest a slight excess of closely inbred relatives in the sample. In relation to the initial question, this indicates that expert knowledge had become a family tradition in the early-modern town of Goslar, in the sense that the people living in the Frankenberg quarter not only shared the same kind of work but also engaged in consanguineous marriages. This would have certainly supported efforts to hold together the substantial profits earned by the smelters.

7.4.3 AB0 Blood groups

The worldwide distributions of the ABO blood groups are well studied; the respective frequencies are known for all continents, all populations and subpopulations (e.g., Vogel and Motulsky 1997). Because of the strongly deviating distributions, the ABO blood groups are known to be related to the immunogenetic response of an individual (Schenkel-Brunner 1995). In general, having blood-group O seems to increase an individual's risk of suffering severely from an infectious agent. Empirical findings show higher morbidity and mortality rates for blood-group O individuals infected with *Escherichia coli* (Van Loon et al. 1991) and *Vibrio cholerae* (e.g., Faruque et al. 1994; Swerdlow et al. 1994). The identification of ABO blood groups also has a long history in the assignment of paternity and the identification of forensic evidence material.

Because of the role of human identification in the work of epidemiology scientists in forensic medicine, human genetics and anthropology, the very first attempts to analyze forensic evidence and archaeological materials at the molecular level was initiated. The earliest analyzed sample materials were mummified soft tissues (e.g., Boyd and Boyd 1933); later attempts were made on keratinous tissues, body fluids and bone (Borgonini Tarli 1982; Borgonini Tarli et al. 1986). However, the serological methodology was ultimately shown to reveal unreliable results when applied to decayed biological material (Berg et al. 1983). This was mainly due to the fact that microorganisms as well as insect larvae can contaminate the samples with pseudoantigenic substances, which interfere with the specificity of the applied serological test.

A PCR-based RFLP analysis of the ABO blood groups (cf. chapter 2.2.5) now allows a reliable method to be used to identify the ABO traits on the phenotypic as well as on the genotypic level (Hummel et al. 2002). Before applying the analysis to ancient skeletal material (Fig. 7.52, Fig. 7.53 and Table 7.12), it was tested on modern samples derived from individuals with known ABO blood-group phenotypes. In order to validate the aDNA results, all aDNA extracts that were used in the amplification of exon 6 and exon 7 sequences determining the ABO blood-group traits underwent STR typing using the AmpFlSTRProfilerPlus (Applied Biosystems), and the results were compared to those obtained in earlier studies (Table 7.13). Using the validation power of autosomal STR typing, the results proved that the newly prepared extracts contained the authentic ancient DNA and had not been subject to contamination.

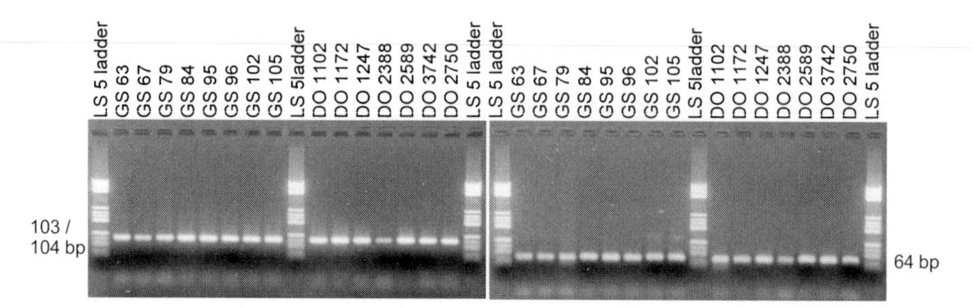

Fig. 7.52. Amplification products (9 µl) of the aDNA extracted from historic skeletal samples. The 103/104 bp products (*left*) are amplified from the exon 6 sequence, the 64 bp products (*right*) from the exon 7 sequence. As for the STR typing, the ABO allele amplifications resulted in amplification products of slightly varying intensity reflecting the different amounts of intact target sequences

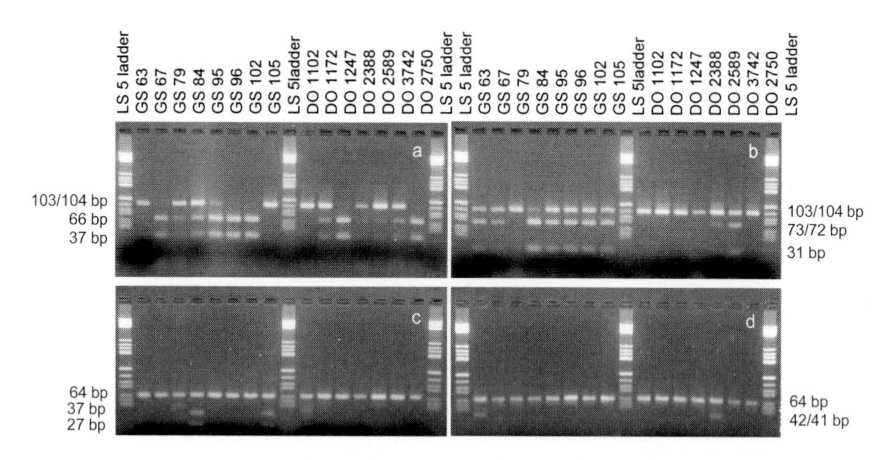

Fig. 7.53. RFLP analysis of the aDNA amplification products by the endonucleases: (a) Rsa I, (b) HpyCH4IV, (c) Nla III and (d) Mnl I. The electrophoresis of the Rsa I and HpyCh4IV restriction analysis are carried out on 3.8% agarose gels, those of the Nla III and Mnl I analysis on 4.4 % agarose gels

The application of this newly established approach for determining ABO blood-group alleles from aDNA extracts will be particularly interesting for analysis of skeletal material from burial sites that are known to have been established for victims of epidemics. There, one should expect an excess of the blood-group O (divided into the subgroups O_1, O_{1v} and O_2), if the hypothesis holds true that O individuals reveal a higher susceptibility to infectious diseases, which is commonly used to explain the worldwide differences in balanced polymorphism.

Moreover, it will be a valuable additional tool in reconstructing kinship. In the case of two individuals from the Lichtenstein cave, the ABO blood-group analysis and the Δccr5 data (cf. section 7.4.4) supported a kinship interpretation, which had remained comparatively weakly founded up to this point (cf. Fig. 7.38). Using these additional findings, DO 183 and DO 3742 may be highly related and are now considered to be siblings (Fig. 7.54).

Table 7.12. ABO genotype determination from ancient bone and teeth samples (-/- PCR product remains undigested, +/- one allele of the PCR product is digested, the other remains undigested, +/+ PCR product is completely digested, (o) carried out, although not necessary for analysis, since the Rsa I/HpyCH4IV patterns are already unique, cf. section 2.2.5)

Sample no.	Rsa I	HpyCH4IV	Nla III	Mnl I	Genotype	Phenotype
GS 63	+/-	+/-	-/-	+/-	O_1O_2	O
GS 67	+/+	+/-	-/- (o)	-/- (o)	O_1O_{1v}	O
GS 79	+/-	-/-	-/- (o)	-/- (o)	AO_1	A
GS 84	+/-	+/-	+/-	-/-	BO_1	B
GS 95	+/-	+/-	-/-	-/-	AO_{1v}	A
GS 96	+/+	+/-	-/- (o)	-/- (o)	O_1O_{1v}	O
GS 102	+/+	+/-	-/- (o)	-/- (o)	O_1O_{1v}	O
GS 105	-/-	+/-	+/-	-/-	AB	AB
DO 1102	-/-	-/-	-/- (o)	-/- (o)	AA	A
DO 1172	+/-	-/-	-/- (o)	-/- (o)	AO_1	A
DO 1247	+/+	-/-	-/- (o)	-/- (o)	O_1O_1	O
DO 2388	+/-	-/-	-/- (o)	-/- (o)	AO_1	A
DO 2589	+/-	+/-	-/-	+/-	O_1O_2	O
DO 3742	+/-	+/-	-/-	-/-	AO_{1v}	A
DO 3750	+/+	-/-	-/- (o)	-/- (o)	O_1O_1	O

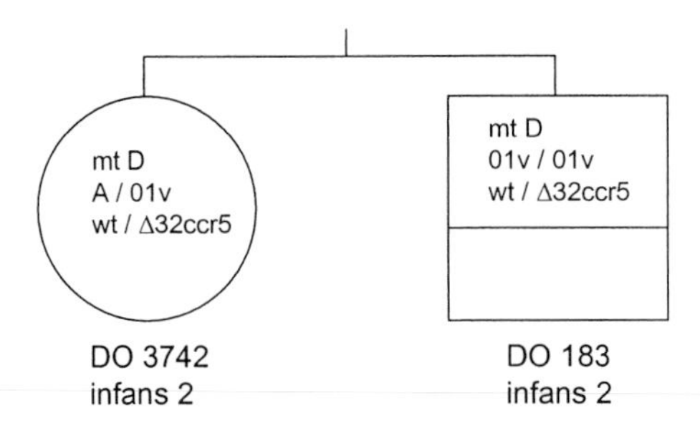

Fig. 7.54. The individuals DO 3742 and Do 183 who had been determined to be siblings (cf. section 7.3.3) reveal ABO blood group alleles and CCR5/Δ32ccr5 genotypes, which strongly support the results derived from mitochondrial haplotyping and statistical likelihood derived from autosomal STR analysis

Table 7.13. Data validation by autosomal STR genotyping (*ps* present study, *es* earlier study) (Bramanti et al. 2000, Schultes 2000). Alleles in parentheses were already determined but were not published in the study of Bramanti et al. (2000) since they were reproduced only once at that time

Sample	Amel	D3S1358	D8S1179	D5S818	vWA	D21S11	D13S317	FGA	D18S51	D7S820
GS 63 ps	XY	14/15	12/13	9/11	16/17	30/30.2	12/14	20/25	12/-	10/-
GS 63 es	XY	14/15	12/13	9/11	(16)/(17)	(30.2)/-	12/(14)	20/(25)	(12)/-	10/10
GS 67 ps	X/Y	15/-	10/13	13/-	15/17	29/33.2	12/-	-/-	-/-	-/-
GS 67 es	-/-	-/-	-/-	-/-	-/-	-/-	-/-	-/-	-/-	-/-
GS 79 ps	XX	16/-	12/15	10/-	15/17	29/31	-/-	19/21	-/-	-/-
GS 79 es	XX	16/16	12/15	10/10	(15)/17	29/31	-/-	(19)/(21)	-/-	-/-
GS 84 ps	XY	15/-	13/-	12/-	15/18	29/30.2	11/12	18/20	16/22	12/-
GS 84 es	XY	15/15	13/13	12/12	15/18	29/30.2	12/-	(18)/20	(22)/-	9/12
GS 95 ps	XY	17/-	13/15	11/-	16/17	29/32.2	10/13	22/24	14/-	11/12
GS 95 es	XY	17/17	13/15	11/11	16/17	29/32.2	10/13	(22)/24	14/14	11/12
GS 96 ps	XY	15/17	13/14	10/12	14/19	30/31.2	13/-	22/24	14/15	9/12
GS 96 es	XY	15/17	13/14	10/12	14/19	30/31.2	13/-	22/(24)	(14)/(15)	(9)/-
GS 102 ps	XY	14/15	11/14	9/12	16/17	29/30	-/-	20/21	15/-	12/-
GS 102 es	XY	14/15	11/14	9/12	17/-	29/30	-/-	20/20	-/-	12/12
GS 105 ps	XY	17/-	11/14	11/-	17/-	30/-	10/-	23.2/25	12/17	8/13
GS 105 es	X(Y)	17/-	11/14	11/11	16/17	28/30	10/-	23.2/25	12/(17)	(8)/-
DO1102 ps	XY	15/16	12/15	9/10	16/17	29/33.2	11/-	20/21	14/18	7/-
DO1102 es	XY	15/16	12/15	9/10	16/16	29/33.2	11/11	20/21	14/18	7/9
DO1172 ps	XY	15/17	12/13/15	10/12	15/16	33.2/-	11/-	21/22/25	-/-	-/-
DO1172 es	XY	15/17	12/13	10/12	15/16	30.2/33.2	11/-	21/25	-/-	8/-
DO1247 ps	XX	15/16	11/13	12/-	17/18	29/32.2	8/-	22/25	-/-	10/-
DO1247 es	XX	15/16	11/13	12/12	17/-	29/-	8/-	22/-	-/-	10/-
DO2388 ps	XX	15/16	13/15	9/11	17/-	30/33.2	-/-	-/-	-/-	-/-
DO2388 es	XX	15/16	13/-	11/-	16/17	30/33.2	11/11	20/21	15/-	10/-
DO2589 ps	XX	15/-	15/-	12/-	-/-	-/-	11/-	22/-	-/-	-/-
DO2589 es	XX	15/16	13/13	11/12	14/15	28/-	11/-	22/24	-/-	-/-
DO3742 ps	XX	14/16	13/-	10/13	13/14	29/-	-/-	24/-	12/-	-/-
DO3742 es	XX	14/16	13/-	10/13	13/14	29/32.2	8/10	24/-	12/15	9/10
DO3750 ps	XX	16/18	12/13	11/12	17/-	29/32.2	12/-	21/23	17/-	8/10
DO3750 es	XX	16/18	12/13	11/12	17/-	29/32.2	9/12	21/23	16/17	8/10

7.4.4 Δ32ccr5 already in the Bronze Age

Δ32ccr5 is a mutation on chromosome 3 that has only recently been discovered (e.g., Dean et al. 1996; Samson et al. 1996) in the context of HIV research. The allele frequencies for Δccr5, which is a 32-bp deletion, range between 2–20% in European populations (Fig. 7.55). However, high allele frequencies are found in the Scandinavian countries, and the lowest rates in Europe are found in Sicily and the Iberian peninsula (e.g., Martinson et al. 1997; Iyer et a. 2001). Allele frequencies in the Middle East, the Arabian peninsula and the Indian subcontinent are considerably lower, and the mutant allele is almost absent in any other populations.

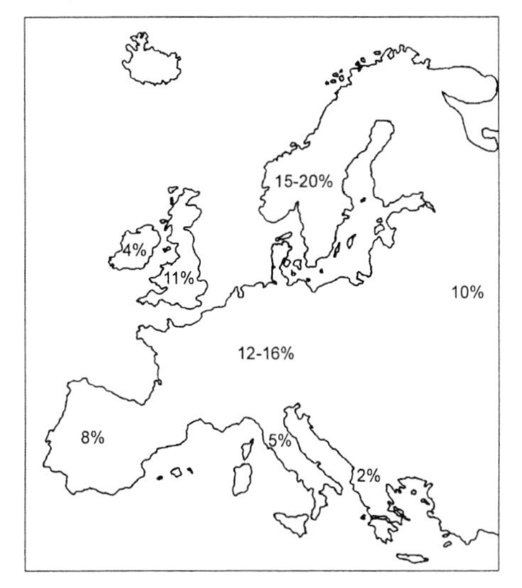

Fig. 7.55. Distribution of Δ32ccr5 in Europe

From the heterozygosity gradient found in Europe and from STR linkage analysis, a single original mutation event in northern Europe is considered to have occurred for the Δ32ccr5. There is general agreement that genetic drift was not responsible, and a heterozygote advantage led to the spreading of the mutant allele throughout Europe. The normal allele CCR5 is a chemokine-receptor gene involved in the immunoreaction in inflammatory diseases. Therefore, heterozygote advantages could be due to chronic rheumatoid arthritis or multiple sclerosis (e.g., Blanpain et al. 2000). Estimates for the time of the first occurrence of Δ32ccr5 range widely between 700–3,500 years ago in northern Europe (Libert et al. 1998; Stephens et al. 1998).

The following study aimed to investigate directly on skeletal material at what point in time this first occured.[38] [39] Therefore, primers were designed that amplify a 130- bp fragment for CCR5 and a 98-bp fragment for Δccr5. The locus is amplified by 35–40 cycles in a multiplex assay consisting of four STRs and the amelogenin gene in order to ensure data validation. The study was carried out on the skeletal material of the early-modern Goslar burial site (cf. sections 7.4.2 and 7.4.3) and on the Bronze Age skeletal material of the Lichtenstein cave, which are 3,000 years old (cf. sections 7.3.3 and 7.4.3).

The results clearly indicate that Δ32ccr5 was already present in central Europe as early as the Bronze Age (Fig. 7.56), long before the huge plague epidemics raged through Europe (Altschuler 2000). The data obtained from the investigation of 38 alleles of individuals from the Goslar series and of the 34 alleles from the Lichtenstein cave samples were calculated to be 18% and 11%, respectively (Tables 7.14a and 7.14b). However, from both burial sites, it is known that close relatives were laid to rest, and therefore, the interpretation of these data in the context of allele frequencies can only be speculative.

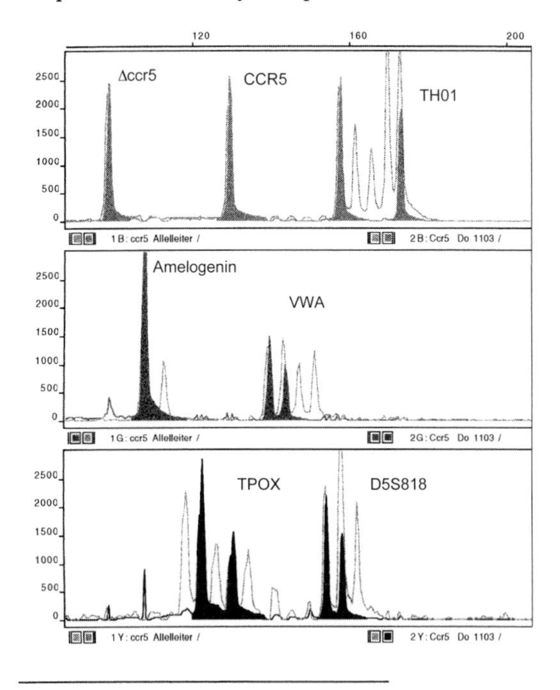

Fig. 7.56. Electropherogram of the multiplex CCR5/Δ32ccr5 (cf. section 2.2.6) amplification results for the Bronze Age sample DO1103. The deleted allele Δ32ccr5 (98 bp) was amplified reproducibly from this sample as well as from a further three samples from the Bronze Age burial site

[38] We are most grateful to Dr. M. Oppermann of the Department of Immunology of the University of Göttingen, who approached us with the idea to investigate the CCR5/Δccr5 directly from ancient skeletons.

[39] The molecular investigations were carried out by Diane Schmidt and the author

This study demonstrates the enormous value of investigating ancient material. The results indicate that statistical modeling of modern population data may reveal incorrect results. It is open to discussion whether this is due to an overestimation of STR linkage analysis, which was the basis for the calculations at the time of the first occurrence, or rather – and this seems much more likely in this case – the full meaning of the Δ32ccr5 mutation and the possible heterozygote advantages connected to the mutation have not yet been recognized.

Table 7.14a. CCR5/Δ32ccr5 in the early-modern population of Goslar (*wt* wild-type, *Δ32* deletion)

Sample no.	CCR5 wt /Δ32	TPOX	VWA	D5S818	TH01	Amelo
GS 30	wt/wt	12/-	-	11/-	-	-
GS 31	wt/wt	10/11	16/18	10/13	6/9.3	XX
GS 39	wt/wt	8/11	17/-	11/11	7/9.3	XY
GS 45.1	wt/wt	11/-	-	10/11	7/-	XY
GS 49	wt/Δ32	8/-	15/17	11/13	9.3/-	XY
GS 52	wt/Δ32	11/-	-	12/13/14	9.3/-	X-
GS 53	wt/wt	8/-	18/19	11/13	8/9	XY
GS 54	wt/wt	8/-	17/-	11/13	8/-	XY
GS 61	wt/Δ32	8/11	15/-	11/12	6/-	XX
GS 67	wt/wt	8/11	15/16	11/-	9/-	XY
GS 68	wt/Δ32	8/11	15/17	11/12	7/9	XY
GS 76	wt/Δ32	9/11	18/19	10/12	7/9.3	XX
GS 79	wt/wt	8/11	16/-	10/11	9/9.3	X-
GS 82	wt/wt	8/11	15/18	12/-	9.3/-	XY
GS 95	wt/wt	8/-	17/-	11/-	9.3/-	XY
GS 96	wt/Δ32	8/11	14/19	10/12	7/9	XY
GS 97	wt/wt	8/-	18/19	12/-	6/9.3	XY
GS 102	wt/Δ32	8/-	16/17	9/12	9/-	XY
GS 105	wt/wt	8/10	16/17	11/-	9/-	X-

Table 7.14b. CCR5/Δ32ccr5 in the Bronze Age population of the Lichtenstein cave (*wt* wild-type, *Δ32* deletion)

Sample no.	CCR5 wt /Δ32	TPOX	VWA	D5S818	TH01	Amelo
DO 26	wt/wt	8/-	16/-	9/11	7/9.3	XX
DO 58:3	wt/wt	8/-	18/-	11/-	6/8	XY
DO 183	wt/Δ32	8/11	14/19	10/-	6/9.3	XY
DO 199	wt/wt	11/-	14/-	12/-	7/9	X-
DO 300	wt/wt	8/11	16/-	-	6/-	XX
DO 902	wt/Δ32	8/-	13/15	12/-	6/9.3	XY
DO 1044	wt/wt	8/(11)	17/-	9/-	9.3/-	XX
DO 1076	wt/wt	8/-	-	12/-	7/9.3	XY
DO 1102	wt/wt	8/-	-	9/10	7/-	XY
DO 1103	wt/Δ32	9/11	14/15	11/12	6/9.3	XX
DO 1500	wt/wt	9/12	-	-	9.3/-	XX
DO 1585	wt/wt	-	-	-	-	XX
DO 1911	wt/wt	8/9	17/19	11/12	9.3/-	XY
DO 1916	wt/wt	8/10	14/16	12/-	6/(7)	XY
DO 2030	wt/wt	8/-	18/-	13/-	7/9	XY
DO 3742	wt/Δ32	8/-	14/-	13/-	9.3/-	XX
DO 3750	wt/wt	9/(12)	17/-	11/-	9.3/-	XX
DO 3756	wt/wt	8/9	17/19	12/-	9/9.3	XX

7.4.5 ΔF 508 in Alia

Alia is a small Sicilian village with a special history. In 1837, a devastating cholera epidemic raged in Alia, causing the deaths of 306 inhabitants (Fig. 7.57). Fearing further infection, the surviving inhabitants buried the dead in a remote place, a natural cave a few kilometers outside of the village, and they covered the bodies with quicklime (Guccione 1991). In 1996, these skeletal remains were retrieved in a good state of preservation.

The question arose as to whether this epidemic led to an increased mortality of those who did not carry the ΔF 508 marker in the historic Alia population. As the most common marker causing cyctic fibrosis, the ΔF 508 mutation is assumed to represent a selective advantage for heterozygote carriers in that they have a higher resistance to electrolyte-wasting diarrhea, which are commonly seen in the course of cholera infections, for example (Gabriel et al. 1994).

Fig. 7.57. In the early 19th century, drinking water fountains were discovered to be the starting points of cholera epidemics in London by the empiric studies of John Snow and Florence Nightingale

The present study intended to investigate the frequency of the ΔF 508 mutation on skeletal samples from the Alia site. This was carried out by including the marker to the AmpFlSTRProfilerPlus kit in order to ensure data validation. In an earlier study on the Alia site, it had already been proven that the aDNA extracts continued to be in a good state of preservation, which enabled us to reliably amplify the ΔF 508 mutation and carry out autosomal STR typing (Bramanti et al. 2000c). To date, 44 samples have been analyzed.[40] The investigations on the ancient skeletal samples have not yet revealed any cases of a heterozygote status for the ΔF 508 locus. All individuals investigated were homozygotic for the normal

[40] The study is a cooperative project between Florence and Göttingen. The experimental work has been carried out in the Göttingen aDNA laboratories by David Caramelli and Angelo Piazza under the supervision of Dr. Barbara Bramanti, who had the idea for the reseach. The final results will be published elsewhere.

allele (Fig. 7.58). Due to frequencies of approximately 50% for ΔF 508 in southern Italy (Rendine et al. 1997), it will be necessary to investigate 75 individuals to validate the hypothesis of a heterozygote advantage at an upper 95% confidence limit.

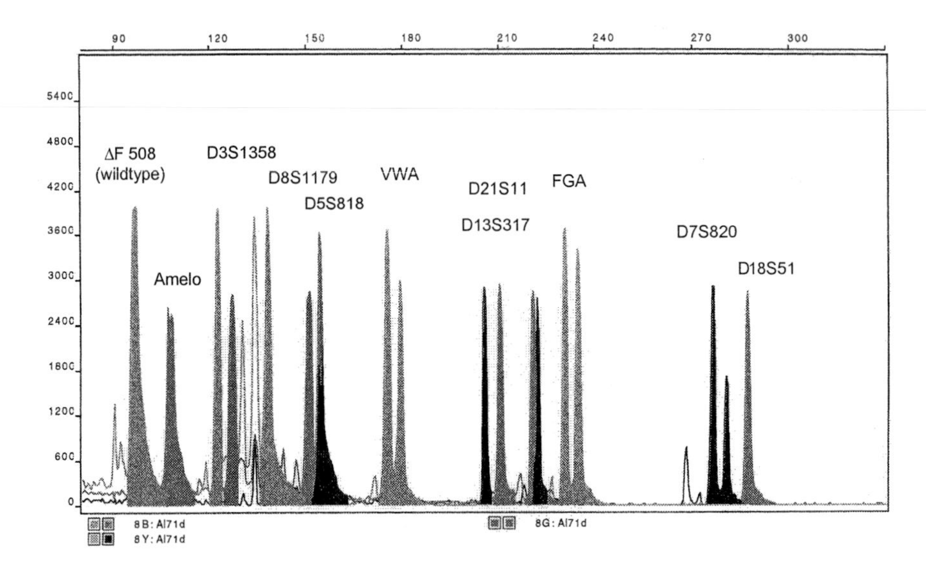

Fig. 7.58. ΔF 508 locus co-amplified with the STR multiplex typing for AL 71

7.4.6 Y-chromosomal haplotypes and the region of origin

Y-haplotyping is the corresponding feature to mitochondrial haplotyping. It discovers paternal lineages (Bosch et al. 1999; Schultes et al. 1999) in applications to either the reconstruction of genealogies or intra-population genetic studies, as shown in sections 7.3.3 and 7.4.1 (e.g., de Knijf et al. 1997; Prinz et al. 2001; Kayser et al. 2001; Lessig et al. 2001). With the help of a newly designed Y-STRs multiplex PCR, it will even be possible to carry out inter-population genetic studies that allow us to determine the most probable regional origin of a single individual or a group of individuals. Although already as few as four Y-chromosomal STRs occasionally may enable this if rare haplotypes are found (cf. section 7.3.4), the number of informative results can be considerably increased by employing more marker systems.

The first applications of the new Y-haplotype multiplex PCR to analyze nine STRs in ancient DNA were successful (Fig. 7.59).[41] They show completely reproducible results from ancient samples derived from different skeletal series.

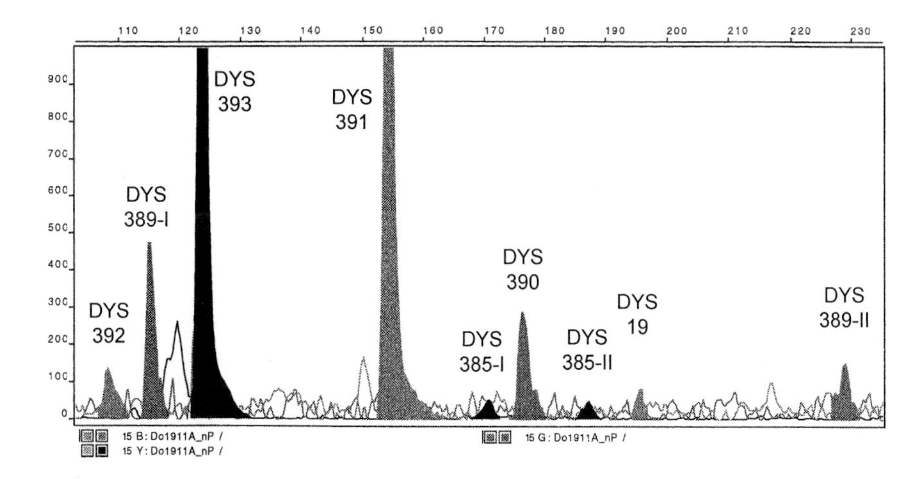

Fig. 7.59. Electropherograms showing Y-haplotyping of DO 1911. As in autosomal STR typing, it is characteristic for ancient DNA samples to have the amplification products of the systems generating large amplicons revealing smaller peak sizes. This is due to fewer intact targets within the extract

References

Allmann M, Höfelein C, Koppel E, Luthy J, Meyer R, Niederhauser C, Wegmuller B, Candrian U (1995) Polymerase chain reaction (PCR) for detection of pathogenic microorganisms in bacteriological monitoring of dairy products. Res Microbiol 146:85–97

Alt K, Vach W (1994) Rekonstruktion biologischer und sozialer Strukturen in ur- und frühgeschichtlichen Bevölkerungen. Innovative Ansätze zur Verwandtschaftsanalyse in der Archäologie. Prähist Z 69:56–91

Alt K, Vach W (1998) Kinship studies in skeletal remains – concepts and examples. In: Alt K, Rösing FW, Teschler-Nicola M (eds) Dental anthropology. Fundamentals, limits and prospects. Springer, Wien, New York

---

[41] The experimental work was carried out by Boris Müller and was part of his diploma thesis.

Althoff G (1983) Der Sachsenherzog Widukind als Mönch auf der Reichenau. Ein Beitrag zur Kritik der Widukind-Mythos. In: Hauck K (ed) Frühmittelalterliche Studien. Jahrbuch des Institutes für Mittelalterforschung der Universität Münster. de Gruyter, Berlin New York. pp. 251-279

Altschuler EL (2000) Plague as HIV vaccine adjuvant. Med Hypotheses 54:1003–1004

Bacher R, Suter PJ, Eggenberger P, Ullrich-Bochsler S, Meyer L Aegerten (1990) Die spätrömischen Anlagen und der Friedhof der Kirche Bürglen. Staatlicher Lehrmittelverlag, Bern

Barriel V, Thuet E, Tassy P (1999) Molecular phylogeny of Elephantidae. Extreme divergence of the extant forest African elephant. C R Acad Sci III 322:447–454

Bartels A (2002) Verwandtschaftsrekonstruktion am Rotfuchs (*Vulpes vulpes*) durch molekulargenetische Analysen an degradierten Gewebeproben. Diplomarbeit, Georg August-Universität, Göttingen

Berg S, Bertozzi B, Meier R, Mendritzki S (1983) Vergleichend-methodologischer Beitrag und kritische Bemerkungen zur Interpretation von Blutgruppenbestimmungen an Mumienrelikten und Skelettfunden. Anthrop Anz 41:1-19

Blanpain C, Lee B, Tackoen M, Puffer B, Boom A, Libert F, Sharron M, Wittamer V, Vassart G, Doms RW, Parmentier M (2000) Multiple nonfunctional alleles of CCR5 are frequent in various human populations. Blood 96:1638–1645

Bollongino R (2000) Bestimmung humanökologisch relevanter Tierarten aus historischen, musealen und forensischen Materialien durch Sequenzierung mitochondrialer Genorte. Diplomarbeit, Georg August-Universität, Göttingen

Borgonini Tarli S, Paoli G (1982) Survey on paleoserological studies. Homo 33: 69–89

Borgonini Tarli S, Paoli G, Francalacci P (1986) Problems and perspectives in palaeoserology. In: Herrmann, B (ed) Innovative trends in prehistoric anthropology. Mitt Berl Ges Anthrop Ethnol Urgesch 7:107–116

Bosch E, Calafell F, Santos FR, Perez-Lezaun A, Comas D, Benchemsi N, Tyler-Smith C, Bertranpetit J (1999) Variation in short tandem repeats is deeply structured by genetic background on the human Y chromosome. Am J Hum Genet 65:1623–1638

Boyd WC, Boyd LC (1933) Blood grouping by means of preserved muscle. Science 78:578

Bramanti B, Hummel S, Schultes T, Herrmann B (2000a) STR allelic frequencies in a German skeleton collection. Anthrop Anz 58:45–49

Bramanti B, Hummel S, Schultes T, Herrmann B (2000b) Genetic characterization of a historical human society by means of aDNA analysis of autosomal STRs. Bienniel Books of EAA 1:147–163

Bramanti B, Sineo L, Vianello M, Caramelli D, Hummel S, Chiarelli B, Herrmann B (2000c) The selective advantage of cystic fibrosis heterocygotes tested by aDNA analysis. Int J Anthropol 15:255–262

Branciari R, Nijman IJ, Plas ME, Di Antonio E, Lenstra JA (2000) Species origin of milk in Italian mozzarella and Greek feta cheese. J Food Prot 63:408–411

Burger J (2000) Sequenzierung, RFLP-Analyse und STR-Genotypisierungen alter DNA aus archäologischen Funden und historischen Werkstoffen. Dissertation Universität Göttingen

Burger J, Hummel S, Pfeiffer I, Herrmann B (2000a) Paleogenetic analysis of (pre)historic artifacts and its significance for anthropology. Anthrop Anz 58:69–76

Burger J, Hummel S, Herrmann B (2000b) Palaeogenetics and cultural heritage. Species determination and STR genotyping from ancient DNA in art and artifacts. Thermochimica Acta 365:141–146

Burger J, Pfeiffer I, Hummel S, Fuchs R, Brenig B, Herrmann B (2001) Mitochondrial and nuclear DNA from (pre)historic hide-derived material. Ancient Biomol 3:227–238

Calvo JH, Zaragoza P, Osta R (2001) Technical note: A quick and more sensitive method to identify pork in processed and unprocessed food by PCR amplification of a new specific DNA fragment. J Anim Sci 79:2108–2112

Dean M, Carrington M, Winkler C, Huttley GA, Smith MW, Allikmets R, Goedert JJ, Buchbinder SP, Vittinghoff E, Gomperts E, Donfield S, Vlahov D, Kaslow R, Saah A, Rinaldo C, Detels R, O'Brien SJ (1996) Genetic restriction of HIV-1 infection and progression to AIDS by a deletion allele of the CKR5 structural gene. Hemophilia growth and development study, Multicenter AIDS cohort study, Multicenter hemophilia cohort study, San Francisco City Cohort, ALIVE Study. Science 273:1856–1862

Eldridge MDB, Browning TL, Close RL (2001a) Provenance of a New Zealand brush-tailed rock-wallaby (Petrogale penicillata) population determined by mitochondrial DNA sequence analysis. Mol Ecol 10:2561-2567

Eldridge MDB, Wilson ACC, Metcalfe CJ, Dollin AE, Bell JN, Johnson PM, Johnston PG, Close RL (2001b) Taxonomy of rock-wallabies, Petrogale (Marsupialia: Macropodidae). III. Molecular data confirms the species status of the purple-necked rock-wallaby Petrogale purpureicollis Le Souef 1924. Austral J Zool 49:323-343.

Faerman M, Kahila G, Smith P, Greenblatt C, Stager L, Oppenheim A, Filon D (1997) DNA analysis reveals the sex of infanticid victims. Nature 385:212–213

Faruque A, Mahalanabis D, Hoque S, Albert M (1994) The relationship between ABO blood groups and suseptibility to diarrhea due to *Vibrio Cholerae* 0139. Clin Infect Dis 18:827–828

Flindt S (1996) Die Lichtensteinhöhle bei Osterode, Landkreis Osterode am Harz. Eine Opferhöhle der jüngeren Bronzezeit im Gipskarst des südwestlichen Harzrandes. Forschungsberichte und erste Grabungsergebnisse. In: Die Kunde N. F.47:435–466

Flindt S (ed) (2001) Höhlen im Westharz und Kyffhäuser. Geologie, Speläologie und Archäologie. Archäologische Schriften des Landkreises Osterode am Harz 3. Mitzkat, Holzminden, pp 62–85

Firestone KB, Elphinstone MS, Sherwin WB, Houlden BA (1999) Phylogeographical population structure of tiger quolls *Dasyurus maculatus* (Dasyuridae: Marsupialia), an endangered carnivorous marsupial. Mol Ecol 8:1613–1625

Gabriel SE, Brigman KN, Koller BH, Boucher RC, Stutts MJ (1994) Cystic fibrosis heterozygote resistance to cholera toxin in the cystic fibrosis mouse model. Science 266:107–109

Gerstenberger J, Hummel S, Herrmann B (1998) Assignment of an isolated skeletal element to the skeleton of Duke Christian II. Ancient Biomol 2: 63–68

Gerstenberger J, Hummel S, Schultes T, Häck B, Herrmann B (1999) Reconstruction of a historical genealogy by means of STR analysis and Y-haplotyping of ancient DNA. Europ J Hum Genet 7:469–477

Gerstenberger J (2002) Analyse alter DNA zur Ermittlung von Heiratsmustern in einer frühmittelalterlichen Bevölkerung. Dissertation, Georg August-Universität, Göttingen

Geschwinde M (1988) Höhlen im Ith. Urgeschichtliche Opferstätten im südniedersächsischen Bergland. August Lax, Hildesheim

Gill P, Ivanov PL, Kimpton C, Piercy R, Benson N, Tully G, Evett I, Hagelberg E, Sullivan K (1994) Identification of the remains of the Romanov family by DNA analysis. Nat Genet 6:130–135

Griep HG (1998) Weltkulturerbe Goslar. Schadach, Goslar

Guccione E (1991) Storia di Alia 1615–1860. Salvatore Sciascia, Caltanissetta, Roma, pp 304–313

Hagelberg E, Gray IC, Jeffreys AJ (1991) Identification of the skeletal remains of a murder victim by DNA analysis. Nature 352:427–429

Hering S, Szibor R (2000) Development of the X-linked tetrameric microsatellite marker DXS9898 for forensic purposes. J Forensic Sci 45:929–931

Hering S, Kuhlisch E, Szibor R (2001) Development of the X-linked tetrameric microsatellite marker HumDXS6789 for forensic purposes. Forensic Sci Int 119:42–46

Herrmann B, Grupe G, Hummel S, Piepenbrink H, Schutkowski H (1990) Prähistorische Anthropolgie. Leitfaden der Feld- und Labormethoden. Springer, Heidelberg

Herrmann B, Hummel S (1997) Genetic analysis of past populations by aDNA studies. Advances in research on DNA polymorphisms, ISFH-Symposion, Toyoshoten, Hakone. pp 33–47

Heyen DW, Beever JE, DA Y, Evert RE, Green C, Bates SR, Ziegele JS, Lewin HA (1997) Exclusion probabilities of 22 bovine microsatellite markers in fluorescent multiplexes for semi-automated parentage testing. Anim Genet 28:21–27

Hobmaier M (1889) Die Grafen von Königsfeld auf Niederaichbach. Verhandlungen des historischen Vereins für Niederbayern 26:163–326

Hopwood AJ, Mannucci A, Sullivan KM (1996) DNA typing from human faeces. Int J Legal Med 108:237–243

Houlden BA, England PR, Taylor AC, Greville WD, Sherwin WB (1996) Low genetic variability of the koala *Phascolarctos cinereus* in south-eastern Australia following a severe population bottleneck. Mol Ecol 5:269–281

Hummel S (1994) DNA aus alten Geweben. In: Hermann B (ed) Archäometrie. Naturwissenschaftliche Analyse von Sachüberresten. Springer, Heidelberg

Hummel S, Herrmann B (1991): Y-chromosome-specific DNA amplified in ancient human bone. Naturwissenschaften 78: 266–267

Hummel S, Nordsiek G, Rameckers J, Lassen C, Zierdt H, Baron H, Herrmann B (1995) aDNA – Ein neuer Zugang zu alten Fragen. Z Morph Anthrop 81:41–65

Hummel S, Bramanti B, Finke T, Herrmann B (2000) Evaluation of morphological sex determinations by molecular analyses. Anthrop Anz 58:9–13

Hummel S, Schmidt D, Kahle M, Herrmann B (2002) ABO blood group genotyping of ancient DNA by PCR-RFLP. Int J Legal Med (in press)

Immel U-D, Hummel S, Herrmann B (1999) DNA profiling of orangutan (Pongo pygmaeus) feces to prove descent and identity in wildlife animals. Electrophoresis 20:1768–1770

Immel U, Hummel S, Herrmann B (2000) Reconstruction of kinship by fecal DNA analysis of Orang Utans. Anthrop Anz 58:63–67

Iyer RK, Kim PS, Bando JM, Lu KV, Gregg JP, Grody WW (2001) A multi-ethnic study of Δ32ccr5 and ccr2b-V64I allele distribution in four Los Angeles populations. Diagn Mol Pathol 10:105–110

Kayser M, Krawczak M, Excoffier L, Dieltjes P, Corach D, Pascali V, Gehrig C, Bernini LF, Jespersen J, Bakker E, Roewer L, de Knijff P (2001) An extensive analysis of Y-chromosomal microsatellite haplotypes in globally dispersed human populations. Am J Hum Genet 68:990–1018

Kimpton C, Walton A, Gill P (1992) A further tetranucleotide repeat polymorphism in the vWF gene. Hum Mol Genet 1:287–287

de Knijff P, Kayser M, Caglia A, Corach D, Fretwell N, Gehrig C, Graziosi G, Heidorn F, Herrmann S, Herzog B, Hidding M, Honda K, Jobling M, Krawczak M, Leim K, Meuser S, Meyer E, Oesterreich W, Pandya A, Parson W, Penacino G, Perez-Lezaun A, Piccinini A, Prinz M,. Schmitt C, Schneider PM, Szibor R, Teifel-Greding J, Weichhold G, Roewer L (1997) Chromosome Y microsatellites: population genetic and evolutionary aspects. Int J Legal Med 110:134–49

Königliche Akademie der Wissenschaften. Historische Kommission (1876) Allgemeine Deutsche Biographie, Vol 4, Leipzig. pp 679–680

Kohn M, Knauer F, Stoffella A, Schröder W, Pääbo S (1995) Conservation genetics of the European brown bear – A study using excremental PCR of nuclear and mitochondrial sequences. Mol Ecol 4:95–103

Kurosaki K, Matsushita T, Ueda S (1993) Individual DNA identification from ancient human remains. Am J Hum Genet 53:638–643

Lassen C, Hummel S, Herrmann B (1996) PCR-based sex determination in ancient human bones by amplification of X- and Y-chromosomal sequences. A comparison. Ancient Biomol 1: 25–33

Lassen C (1998) Molekulare Geschlechtsdetermination der Traufkinder des Gräberfeldes Aegerten (Schweiz). Dissertation, Georg August-Universität, Göttingen. Cuvillier, Göttingen

Lassen C, Hummel S, Herrmann B (2000) Molecular sex identification of stillborn and neonate individuals (Traufkinder) from the burial site Aegerten. Anthrop Anz 58:1–8

Launhardt K, Epplen C, Epplen JT, Winkler P (1998) Amplification of microsatellites adapted from human systems in faecal DNA of wild Hanuman langurs (*Presbytis entellus*). Electrophoresis 19:1356–1361

Lessig R, Edelmann J, Krawczak M (2001) Population genetics of Y-chromosomal microsatellites in Baltic males. Forensic Sci Int 118:153–157
Levi-Strauss C (ed) (1981) Die elementaren Strukturen der Verwandtschaft. Suhrkamp, Frankfurt
Libert F, Cochaux P, Beckman G, Samson M, Aksenova M, Cao A, Czeizel A, Claustres M, de la Rua C, Ferrari M, Ferrec C, Glover G, Grinde B, Guran S, Kucinskas V, Lavinha J, Mercier B, Ogur G, Peltonen L, Rosatelli C, Schwartz M, Spitsyn V, Timar L, Beckman L, Parmentier M, Vassart G (1998) The Δ ccr5 mutation conferring protection against HIV-1 in Caucasian populations has a single and recent origin in Northeastern Europe. Hum Mol Genet 7:399–406
Loreille O, Roumat E, Verneau O, Bouchet F, Hanni C (2001) Ancient DNA from Ascaris: extraction amplification and sequences from eggs collected in coprolites. Int J Parasitol 31:1101–1106
Martinson JJ, Chapman NH, Rees DC, Liu YT, Clegg JB (1997) Global distribution of the CCR5 gene 32-base pair deletion. Nat Genet 16:100–103
Mayer, H. (1996) Christian der Jüngere, Herzog von Braunschweig-Lüneburg-Wolfenbüttel (1599-1626), Braunschweigisches Jahrbuch für Landesgeschichte 77:181–201
Meyer R, Höfelein C, Luthy J, Candrian U (1995) Polymerase chain reaction-restriction fragment length polymorphism analysis: a simple method for species identification in food. J AOAC Int 78:1542–1551
Poinar HN, Hofreiter M, Spaulding WG, Martin PS, Stankiewicz BA, Bland H, Evershed RP, Possnert G, Pääbo S (1998) Molecular coproscopy: dung and diet of the extinct ground sloth *Nothrotheriops shastensis*. Science 281:402–406
Prinz M, Ishii A, Coleman A, Baum HJ, Shaler RC (2001) Validation and casework application of a Y chromosome specific STR multiplex. Forensic Sci Int 120:177–188
Rendine S, Calafell F, Cappello N, Gagliardini R, Caramia G, Rigillo N, Silvetti M, Zanda M, Miano A, Battistini F, Marianelli L, Taccetti G, Diana MC, Romano L, Romano C, Giunta A, Padoan R, Pianaroli A, Raia V, De Ritis G, Battistini A, Grzincich G, Japichino L, Pardo F, Piazza A (1997) Genetic history of cystic fibrosis mutations in Italy. I. Regional distribution. Ann Hum Genet 61:411–424
Rösing FW (1986) Kith or kin? On the feasability of kinship reconstruction in skeletons. In: David AR (ed) Science in Egyptology. Manchester University Press, Manchester, pp 223–237
Roth H, Theune C (1988) Zur Chronologie merowingerzeitlicher Frauengräber in Südwestdeutschland. Ein Vorbericht zum Gräberfeld von Weingarten, Kr. Ravensburg. Archäolog Informat Baden-Württemburg 6:7–18
Samson M, Labbe O, Mollereau C, Vassart G, Parmentier M (1996) Molecular cloning and functional expression of a new human CC-chemokine receptor gene. Biochemistry 35:3362–3367
Schenkel-Brunner H (1995) Human blood groups. Springer, Wien, New York
Schmidt D (2001) DNA-Extraktion aus konservierten Lebensmitteln. Staatsexamensarbeit, Georg August-Universität, Göttingen

Schmid K (1964) Die Nachfahren Widukinds. Deutsche Archiv zur Erforschung des Mittelalters 20:1–4

Schultes T (2000) Typisierungen alter DNA zur Rekonstruktion von Verwandtschaft in einem bronzezeitlichen Skelettkollektiv. Dissertation, Georg August-Universität Göttingen. Cuvillier, Göttingen

Schultes T, Hummel S, Herrmann B (1997a) Zuordnung isolierter Skelettelemente mittels aDNA-typing. Anthrop Anz 55:207–216

Schultes T, Hummel S, Herrmann B (1997b) Recognizing and overcoming inconsistencies in microsatellite typing of ancient DNA samples. Ancient Biomol 1:227–233

Schultes T, Hummel S, Herrmann B (1999) Amplification of Y-chromosomal STRs from ancient skeletal remains. Hum Genet 104:164–166

Schutkowski H (1990) Zur Geschlechtsdiagnose an Kinderskeletten. Morphognostische, metrische und diskriminanzanalytische Untersuchungen. Dissertation, Georg August-Universität Göttingen

Schutkowski H (1993) Sex determination of infant and juvenile skeletons: I. Morphognostic features. Am J Phys Anthrop 90:199–205

Sebastio P, Zanelli P, Neri TM (2001) Identification of anchovy (*Engraulis encrasicholus L.*) and gilt sardine (*Sardinella aurita*) by polymerase chain reaction, sequence of their mitochondrial cytochrome b gene, and restriction analysis of polymerase chain reaction products in semipreserves. J Agric Food Chem 49:1194–1199

Siegmund F (1998) Social structure and relations. In: Ausenda G, Wood I (eds) Franks and Alamanni in the Merovingian period. An ethnographic perspective. Boydell Press, Woodbridge, pp 177–199

Sinclair AR (1997) Fertility control of mammal pests and the conservation of endangered marsupials. Reprod Fertil Dev 9:1–16

Spencer PB, Odorico DM, Jones SJ, Marsh HD, Miller DJ (1995) Highly variable microsatellites in isolated colonies of the rock-wallaby (*Petrogale assimilis*). Mol Ecol 4:523–525

Spencer PB, Bryant KA (2000) Characterization of highly polymorphic microsatellite markers in the marsupial honey possum (*Tarsipes rostratus*). Mol Ecol 9:492–494

Stephens JC, Reich DE, Goldstein DB, Shin HD, Smith MW, Carrington M, Winkler C, Huttley GA, Allikmets R, Schriml L, Gerrard B, Malasky M, Ramos MD, Morlot S, Tzetis M, Oddoux C, di Giovine FS, Nasioulas G, Chandler D, Aseev M, Hanson M, Kalaydjieva L, Glavac D, Gasparini P, Kanavakis E, Claustres M, Kambouris M, Ostrer H, Duff G, Baranov V, Sibul H, Metspalu A, Goldman D, Martin M, Duffy D, Schmidtke J, Estivill X, O'Brien S, Dean M (1998) Dating the origin of the CCR5-Δ32 AIDS-resistance allele by the coalescence of haplotypes. Am J Hum Genet 62:1507–1515

Steuer H (1982) Frühgeschichtliche Sozialstrukturen in Mitteleuropa: Eine Analyse der Auswertung des archäologischen Quellenmaterials. Vandenhoeck und Ruprecht, Göttingen

Stone AC, Milner GR, Pääbo S, Stoneking M (1996) Sex determination of ancient human skeletons using DNA. Am J Phys Anthropol 99:231–238

Swerdlow D, Mintz E, Rodriguez M, Tejada E, Okampo C, Espejo L, Barrett T, Petzelt J, Bean N, Seminario L, Tauxe R (1994) Severe life-threatening cholera associated with blood group O in Peru: implications for the Latin American epidemic. J Infect Dis 170: 468–472

Taberlet P, Camarra JJ, Griffin S, Uhres E, Hanotte O, Waits LP, Dubois-Paganon C, Burke T, Bouvet J (1997) Noninvasive genetic tracking of the endangered Pyrenean brown bear population. Mol Ecol 6:869–876

Thomas MG, Hagelberg E, Jone HB, Yang Z, Lister AM (2000) Molecular and morphological evidence on the phylogeny of the Elephantidae. Proc R Soc Lond B Biol Sci 267:2493–500

Ullrich H (1972) Anthropologische Untersuchungen zur Frage nach der Entstehung und Verwandtschaft der thüringischen, böhmischen und mährischen Aunjetitzer. Veröffentlichungen des Museums für Ur- und Frühgeschichte Thüringen 3:9–147

Ullrich-Bochsler S (1997) Anthropologische Befunde zur Stellung von Frau und Kind in Mittelalter und Neuzeit. Berner Lehrmittel- und Medienverlag, Bern

Uytterschaut H (1993) Human bone remodelling and aging. In: Grupe G, Garland N (eds) Histology of ancient human bone. Springer, Berlin, Heidelberg

Van Loon F, Clemens J, Sack D, Rao M, Faruque A, Chowdhury S, Harris J, Ali M, Chakraborty J, Khan M, Neogy P, Svennerholm A, Holmgren J (1991) ABO blood groups and the risk of diarrhea due to enterotoxigenic *Escherichia coli*. J Infect Dis 163: 1243–1246

Vogel F, Motulski AG (1979) Human genetics. Problems and approaches (3rd edn). Springer, Berlin, Heidelberg, New York

Voland E (1984) Human sex ratio manipulation: historical data from a German parish. J Hum Evol 13:99–107

Wasser SK, Houston CS, Koehler GM, Cadd GG, Fain SR (1997) Techniques for application of faecal DNA methods to field studies of Ursids. Mol Ecol 6:1091–1097

Whitehouse AM, Harley EH (2001) Post-bottleneck genetic diversity of elephant populations in South Africa, revealed using microsatellite analysis. Mol Ecol 10:2139–2149

Zierdt H, Hummel S, Herrmann B (1996) Amplification of human short tandem repeats from medieval teeth and bone samples. Hum Biol 68:185–199

8 Protocols

The collection of protocols that is presented in this chapter does not, and is not intended to, represent a complete summary of all protocols available in the ancient DNA literature (also cf. section 3.1). Obviously, as with all protocols, there can be no guarantee that any protocol presented here describes the optimal way to perform a certain task. Nonetheless, because of the extensive experience of our laboratory with ancient DNA analysis, I am confident that all of them are consistent and reliable. Primarily, this is thanks to the many graduate and Ph.D. students, the post-doctoral scientists and all the guests who have carried out research work in our laboratory, all of them putting a great deal of effort into optimizing the procedures, primers and protocols in order to improve aDNA analysis.

The collection of protocols and procedures is comprised of instructions for basic chemistry (cf. section 8.10, which lists all the necessary ingredients), although many of these can also be found in any good laboratory manual (e.g., Sambrook et al. 1989). There were two reasons for listing them here: firstly, it is convenient to use a single source; secondly and importantly, it is our experience that even minor inconsistencies in procedures or parameters will have an enormous influence on the final results (see chapter 3). For example, in the case of unspecific results or amplification failure, anyone who wants to perform a PCR amplification with one of the primer sets described here (section 8.5) which have been specifically designed for aDNA use, will have the opportunity to compare their own protocols with the ones we use.

8.1 Collection and storage of samples

Materials: disposable gloves, face mask, head-dress gown, lab coat, glasses, saw (for hard tissues), forceps or scalpel (for soft tissues), sterile tubes or plastic bags, tissues, permanent pencil
Chemicals and reagents: liquid soap, doubly distilled and deionized H_2O (dH_2O), ethanol (99.5%)

Whenever possible, sample collection should be done at the original (archaeological) site. This minimizes the risk of the samples becoming contaminated by modern DNA and enables the immediate and appropriate storage of the samples. Storage of sample material should be as cool and dry as possible, preferably at -20°C. If samples are taken from museum collections where they generally have been stored at approximately room temperature, they should also be stored frozen, since it has been shown that periods at room temperature can cause further DNA damage (cf. section 3.3.2). If electric or mechanical saws, forceps or scalpels are necessary for sampling, it is important to ensure proper cleaning of the blades between the sampling of two different individuals. This is best done by using successively concentrated soap or bleach, distilled water and a final wash in absolute ethanol in order to prevent cross-contaminations. It is also necessary to change the disposable gloves between different samples.

If it cannot be excluded that contamination of the samples by members of the excavation team or by museum or any other personnel who have handled the material has occurred, it is recommended that a saliva sample from each person is taken as a control sample in order to enable genetic typing and the comparison of results. This is important in particular if only a single sample is going to be investigated (cf. section 6.2.2). In cases where it is not possible to obtain saliva (or any other tissue) of a person suspected to be a potential contaminator, then some other material that is known to have been in contact with the person in question should be sampled and analyzed. Preferably the material should be of non-human origin. This sample is an important control and should not undergo the decontamination measures that are carried out for the sample that is the focus of interest.

8.2 Sample preparation

Materials: disposable gloves, face mask, head-dress gown, lab coat, glasses, scalpel, mortar, grinding mill, spatula, 2-ml Eppendorf tubes, tube rack, tissues, permanent pencil
Chemicals and reagents: liquid soap, dH_2O, ethanol (70%)

Any surfaces of the sample (except particular types of control samples mentioned previously) that have potentially been cross-contaminated or may have been touched in any other way by cellular material of another individual (e.g., through sneezing, coughing, talking, eyelid blinking) must be removed. In general, a disposable scalpel is sufficient for this task.

In order to prepare the sample for DNA extraction, it should be homogenized. For mummified soft tissues, hair, fabric or soft or brittle material, this can be carried out using a mortar and pestle. Homogenizing bones and teeth usually requires a grinding mill similar to those used by geologists and mineralogists. The vessels of such mills are easy to clean, again by using a series of cleaning steps involving liquid soap, double-distilled water and ethanol.[42] The bone samples or teeth are initially broken down with the help of a resistant mortar (metal) or a hammer before milling them in order to reduce the milling time. If milling times of more than 3 min are necessary, the processes should be interrupted for a couple of minutes in order to allow the samples to cool down.

Although there is no hard data available, the experience from our laboratory indicates that homogenizing the samples should be done as shortly prior to DNA extraction. The long-term storage of powdered bone and teeth appears to result in the reduction of DNA preservation. This degradation can be best explained by an increase in oxidative damage even at the low temperature of -20°C as a result of the enormous increase in exposed surface area.

8.3 Phenol-based DNA extraction

***Material*:** disposable gloves, face mask, glasses, lab coat, head-dress gown, 2-ml Eppendorf tubes, tube rack, 15-ml falcon tubes, pipettes and pipette tips, organic waste bottle, over-head rotator, bench-top centrifuge, thermostatic mixer, incubator
Chemicals and reagents: 0.5 M EDTA (pH 8.3), 2×Lysis buffer, proteinase K (100 U/ml), phenol, chloroform, 2 M NaAc (pH 4.5), 2-propanol (100%), silica-particle solution, ethanol (80%), DNA-free H_2O

The phenol-based method of DNA extraction is regularly carried out in our laboratory on sample material such as archaeological and forensic evidence, bone, teeth and feces. Additionally, this method has also been successful on many other types of source material that are thought to contain significant amounts of inhibitors (humic acids, tannins) because of contact with soil or natural oxidative aging processes. In general, it seems advisable to carry out a phenol-based organic extraction in all cases of

[42] Vessels of such mills should regularly be cleaned thoroughly by running them with sterile, fine grade sand (e.g., from aquarium merchants).

ancient material where chromosomal DNA is the intended target (cf. section 3.1). In our laboratory the organic extraction is carried out using an automated system (nucleic acids extractors, types 340a and 341, Applied Biosystems). This prevents the samples from being handled extensively and, as a result, minimizes the contamination risk. Furthermore, it reduces the exposure of the investigators to the organic solvents, many of which are hazardous. Unfortunately, these machines are no longer available. Obviously, these machines that were originally manufactured for automated modern DNA extraction have not been economically viable to produce. This is almost certainly due their low sample capacity (only eight samples can be processed at any one time), and they therefore cannot compete with the many kits that are available on the market. This also explains why there are no alternative retailers for similar machines at present. Nevertheless, the phenol-based extraction method can also be carried out manually, although it is of course far less convenient than other techniques and must be carried out under a fume hood.

The volumes given in the following protocol are those we use in the automated extractors. They may be varied as long as volumes are proportional and appropriate handling is ensured. It is important to note that the machines do not use centrifugation for the separation of organic and aqueous phases; they carry out this step by simply keeping the vessels in a horizontal position and heating the fluids to 56°C, which promotes the separation process. This can also be done in the same way using the manual extraction method by following the same protocol instructions. We have tested that this mode of phase separation can be replaced by centrifugation (which is usually described by standard laboratory manual protocols). Furthermore, the pelleting and washing steps of the DNA-containing pellet are carried out somewhat differently from those described in the automated protocol. However, we have tested and found no reduction in the quantity and quality of the DNA in the procedure carried out in the machine compared to those described in the following protocol that are adopted for manual extraction.

Protocol

- 0.3 g of homogenized sample
- + 1.7 ml 0.5 M EDTA (pH 8.3)
- rotate for 18–24 h
- centrifuge gently (≤6,000 rpm) 3 min
- transfer supernatant (ca. 1.5 ml) to falcon tube
- + 1.5 ml of 2×Lysis buffer
- + 0.5 ml proteinase K
- incubate at constant movement at 56°C for 1 h

- + 3.0 ml phenol[43]
- shake at RT for 6 min
- allow organic and aqueous phases to separate (56°C, 10 min)
- transfer aqueous phase (upper[44]) to fresh tube
- + 4.5 ml chloroform
- shake at RT for 6 min
- allow organic and aqueous phases to separate (56°C, 10 min)
- transfer aqueous phase (upper) to a fresh tube[45]
- + 100 µl 2 M NaAc (pH 4.5)
- + 3.3 ml 2-propanol (100%)
- + 5 µl silica particle solution[46]
- shake gently at RT for 10 min
- allow silica particles to pellet (5 min), (or centrifuge 1,000 rpm 1 min) [47]
- discard supernatant
- + 3.3 ml ethanol (80%), shake gently
- let silica particles pellet (5 min), (or centrifuge 1,000 rpm 1 min)
- discard supernatant
- + 3.3 ml ethanol (80%), shake gently
- allow silica particles to pellet (5 min), (or centrifuge 1,000 rpm 1 min)
- discard supernatant
- dry pellet at room temperature (30–60 min)[48]
- + 50 µl Ampuwa water
- the sample is now ready for use in PCR[49] [50]
- store at -20°C

---

[43] This step and all following steps involving organic solvents and must be carried out under a fumehood.

[44] If it is uncertain which is the aqueous phase, test by adding a droplet of water.

[45] In case the aqueous phase still reveals intense yellowish to brownish color, repeat phenol and chloroform steps.

[46] If only small amounts of DNA are present in the extract, use less silica, since certain amounts of DNA bind irreversibly to the silica particles and cannot be eluted.

[47] Do not centrifuge the sample at higher speed, since shear forces may cause extensive DNA damage while the DNA is bound to the silica particles.

[48] The ethanol should be completely evaporated, but the pellet should not be over-dried, since amplification of DNA suffers from the irreversible binding of the DNA to the silica particles.

[49] Leave silica particles within aDNA extract. They apparently improve intact storage of aDNA extracts. In contrast to what the suppliers indicate, they do not disturb the amplification process (cf. section 3.3.4). Make sure to homogenize extracts containing silica particles before aliquots are transferred to the PCR reaction mix, since major parts of the aDNA adhere to silica particles at room temperature.

[50] If the extract still contains inhibitors a further clean-up may be carried out by, e.g., Wizard Prep Purification System (Promega).

8.4 DNA extraction from saliva

Material: disposable gloves, face mask, glasses, lab coat, head-dress gown, cotton tips, 2-ml Eppendorf tubes, tube rack, pipettes and pipette tips, bench-top centrifuge, thermostatic mixer, vortex machine, (if available, a second thermostatic mixer or water bath)
***Chemicals and reagents*:** Chelex100 (5%), proteinase K (100 U/ml)

Saliva samples serve as convenient control samples in ancient DNA analysis. They can easily be collected by rubbing a cotton tip along the insides of the cheeks. If the sample is not processed immediately, the tips should be dried by placing them between two layers of tissues. Stored dry at room temperature, the DNA can successfully be recovered from the samples for at least a couple of weeks. If long-term storage is necessary, then it is advisable to store them frozen.

DNA extraction from saliva can be done either using kits or, for example, by Chelex100. Since DNA extractions by Chelex100 reveal good results and are significantly less expensive than kits, the protocol, which is a variation of the original protocol published by Walsh et al. 1991, is given here.

Protocol

- remove cotton part of the tip
- place in 2-ml Eppendorf tube
- add 200 µl Chelex100 (5%)
- shake at RT for 15 min
- squeeze cotton with pipette tip inside cup and remove cotton
- add 2 µl proteinase K
- shake at 56°C for 30 min

(if there is not a second thermomixer or water bath available, leave the sample at room temperature until the thermomixer is heated up to 95°C)

- vortex briefly
- 95°C for 8 min
- centrifuge at 12,000 rpm for 3 min
- transfer supernatant to a fresh 2-ml Eppendorf tube
- ready for use in PCR
- store at -20°C[51]

---

[51] DNA purified by Chelex100 (Walsh et al. 1991) will show no signals under UV light (λ 254 nm) when the extract is electrophoresed on gels that are stained by ethidium bromide. This is due to the fact that Chelex100 denatures the DNA

8.5 PCR primers and protocols

Material: disposable gloves, face mask, glasses, lab coat, head-dress gown, 0.5-ml Eppendorf tubes, tube rack, pipettes and pipette tips, thermocycler
Chemicals and reagents: reaction buffer II, $MgCl_2$ (25 mM), dNTPs (1.25 mM each), primer set, AmpliTaqGold DNA polymerase, DNA-free H_2O, mineral oil

In general, amplifications of ancient DNA should be carried out in larger (50 µl) reaction volumes than are commonly used in "routine" amplifications. This makes a maximum number of intact DNA targets available while also minimizing the concentration of residual inhibitors. If the same volume of aDNA extract were to be subjected to a lower total reaction volume, the amplification could be inhibited, or at least made less efficient, requiring more amplification cycles. Reducing the amount of aDNA extract at this point would indeed lower the concentration of inhibitors but also the total numbers of intact targets, making large-volume approaches much more appealing. Of course, smaller reaction volumes are possible if the aDNA extracts contain more than average intact DNA or if the extract is thought to be very pure.

A PCR reaction mix is generally prepared as a master mix, which has several advantages in sample preparation. This ensures uniform concentrations of all the components in all samples. A master mix consists of all ingredients for amplification except the DNA extract and mineral oil. If different variations of the total volume are required, for example, the volume of DNA extracts in the reaction vary, only a partial volume of Ampuwa water is pipetted to the master mix. The remaining volume of Ampuwa water is added with the DNA extract.

All cycling parameters given in this section are adapted for DNA Thermal Cycler models TC480 and TC1 (Perkin Elmer). Both types of cyclers are known to fall below the temperatures by 1–2°C after long cool-down phases. The annealing temperatures given in the protocols are the displayed values. Therefore, in cycling machines that have more accurate cool-down phases, the annealing temperatures given in the protocols may be a little too high. Due to the system employed in the TC1 and TC480, all reactions were covered by mineral oil.

---

into single strands. It has not been tested whether this causes DNA degradation over long-term storage of extracts. Respective DNA extracts that have been kept at -20°C for up to 2 years were amplified successfully without a noticeable decrease of amplification product yields.

All protocols are calculated on the use of AmpliTaqGold (Applied Biosystems). This enzyme needs an initial phase of 94°C for 11 min in order to overcome the inhibition that keeps the enzyme inactive at room temperature. This enables the polymerase to be added directly to the master mix. If more than about 45 cycles are used for amplification (e.g., because of very few intact targets), it may be useful to decrease the initial activation time to 7–9 min. This activates fewer polymerase fragments in the beginning and conserves more polymerase activity for late cycles. AmpliTaqGold is provided with reaction buffer and $MgCl_2$ solution.

The protocols will give no values for particular numbers of amplification cycles. This decision depends on the aDNA preservation of the individual sample or the sample series, respectively. In the experience of our laboratory, aDNA amplifications are usually successful using 35–45 PCR cycles, provided that an effective DNA extraction procedure has been performed (cf. section 3.1). However, the actual number of cycles that are necessary to generate detectable amounts of amplification product also depends on the reaction efficiency, which is influenced by various factors (cf. section 4.1). These may be factors affecting the quality of the aDNA extract (e.g., inhibitors) or factors that are characteristic for the particular PCR assay. A most typical example for the latter case is the influence of the particular primer characteristics on the reaction efficiency (cf. section 4.4). As a consequence, a particular PCR assay may need many more amplification cycles than another PCR assay, even when aliquots of the same aDNA extract are investigated. In our experience this applies, for example, to the amplification of Y-chromosomal STRs, which always require approximately ten more amplification cycles than the amplification of autosomal STRs. This is most probably due to the AT-rich, low energy sequences neighboring the Y-chromosomal STRs, which induce ineffective primer binding and lead to comparatively inefficient amplifications.

Before the actual protocols are given, the following table (Table 8.1) provides an overview of the sequences amplified.

Table 8.1. Overview of amplification protocols

Name of protocol	Product [bp]	Purpose	Refers to sections:
ABO Ex6	103/104	ABO blood-group typing by RFLP	2.2.5, 7.4.3
ABO Ex7	64	ABO blood-group typing by RFLP	2.2.5, 4.4.6, 7.4.3
Amelogenin	106/112	Sex determination	2.2.1, 7, 7.1.1
AmpFlSTRProfilerPlus	106–348	Identification, kinship, population genetics	2.2.2, 2.2.7, 3.1, 5.1, 6.2, 7.3, 7.4
CCR5 and STRs	98–178	Identification, kinship, population genetics	2.2.6, 4.4.2, 7.4.4
ΔF 508 and ProfilerPlus	92–348	Identification, kinship, population genetics	2.2.7, 7.4.5
D-loop	121/112	Population genetics	2.1.1, 6.1.2
Hypervariable region I+II	131–217	Maternal lineages, population genetics	2.1.1, 3.1, 5.2, 7.3
Sex multiplex	91–166	Sex determination, kinship, identification	2.2.4, 7.1.2
Species I (sheep, goats, etc.)	195	Species determination by RFLP	2.1.2, 7.2.4
Species II (food)	248	Species determination by RFLP	2.1.2, 7.2.8
Species III (sheep, goats, etc.)	115–195	Species determination and sexing	2.1.2, 7.2.5
Species IV (elephants)	152	Identification of elephant species	2.1.2, 7.2.7
Species V (rhinoceroses)	192	Identification of rhinoceros subspecies	2.1.2, 7.2.7
Species VI (foxes, dogs)	187	Maternal lineages, subspecies determination	2.1.2, 7.2.6
Species VII (marsupials)	156–209	Maternal lineages, species determination	7.2.9
STR singleplex D1S1656	122–160	Identification, kinship, population genetics	2.2.2, 7.2.2, 7.3.2
STR singleplex D3S366	138–162	Identification, kinship, population genetics	2.2.2, 7.2.1
STR triplex	126–235	Identification, kinship, population genetics	2.2.2, 7.2, 7.3
STR duplex	86–178	Identification, kinship, population genetics	2.2.2, 7.2.1, 7.3
STR singleplex VWA	126–170	Identification, kinship, population genetics	2.2.2, 7.4.1
Stock marks (STR kit for cattle)	64–265	Identification, kinship, population genetics	2.2.2, 7.2.3
Y-STR quadruplex	145–227	Paternal lineages, population genetics	2.2.3, 7.3, 7.4.1
Y-STR multiplex	91–210	Paternal lineages, population genetics	2.2.3, 7.4.6
Y-chromosomal repeat	154	Sex determination	

ABO Ex6 [52] (ABO blood-group typing, population genetics, human identification)

103/104-bp amplification products
AB0-exon 6-upper: 5' – CTC TCT CCA TGT GCA GTA GGA AGG
AB0-exon 6-lower: 5' – GAA CTG CTC GTT GAG GAT GTC G

- 5 μl buffer II (50 mM KCl, 10 mM Tris-HCl)
- + 3 μl $MgCl_2$ (final concentration of 1.5 mM)
- + 8 μl dNTPs (200 μM each dNTP)
- + 0.12 μM of each primer
- + 2.5 U AmpliTaqGold
- + 2–10 μl aDNA extract
- + make up reaction volume to 50 μl using Ampuwa water

initial: 94°C for 11 min.
cycling: 94°C for 1 min, 56°C for 1 min, 72°C for 1 min

ABO Ex7 [53](ABO blood-group typing, population genetics, human identification)

64-bp amplification product
ABO-exon 7-upper: 5' – AAG GAC GAG GGC GAT TTC TA
ABO-exon 7-lower: 5' – GCT GCA ACT CTT GCA CCG ACC

- 5 μl buffer II (50 mM KCl, 10 mM Tris-HCl)
- + 3 μl $MgCl_2$ (final concentration of 1.5 mM)
- + 8 μl dNTPs (200 μM each dNTP)
- + 0.12 μM of each primer
- + 2.5 U AmpliTaqGold
- + 2–10 μl aDNA extract
- + make up reaction volume to 50 μl using Ampuwa water

initial: 94°C for 11 min.
cycling: 94°C for 1 min, 53°C for 1 min, 72°C for 1 min

---

[52] cf. Hummel et al. (2002)
[53] cf. Hummel et al. (2002)

Amelogenin [54] *(sex determination)*

106/112-bp amplification products
Amel-upper: 5' – CCC TGG GCT CTG TAA AGA ATA GTG
Amel-lower: 5' – ATC AGA GCT TAA ACT GGG AAG CTG

- 5 µl buffer II (50 mM KCl, 10 mM Tris-HCl)
- + 5 µl $MgCl_2$ (final concentration of 2.5 mM)
- + 7 µl dNTP (175 µM each dNTP)
- + 0.1 µM of each primer
- + 2.5 U AmpliTaqGold
- + 5–10 µl aDNA extract
- + make up reaction volume to 50 µl using Ampuwa water

initial: 94°C for 11 min.
cycling: 94°C for 1 min, 56°C for 1 min, 72°C for 1 min

AmpFlSTRProfilerPlus [55] *(human identification, kinship, population genetics)*

For product lengths and labels of each STR, see the manufacturer's instructions (Applied Biosystems). The primer sequences are not published.

- 10 µl Ready Reaction Mix
- + 5 µl Primer Set
- + 5–10 µl aDNA extract
- fill up reaction volume to 25 µl with Ampuwa water

initial: 94°C for 11 min.
cycling: 94°C for 1 min, 59°C for 1 min, 72°C for 1 min
delay: 60°C for 45 min

[54] Primer sequences following Mannucci et al. (1994)
[55] cf. User Manual (Applied Biosystems)

CCR5 and STRs (population genetics, kinship, human identification)

CCR5: 130/98-bp amplification products
CCR5-upper: 5' – CCA GGA ATC ATC TTT ACC AGA TCT C – *6Fam
CCR5-lower: 5' – GGA CCA GCC CCA AGA TGA CTA
Amelogenin: 106/112-bp amplification products
Amel-upper: 5' – CCC TGG GCT CTG TAA AGA ATA GT – *Hex
Amel-lower[56]: 5' – ATC AGA GCT TAC ACT GGG AAG CTG
D5S818: 145–177-bp amplification products
D5S818-upper: 5' – CAA GGG TGA TTT TCC TCT TTG GTA T – *Ned
D5S818-lower: 5' – GTG ATT CCA ATC ATA GCC ACA GTT
TPOX: 110–142-bp amplification products
TPOX -upper: 5' – GAA CCG TCG ACT GGC ACA G – *Ned
TPOX -lower: 5' – CGC TAG GCC CTT CTG TCC T
VWA: 126–170-bp amplification products
VWA-upper: 5'–CCC TAG TGG ATG ATA AGA ATA ATC AGT ATG *Hex
VWA-lower: 5' – GGA CAG ATG ATA AAT ACA TAG GAT GGA TGG
TH01: 154–178 bp amplification products
TH01-upper: 5' – GTG GGC TGA AAA GCT CCC GAT TAT – *6-Fam
TH01-lower: 5' – GTG ATT CCC ATT GGC CTG TTC CTC

- 5 µl buffer II (50 mM KCl, 10 mM Tris-HCl)
- + 5 µl $MgCl_2$ (final concentration of 2.5 mM)
- + 8 µl dNTP (200 µM each dNTP)
- + 0.28 µM of each primer CCR5
- + 0.25 µM of each primer amelogenin
- + 0.22 µM of each primer D5S818
- + 0.24 µM of each primer TPOX
- + 0.2 µM of each primer VWA
- + 0.2 µM of each primer TH01
- + 2.5 U AmpliTaqGold
- + 2–10 µl DNA extract
- + make up reaction volume to 50 µl using Ampuwa water

initial: 94°C for 11 min.
cycling: 94°C for 1 min, 54°C for 1 min, 72°C for 1 min
delay: 60°C for 30 min

[56] a mismatch was built in to this primer in order to enable the multiplex amplification at a low annealing temperature

Δ F 508 and AmpFlSTRProfilerPlus [57] *(population genetics, kinship, human identification)*

Δ F 508: 95/92-bp amplification products
ΔF 508-upper: 5' – GGA TTA TGC CTG GCA CCA TTA – *6-Fam
ΔF 508-lower: 5' – TTC TAG TTG GCA TGC TTT GAT GA

- 5 μl Ready Reaction Mix (AmpFlSTRProfiler Plus)
- + 2.5 μl Primer Mix (AmpFlSTRProfiler Plus)
- + 0.16 μM of primer *Δ F508-upper*
- + 0.24 μM of primer *Δ F508-lower*
- + 2.5 U AmpliTaqGold
- + 1–4 μl DNA extract
- fill up reaction volume to 12.5 μl with Ampuwa water

initial: 94°C for 11 min.
cycling: 94°C for 1 min, 59°C for 1 min, 72°C for 1 min
delay: 60°C for 45 min

D-loop [58] *(population genetics)*

121/112-bp amplification products
H8316-upper: 5' – ATG CTA AGT TAG CTT TAC AG
L8196-lower: 5' – ACA GTT TCA TGC CCA TCG TC

- 5 μl buffer II (50 mM KCl, 10 mM Tris-HCl)
- + 3 μl $MgCl_2$ (final concentration of 1.5 mM)
- + 8 μl dNTP (200 μM each dNTP)
- + 0.2 μM of each primer
- + 2.5 U AmpliTaqGold
- + 5–10 μl aDNA extract
- + make up reaction volume to 50 μl using Ampuwa water

initial: 94°C for 11 min.
cycling: 94°C for 1 min, 59°C for 1 min, 72°C for 1 min

---

[57] The primer sequences are different from those published by Bramanti et al. (2000). They were redesigned by Dr. B. Bramanti in order to enable the simultaneous use with the AmpFlSTRProfilerPlus.
[58] Primer sequences following Wrischnik et al. (1987)

Hypervariable regions I + II[59] *(maternal lineages, kinship, population genetics)*

primer set 1: 131-bp amplification product
H16210: 5' – ACA GCA ATC AAC CCT CAA CTA TCA
L16340: 5' – TGT GCT ATG TAC GGT AAA TGG CTT

primer set 2: 168-bp amplification product
H16028: 5' – TTC ATG GGG AAG CAG ATT TGG
L16195: 5' – ATG GGG AGG GGG TTT TGA TGT GG

primer set 3: 217-bp amplification product
H128: 5' – CTG TCT TTG ATT CCT GCC TCA T
L344: 5' – AGA TGT GTT TAA GTG CTG TGG C

- 5 µl buffer II (50 mM KCl, 10 mM Tris-HCl)
- + 4 µl $MgCl_2$ (final conc. of 2 mM)
- + 7 µl dNTP (175 µM each dNTP)
- + 0.14 µM of each primer set 1
- + 0.18 µM of each primer set 2
- + 0.22 µM of each primer set 3
- + 2.5 U AmpliTaqGold
- + 5–10 µl aDNA extract
- + make up reaction volume to 50 µl using Ampuwa water

initial: 94°C for 11 min.
cycling: 94°C for 1 min, 56°C for 1 min, 72°C for 1 min

[59] cf. Schultes (2000)

Sex-multiplex (positive identification of females, human identification, kinship)

Amelogenin: 106/112-bp amplification products
Amel-upper: 5' – CCC TGG GCT CTG TAA AGA ATA GT - *Hex
Amel-lower[60]: 5' – ATC AGA GCT TAC ACT GGG AAG CTG

DXS 6789: 120–164 bp amplification products (tetra)
DXS6789-upper: 5' – GTT GGT ACT TAA TAA ACC CTC TTT T –*Ned
DXS6789-lower: 5' – GGA TCC CTA GAG GGA CAG AA

DXS 9898: 130–155-bp amplification products (tetra)
DXS9898-upper: 5' – CAC ACC TAC AAA AGC TGA GAT ATA T – *Hex
DXS9898-lower: 5' – CAT CCA GAT AGA CAG ATC AAT AGA TT

DYS391: 138–166-bp amplification product (tetra)
DYS391-upper: 5' – TTG TGT ATC TAT TCA TTC AAT CAT A
DYS391-lower: 5' – GGA ATA AAA TCT CCC TGG T – *6-Fam

DYS392: 91–121-bp amplification product (tri)
DYS392-upper: 5' – CAA GAA GGA AAA CAA ATT TTT T
DYS392-lower: 5' – GGA TCA TTA AAC CTA CCA ATC – *6-Fam

- 5 µl buffer II (50 mM KCl, 10 mM Tris-HCl)
- + 4 µl $MgCl_2$ (final concentration of 2.0 mM)
- + 8 µl dNTP (200 µM each dNTP)
- + 0.25 µM of each primer amelogenin
- + 0.2 µM of each primer DXS6789
- + 0.2 µM of each primer DXS9898
- + 0.32 µM of each primer DYS391
- + 0.24 µM of each primer DYS392
- + 2.5 U AmpliTaqGold
- + 2–10 µl aDNA extract
- + make up reaction volume to 50 µl using Ampuwa water

initial: 94°C for 11 min.
cycling: 94°C for 1 min, 50°C for 1 min, 72°C for 1 min
delay: 60°C for 30 min

---

[60] a mismatch was built in to this primer in order to enable the multiplex amplification at a low annealing temperature

Species I [61] *(identification of sheep, goats, cattle, red deer, deer by RFLP)*

195-bp amplification product
H15342: 5' – GCG TAC GCA ATC TTA CGA TCA A
L15536: 5' – CTG GCC TCC AAT TCA TGT GAG

- 6 μl buffer II (60 mM KCl, 12 mM Tris-HCl)
- + 5 μl $MgCl_2$ (final concentration of 2.5 mM)
- + 6 μl dNTP (150 μM each dNTP)
- + 0.18 μM of each primer
- + 2 U AmpliTaqGold
- + 5–10 μl aDNA extract
- + make up reaction volume to 50 μl using Ampuwa water

initial: 94°C for 11 min.
cycling: 94°C for 1 min, 62°C for 1 min, 72°C for 1 min

Species II [62] *(identification of cattle, pig, sheep, goats, chicken, turkeys by RFLP)*

248-bp amplification product
HCytb526: 5' – CCC GAT TTT TCG CCT TCC ACT T
LCytb777: 5' – TGG GGT GTA GTT ATC TGG GTC T

- 5 μl buffer II (50 mM KCl, 10 mM Tris-HCl)
- + 4 μl $MgCl_2$ (final concentration of 2 mM)
- + 8 μl dNTP (200 μM each dNTP)
- + 0.4 μM of each primer
- + 2 U AmpliTaqGold
- + 10 ng DNA extract (≅1–10 μl aDNA)
- + make up reaction volume to 50 μl using Ampuwa water

initial: 94°C for 11 min.
cycling: 94°C for 1 min, 60°C for 1 min, 72°C for 1 min

---

[61] cf. Burger (2000)
[62] cf. Schmidt (2001)

Species III[63] *(species identification by sequencing, sex determination)*

primer set 1: 195-bp amplification product
H15342: 5' – GCG TAC GCA ATC TTA CGA TCA A
L15536: 5' – CTG GCC TCC AAT TCA TGT GAG
primer set 2: 149-bp amplification product
SRYupper: 5' – ATT GTG TGG TCT CGT GAA CGA
SRY lower: 5' – AGT AGT CTC TGT GCC TCC TCA AAG
primer set 3: 115-bp amplification product
ZFXYupper: 5' – GGG TTA CTG AAT CGC CAC CTT
ZFXYlower: 5' – GGA TTC GCA TGT GCT TTT TGA

- 5 μl buffer II (50 mM KCl, 10 mM Tris-HCl)
- + 5 μl $MgCl_2$ (final concentration of 2.5 mM)
- + 7 μl dNTP (175 μM each dNTP)
- + 0.1 μM of each primer H15342/L15536
- + 0.5 μM of each primer SRY
- + 0.4 μM of each primer ZFXY
- + 3 U AmpliTaqGold
- + 5–10 μl aDNA extract
- + make up reaction volume to 50 μl using Ampuwa water

initial: 94°C for 10 min
cycling: 94°C for 30 s, 58°C for 30 s, 72°C for 45 s

[63] cf. Burger (2000)

Species IV[64] (identification of Loxodonta africana and Elephas maximus)

152-bp amplification product
CBelephant-upper Cyt238: 5' – CGA CAA CTA CAC TTA AAC GGA GCA
CBelephant-lower Cyt389: 5 – CCT ATG AAG GCG GTG GTT ATG

- 6 µl buffer II (60 mM KCl, 12 mM Tris-HCl)
- + 5 µl $MgCl_2$ (final concentration of 2.5 mM)
- + 6 µl dNTP (150 µM each dNTP)
- + 0.2 µM of each primer
- + 2 U AmpliTaqGold
- + 1–10 µl aDNA
- + make up reaction volume to 50 µl using Ampuwa water

initial: 94°C for 10 min
cycling: 94°C for 1 min, 62°C for 1 min, 72°C for 1 min

Species V[65] (identification of subspecies of rhinoceroses)

192-bp amplification product
CBrhino-upper Cyt413: 5' – TAT CCT TCT GAG GGG CAA CAG TC
CBrhino-lower Cyt604: 5 – CGT GTA GGA ATA GTA GGT GGG TGA

- 6 µl buffer II (60 mM KCl, 12 mM Tris-HCl)
- + 5 µl $MgCl_2$ (final concentration of 2.5 mM)
- + 6 µl dNTP (150 µM each dNTP)
- + 0.2 µM of each primer
- + 2 U AmpliTaqGold
- + 1–10 µl aDNA
- + make up reaction volume to 50 µl using Ampuwa water

initial: 94°C for 10 min.
cycling: 94°C for 1 min, 60°C for 1 min, 72°C for 1 min

---

[64] cf. Bollongino (2000)
[65] cf. Bollongino (2000)

Species VI[66] *(hypervariable region of Vulpes vulpes and Canis familiaris)*

187-bp amplification product
V.vulpes-upper: 5' – GCC CTA TGT ACG TCG TGC ATT A
V.vulpes-lower: 5' – GCA AGG ATT GAT GGT TTC TCG

- 5 µl buffer II (50 mM KCl, 10 mM Tris-HCl)
- + 4 µl $MgCl_2$ (final concentration of 2.0 mM)
- + 7 µl dNTP (175 µM each dNTP)
- + 0.4 µM of each primer
- + 1.5 U AmpliTaqGold
- + 1–10 µl aDNA
- + make up reaction volume to 50 µl using Ampuwa water

initial: 94°C for 11 min
cycling: 94°C for 1 min, 58°C for 1 min, 72°C for 1 min

Species VII[67] *(hypervariable region of Petrogale lateralis, marsupials)*

primer set 1: 103-bp amplification product
MarsDloo1F: 5' – AAA GTA CAT AAC ACC TTA AAA CAC TAG
MarsDloo1R: 5' – GAG GAT TTA ACT TRT TAT GTA AAG TT
primer set 2: 156-bp amplification product
MarsDloo2F: 5' – CCA CAA CAC ATC AAC TYA TTT G
MarsDloo2R: 5' – ATT CAT TTT ATG TAT TAC TAG AAT TAT GTA
primer set 3: 209-bp amplification product
MarsDloo3F: 5' – CAT AMA CTC ATA YAT TAC TAA ATA CAT TAH AC
MarsDloo3R: 5' – TGG RTT TTT ATG TAG CTA GTA ATA TRT

- 5 µl buffer II (50 mM KCl, 10 mM Tris-HCl)
- + 4 µl $MgCl_2$ (final concentration of 2.0 mM)
- + 8 µl dNTP (200 µM each dNTP)
- + 0.25 µM of each primer
- + 2 U AmpliTaqGold
- + 1–10 µl aDNA
- + make up reaction volume to 50 µl using Ampuwa water

initial: 94°C for 11 min
cycling: 94°C for 1 min, 48°C for 1 min, 72°C for 1 min

---

[66] cf. Bartels (2002)
[67] IUB-code for primer sequences: R=A or G; Y=C or T; M=A or C; H=A,C or T

STR singleplex D1S1656 [68] *(human identification, kinship, population genetics)*

122–160 bp amplification products (tetra)
D1S1656-upper: 5' – GTG TTG CTC AAG GGT CAA CT
D1S1656-lower: 5' – GAG AAA TAG AAT CAC TAG GGA ACC – *6-Fam

- 5 μl buffer II (50 mM KCl, 10 mM Tris-HCl)
- + 4 μl $MgCl_2$ (final concentration of 2.0 mM)
- + 7 μl dNTP (175 μM each dNTP)
- + 0.08 μM of each primer D1S1656
- + 2 U AmpliTaqGold
- + 5–10 μl aDNA extract
- + make up reaction volume to 50 μl using Ampuwa water

initial: 94°C for 11 min.
cycling: 94°C for 1 min, 56°C for 1 min, 72°C for 1 min

STR singleplex D6S366 [69] *(human identification, kinship, population genetics)*

138–162 bp amplification products (tri)
D6S366-upper: 5' – AGA GGT TAC AGT GAG CCG AGA TTG – *Ned
D6S366-upper: 5' – GAA GTC CTA ACA GAA TGG AAG GTC C

- 5 μl buffer II (50 mM KCl, 10 mM Tris-HCl)
- + 3 μl $MgCl_2$ (1.5 mM)
- + 7 μl dNTP (150 μM each dNTP)
- + 0.14 μM of each primer
- + 2.5 U AmpliTaqGold
- + 1–10 μl aDNA
- + make up reaction volume to 50 μl using Ampuwa water

initial: 94°C for 11 min.
cycling: 94°C for 1 min, 54°C for 1 min, 72°C for 1 min

[68] cf. Gerstenberger et al. (1998). The primer sequences are following Lareu et al. 1998).
[69] The primer sequences are following Panzer et al. (1993).

STR triplex[70] *(human identification, kinship, population genetics)*

VWA: 126–170 bp amplification products (tetra)
VWA-upper: 5'–CCC TAG TGG ATG ATA AGA ATA ATC AGT ATG *Hex
VWA-lower: 5'–GGA CAG ATG ATA AAT ACA TAG GAT GGA TGG

FES/FPS: 213–237 bp amplification product (tetra)
FES/FPS -upper: 5' – GGG ATT TCC CTA TGG ATT GG
FES/FPS -lower: 5' – GCG AAA GAA TGA GAC TAC AT – *6-Fam

F13A1: 181–235 bp amplification product (tetra)
F13A1-upper: 5' – ATG CCA TGC AGA TTA GAA A – *Hex
F13A1-lower: 5' – GAG GTT GCA CTC CAG CCT TT

- 5 µl buffer II (50 mM KCl, 10 mM Tris-HCl)
- + 4 µl $MgCl_2$ (final concentration of 2.0 mM)
- + 7 µl dNTP (175 µM each dNTP)
- + 0.04 µM of each primer VWA
- + 0.1 µM of each primer FES/FPS
- + 0.1 µM of each primer CD4
- + 2 U AmpliTaqGold
- + 5–10 µl aDNA extract
- + make up reaction volume to 50 µl using Ampuwa water

initial: 94°C for 11 min
cycling: 94°C for 1 min, 50°C for 1 min, 72°C for 1 min

[70] cf. Gerstenberger et al. (1999). The primer sequences are following Kimpton et al. (1992), Polymeropoulos et al. (1991a), Polymeropoulos et al. (1991b).

STR duplex[71] *(human identification, kinship, population genetics)*

TH01: 154–178 bp amplification products (tetra)
TH01-upper: 5' – GTG GGC TGA AAA GCT CCC GAT TAT – *6-Fam
TH01-lower: 5' – GTG ATT CCC ATT GGC CTG TTC CTC
CD4: 86–121 bp amplification products (penta)
CD4-upper: 5' – TTG GAG TCG CAA GCT GAA CTA GC – *Hex
CD4-lower: 5' – GCC TGA GTG ACA GAG TGA GAA CC

- 5 µl buffer II (50 mM KCl, 10 mM Tris-HCl)
- + 4 µl $MgCl_2$ (final concentration of 2.0 mM)
- + 7 µl dNTP (=175 µM each dNTP)
- + 0.12 µM of each primer TH01
- + 0.12 µM of each primer CD4
- + 2 U AmpliTaqGold
- + 5–10 µl aDNA extract
- + make up reaction volume to 50 µl using Ampuwa water

initial: 94°C for 11 min.
cycling: 94°C for 1 min, 53°C for 1 min, 72°C for 1 min

STR singleplex VWA[72] *(human identification, kinship, population genetics)*

126–170 bp amplification products (tetra)
VWA-upper: 5' – CCC TAG TGG ATG ATA AGA ATA ATC
VWA-lower: 5' – GGA CAG ATG ATA AAT ACA TAG GAT GGA TGG

- 5 µl buffer II (50 mM KCl, 10 mM Tris-HCl)
- + 3 µl $MgCl_2$ (final concentration of 1.5 mM)
- + 5 µl dNTP (125 µM each dNTP)
- + 0.06 µM of each primer
- + 2 U AmpliTaqGold
- + 5–10 µl aDNA extract
- + make up reaction volume to 50 µl using Ampuwa water

initial: 94°C for 11 min
cycling: 94°C for 1 min, 52°C for 1 min, 72°C for 1 min

---

[71] cf. Gerstenberger et al. (1999). The primer sequences are following Kimpton et al. (1994) and Urquhart et al. (1995).
[72] The primer sequences following Kimpton et al. (1992) are suboptimal for ancient DNA use. Preferably use VWA primers of STR triplex amplification.

Stock Marks (STR kit for cattle) [73] *(identification, kinship, population genetics)*

For product lengths and labels of each STR, see the manufacturer's instructions. (Applied Biosystems). The primer sequences are not published.

- 3 µl StockMarks PCR buffer
- + 4 µl dNTP mix
- + 5.5 µl Amplification primer mix
- + 0.5 µl AmpliTaq Gold (2.5 U)
- + 2 µl aDNA extract (10-100 ng)
- fill up reaction volume to 15 µl with Ampuwa water

initial: 94°C for 11 min.
cycling: 94°C for 45 s, 61°C for 45 s, 72°C for 1 min
1. delay: 72°C for 1 h
2. delay: 25°C for 2 h

Y-chromosomal repeat [74] *(sex determination)*

154-bp amplification product
Yrep-upper: 5' – TCC ACT TTA TTC CAG GCC TGT CC
Yrep-lower: 5' – TTG AAT GGA ATG GGA ACG AAT GG

- 5 µl buffer II (50 mM KCl, 10 mM Tris-HCl)
- + 3 µl $MgCl_2$ (final concentration of 1.5 mM)
- + 6 µl dNTP (150 µM each dNTP)
- + 0.3 µM of each primer
- + 2 U AmpliTaqGold
- + 1–5 aDNA extract
- + make up reaction volume to 50 µl using Ampuwa water

initial: 94°C for 11 min.
cycling: 94°C for 1 min, 58°C for 1 min, 72°C for 1 min

---

[73] cf. Heyen et al. (1997)
[74] Primer sequences are following Kogan et al. (1987).

Y- chromosomal STR quadruplex [75] *(paternal lineages, population genetics)*

DYS 19: 174–210 bp amplification product (tetra)
DYS19-upper: 5' – CTA CTG AGT TTC TGT TAT AGT – *6-Fam
DYS19-lower: 5' – ATG GCA TGT AGT GAG GAC A
DYS 389I + II: 145–169 bp and 261–293 bp amplification products (tetra)
DYS389-upper: 5' – CCA ACT CTC ATC TGT ATT ATC TAT – *Hex
DYS389-lower: 5' – TTA TCC CTG AGT AGT AGA AGA AT
DYS 390: 191–227 bp amplification product (tetra)
DYS 390-upper: 5' – TAT ATT TTA CAC ATT TTT GGG CC – *Hex
DYS 390-lower: 5' – TGA CAG TAA AAT GAA CAC ATT GC

- 5 µl buffer II (50 mM KCl, 10 mM Tris-HCl)
- + 3 µl $MgCl_2$ (final concentration of 1.5 mM)
- + 7 µl dNTP (175 µM each dNTP)
- + 0.14 µM of each primer DYS19
- + 0.1 µM of each primer DYS389I+II
- + 0.08 µM of each primer DYS390
- + 2.5 U AmpliTaqGold
- + 5–10 µl aDNA extract
- + make up reaction volume to 50 µl using Ampuwa water

initial: 94°C for 11 min.
cycling: 94°C for 1 min, 50°C for 1 min, 72°C for 1 min

[75] cf. Schultes et al. (1999). The primer sequences are following Roewer et al. (1992), Schultes et al. (1999), Kayser et al. (1997).

Y-chromosomal STR multiplex[76] *(paternal lineages, population genetics)*

DYS 19: 174–210 bp amplification product (tetra)
DYS19-upper: 5' – CTA CTG AGT TTC TGT TAT AGT – *6-Fam
DYS19-lower: 5' – ATG GCA TGT AGT GAG GAC A
DYS385: 146–210 bp amplification products (tetra)
DYS385-upper: 5' – AAA GAA AGA GAA GAA AGA GAA AGA
DYS385-lower: 5' – AAT ATT TTA AAA AAT AAT CTA TCT ATT C *Ned
DYS389: 103–131 bp and 223-251 bp amplification products (tetra)
DYS389-upper: 5' – CCA ACT CTC ATC TGT ATT ATC TAT – *Hex
DYS389-lower: 5' – TGG ACT GCT AGA TAA ATA GAT AGA T
DYS390: 150–186 bp amplification product (tetra)
DYS390-upper: 5' – TGC ATT TTG GTA CCN CAT AA
DYS390-lower: 5' – ATT GCT ATG TGT ATA CTC AGA AAC – *6-Hex
DYS391: 138–166 bp amplification product (tetra)
DYS391-upper: 5' – TTG TGT ATC TAT TCA TTC AAT CAT A
DYS391-lower: 5' – GGA ATA AAA TCT CCC TGG T – *6-Fam
DYS392: 91–121 bp amplification product (tri)
DYS392-upper: 5' – CAA GAA GGA AAA CAA ATT TTT T
DYS392-lower: 5' – GGA TCA TTA AAC CTA CCA ATC – *6-Fam
DYS393: 97–125 bp amplification product (tetra)
DYS393-upper: 5' – GTG GTC TTC TAC TTG TGT CAA TAC – *Ned
DYS393-lower: 5' – CAA AAA ATG AGG TAT GTC TCA T

- 5 μl buffer II (50 mM KCl, 10 mM Tris-HCl)
- + 6 μl $MgCl_2$ (final concentration of 3.0 mM)
- + 14 μl dNTP (350 μM each dNTP)
- + 0.6 μM of each primer DYS19
- + 2.4 μM of each primer DYS385
- + 0.3 μM of each primer DYS389
- + 0.6 μM of each primer DYS390
- + 0.28 μM of each primer DYS391
- + 0.8 μM of each primer DYS392
- + 0.4 μM of each primer DYS393
- + 2.5 U AmpliTaqGold
- + 5–10 μl aDNA extract
- + make up reaction volume to 50 μl using Ampuwa water

initial: 94°C for 11 min.
cycling: 94°C for 1 min, 50°C for 1 min, 72°C for 1 min

[76] cf. Müller (2002)

8.6 Agarose gel electrophoresis

Material: 2 Erlenmeyer flasks (200 ml), watch glass dish, pipettes and pipette tips, heatable magnetic stirrer,[77] magnetic stirring bar, balance, graduated cylinders (100 ml and 1L), horizontal electrophoresis unit (ca. 11×16 cm), power supply (150V, 300mA), photographic documentation unit (either a Polaroid or a digital camera system), UV transilluminator (254 nm), UV-protection face shield, disposable gloves, ethidium bromide waste bottle, titration plates, tube rack
Chemicals and reagents: 1×TBE buffer, agarose, ethidium bromide (1%),[78] bromphenol blue/sucrose, DNA marker standard

Agarose gels are used for the electrophoresis of DNA extracts, PCR products and RFLP analysis of PCR products. In the case of DNA extracts, it allows a general estimate of the total amount of DNA that was extracted. In addition, it also gives insight into the distribution of the molecular weights of the fragments and allows a visual inspection of whether residues of humic acids or tannins are present in the extract. The electrophoresis of PCR products enables an evaluation of the overall amplification success. If presence/absence results are expected, this will already be the final analysis step. Furthermore, agarose gel electrophoresis allows for the determination of the optimal volumes of PCR products that may be subjected to (1) fragment length detection by PAA electrophoresis, (2) a Taq-cycle sequencing reaction or (3) RFLP digestion. In the latter case, a second high-percentage agarose gel will be used in the final analysis step.

Preparation of a 2.5% agarose gel (11×16 cm)

- 2 g agarose to Erlenmeyer flask with magnetic stirring bar
- add 80 ml 1×TBE
- cover with watch glass dish
- boil and stir until gel remains clear
- pour gel into second flask[79]

---

[77] Agarose gels can also be prepared in a microwave oven. However, in our experience this often leads to a combustion because of boiling delay. Indeed, this does not happen if a heatable magnetic stirrer is used. This device is also a necessary piece of equipment for preparation of PAA gels.

[78] Always wear double disposable gloves while handling ethidium bromide (EtBr), which is a powerful mutagen and migrates quickly even through gloves

- add 3 µl of ethidium bromide
- mix thoroughly
- pour into the tray of the electrophoresis unit
- place comb
- draw possible bubbles away with a pipette tip
- cover with 1×TBE buffer after gel is solid (ca. 30 min)

For the preparation of high percentage agarose gels (3–4.4%), the procedure is basically the same, except all steps must by carried out very quickly in order to prevent the gel from solidifying before being poured into the tray. This is of particular concern when ethidium bromide has been incorporated in the gel solution, since there will be no time for thoroughly mixing the ethidium bromide in the second flask. In addition, some bubbles may also remain in the gel. Draw them as quickly as possible to the anodic end of the gel with the help of a pipette tip. If the bubbles are small and do not clump together, they will not interfere with a good electrophoresis result.

Loading the gel

As a general rule of thumb, about 10–20% of the total PCR volume (5–10 µl of the sample) should be loaded to a gel depending on the yields expected. In case of small PCR products or RFLP restriction products (<100 bp), this amount should be increased to about 10–15 µl, since the intensity of the fluorescence under UV light is dependent not only on the amount but also on the fragment length and therefore the amount of ethidium bromide that has been incorporated. Bromphenol blue/sucrose should be added with the sample aliquots before pipetting them to the gel in order to increase the weight, which makes them stay in the wells of the gel in the presence of 1×TBE buffer. Also, it helps to visualize the distance the smaller molecules have already covered during the electrophoresis process. The sample aliquots are best mixed with about 1 µl of bromphenol blue/sucrose in a titration plate. Choose a suitable DNA marker standard (Fig. 8.1). These should be loaded at either side of the gel.

[79] The second Erlenmeyer flask, which is contaminated by ethidium bromide, should never be used for boiling a gel because of the evaporation of EtBr during the boiling process.

Running the gel

The electrophoresis time depends on the intended separation result. The 2.5–4.4% agarose gels can withstand 100–120 V/300mA without starting to melt.

Visualizing the DNA

In order to visualize the DNA molecules with the incorporated ethidium bromide, the gel must be transilluminated by ultra-short-wave UV light (λ 254 nm).[80] DNA molecules will glow pinkish to orange at this wavelength. Residues of humic acids and tannin appear turquoise in color.

When very small RFLP products (ca. 30 bp) are separated on high percentage gels, the bromphenol blue travels only slightly faster through the gel than the actual DNA molecules. Therefore, for quite some time the bromphenol blue totally obscures the RFLP products, making them almost invisible in UV light. However, by allowing the gel to run for a longer time (i.e., longer separation distance), this effect will be overcome.

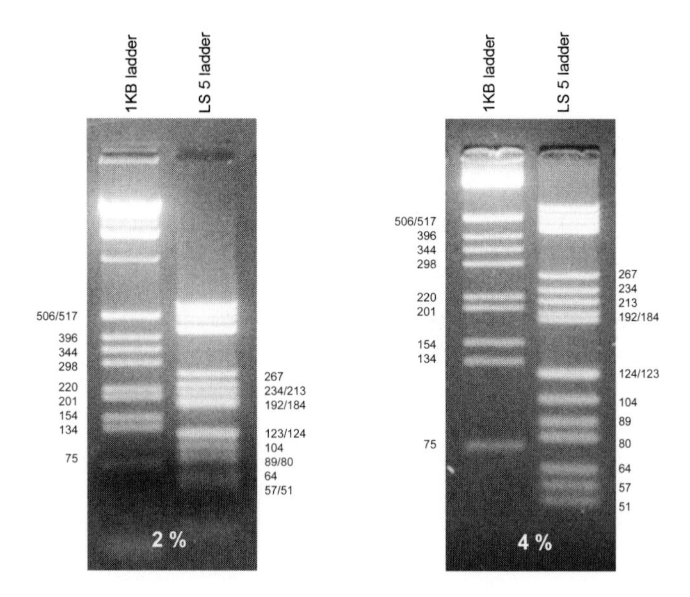

Fig. 8.1. DNA marker standards separated on 2.0% (*left*) and 4.0% (*right*) agarose gels

[80] Make sure to wear UV protective glasses or rather a face shield when working at a transilluminator generating 254 nm UV light. Short exposure for a few seconds may cause irreversible damage to eyes and skin.

8.7 Taq-cycle sequencing

Material: disposable gloves, 0.5-ml Eppendorf tubes, 2-ml Eppendorf tubes, tube rack, pipettes and pipette tips, thermocycler
Chemicals and reagents: QIAquick PCR purification kit, H_2O (HPLC grade), sequencing primer, BigDye Terminator Cycle Sequencing Reaction kit, mineral oil, ethanol (70%), ethanol (99.5%), 3 M NaAc (pH 4.6)

Before the cycle sequencing reaction is started, unused primers and dNTPs must be removed from the amplification mix. This is best carried out using spin column purification.

For the amplification, one of the two original PCR primers can be used as so-called "sequencing primer." The annealing temperature (AT) depends on the optimum for the particular primer that is used. When the sequencing reaction is finished the single stranded products are precipitated and washed in order to remove free-labeled ddNTPs and salts from the extract.

If the cycle sequencing reaction should be performed for a multiplex PCR product, a separate cycle sequencing reaction must be prepared for each sequence.

Protocol for spin column purification [81]

- 40 µl PCR product
- + 200 µl buffer PB
- vortex briefly
- place spin column in collection tube (2 ml)
- transfer product/buffer mix to spin column
- centrifuge 4,000 rpm 1 min
- discard flow-through
- + 750 µl buffer PE to spin column
- centrifuge 4,000 rpm 1 min
- discard flow-through
- centrifuge 13,000 rpm 1 min
- place spin column to fresh 2-ml Eppendorf tube
- + 50 µl H_2O (HPLC grade)82
- centrifuge 14,000 rpm 1 min
- ready for cycle sequencing reaction

---

[81] This protocol uses components of the QIAquick purification kit but varies from the instructions given by the manufacturer.
[82] Do not use elution buffer EB, since H_2O (HPLC grade) ensures improved Taq cycle sequencing

Protocol for cycle sequencing reaction

- 4 µl BDT Ready Reaction mix
- + 3 pmol sequencing primer
- + 2–6 µl PCR product
- fill to 20 µl with H_2O (HPLC grade)
- mineral oil cover

25 cycles: 96°C for 30 sec, AT[83] for 15 sec, 60°C for 4 min

Protocol for precipitation

- 50 µl ethanol (99.5%)
- + 2 µl 3 M NaAc (pH 4.6)
- + 18 µl cycle sequencing product (avoid any mineral oil)
- mix gently
- give 10 min at RT to precipitate
- centrifuge at 14,000 rpm 30 min
- cautiously remove supernatant (the pellet is invisible)
- + 250 µl ethanol (70%)
- vortex briefly
- centrifuge at 14,000 rpm 15 min
- remove cautiously supernatant
- dry pellet at 37°C for15 min
- resolve pellet in 10 µl of H_2O (HPLC grade)
- ready for sample preparation for PAA electrophoresis

8.8 PAA gel electrophoresis

If the so-called "denaturing electrophoresis conditions" are chosen, the most appropriate approach would be to use polyacrylamide[84] (PAA) gel electrophoresis, which allows the separation of length differences of DNA fragments down to a single base pair. PAA gel electrophoresis can be carried out either as slab gel electrophoresis or capillary gel electrophoresis. The following protocols refer to both, slab gels on a 373 stretch system (Applied Biosystems) and on a 310 capillary system (Applied Biosystems).

---

[83] AT = annealing temperature of the respective primer

[84] All acrylamides are neurotoxic, i.e., make sure to always wear gloves when handling acrylamide and to avoid any contact with the skin.

Slab gel electrophoresis

Material: disposable gloves, 100-ml beaker, heatable magnetic stirrer, magnetic stirrer bar, 0.5-ml Eppendorf tubes, tube racks, pipettes, regular and gel loading pipette tips, a pair of glass plates (notched and plain),[85] including spacers, comb and clamps,[86] ice box, syringe, waste bin for acrylamide
Chemicals and reagents: 40% acrylamide-bisacrylamide (29:1), urea, 1×TBE buffer, ammonium persulfate (10%), TEMED, formamide, DNA fragment length standard,[87] loading buffer,[88] sequencing loading mix, 2-propanol (100%), H_2O (HPLC grade), liquid soap

Preparing a 6% denaturing gel (12 cm plates)

- set up glass plates with spacers and clamps
- place the set horizontally
- set up gel mix in 100-ml beaker including magnetic stirring bar with:
- 15 g urea
- + 3 ml 10×TBE buffer
- + 4.5 ml 40% acrylamide-bisacrylamide (29:1)
- + 12 ml H_2O (HPLC grade)
- mix, heat to 50°C
- let the solution cool down to about RT
- + 90 µl ammonium persulfate (10%)
- + 12 µl TEMED
- mix gently
- pour to plates, avoid bubbles
- place comb
- gel needs ca. 2 h to polymerize

[85] Plates are cleaned with liquid soap, H_2O (HPLC grade), 2-propanol (100%). Results may be strongly affected in quality (high background, loss of detection sensitivity) if plates were in contact with material that contains fluorescent particles (e.g., many paper towels, permanent pencils). If the contamination of the plates was intensive, it may be irreversible.

[86] For separation of 2–6 bp differences into short total product length (e.g., STRs), 12-cm glass plates are suitable in combination with a 6% denaturing PAA gel. Sequence analysis (1-bp differences) require at least 24 cm glass plates.

[87] GS 500Rox or GS 350Rox (Applied Biosystems) are suitable for aDNA analysis.

[88] This is supplied with the GS 500Rox or GS 350Rox.

The preparation of all components except for the PCR product are done in a master mix, i.e., all the respective volumes must be multiplied by the number of samples that are to be analyzed.

Preparation of samples for fragment length analysis

- 2 µl formamide
- + 0.5 µl loading buffer
- + 0.25 µl GS 500Rox (or GS 350Rox)
- + 0.1–2.5µl PCR product (dependent on product yield)
- 95°C for 2 min
- store on ice

Preparation of samples for sequence analysis

- 4 µl sequencing loading mix
- 10 µl cycle sequencing product (cf. end of precipitation, section 10.7)
- 95°C for 2 min
- store on ice

The samples are stored on ice until loading them to the gel in order to keep the DNA strands denatured. The storage may be done overnight at -18°C. Also, the polymerized gels can be kept over night in a refrigerator (ca. 8°C). In this case, the plates should then be enclosed in plastic wrap to prevent the polyacrylamide from drying out.

The slab gels and the buffer chambers are placed in the machine following the manufacturer's instructions. Before samples can be loaded, the wells should be rinsed repeatedly with the 1×TBE electrophoresis buffer using a syringe.

At cool room temperature (<20°C), slab gels can often be used two or even three times. However, prerequisites are that no unspecific high molecular-weight products in the samples of the earlier run are present and that sufficient electrophoresis time has been given for DNA fragments (PCR products and lane-internal DNA standards) to have migrated off the end of the gel. This can easily be monitored, based on the previous gel image.

Capillary gel electrophoresis

Materials: disposable gloves, 100-ml beaker, pipettes and pipette tips, capillary, valve block (including syringe, vents, Buver vials), 0.5-ml analyzer tubes, analyzer septa, waste vial, 2 buffer vials, autosampler tray
Chemicals and reagents: loading buffer,[89] formamide (purity 99.5%), DNA fragment length standard,[90] POP6,[91] TSR buffer

The preparation of all components except for the PCR product is done using a master mix, i.e., all the respective volumes must be multiplied by the number of samples that are to be analyzed. In capillary electrophoresis, the amount of PCR product to be analyzed is controlled with the injection time.

Preparation of samples for fragment length analysis

- 24 µl formamide
- \+ 1 µl GS 500Rox
- \+ 2 µl PCR product
- 95°C for 2 min
- store on ice

Preparation of samples for sequence analysis

- 16 µl TSR buffer
- \+ 4 µl cycle sequencing product (see precipitation, section 8.7)
- 95°C for 2 min
- store on ice

The samples are now placed on the autosampler tray. If they are to be stored prior to any analysis, they should be kept frozen (-18°C). The capillaries are placed in the machine following the manufacturer's instructions.

---

[89] This is supplied with the GS 500Rox or GS 350Rox.
[90] GS 500Rox or GS 350Rox
[91] This is a readily prepared polyacrylamide solution for use in capillaries and is suitable for fragment length and sequence analysis.

8.9 RFLP digestion

Material: disposable gloves, 0.5-ml Eppendorf tubes, tube rack, pipettes and pipette tips, thermostatic incubator
Chemicals: restriction endonuclease, digestion buffer,[92] BSA (10mg/ml), mineral oil

Restriction fragment length polymorphism (RFLP) analysis can be carried out on PCR products if the detection and identification of a single polymorphic site is the goal. As such, RFLP analysis is more convenient, faster and less expensive than sequence analysis. In case the RFLP products reveal sticky ends (cf. section 5.3), loading of the samples to the gel should be carried out as fast as possible.

Preparation of samples

The preparation of all components, except for the PCR product and mineral oil, is done using a master mix. A master mix is made up by multiplying all the volumes given here by the number of samples that are to be analyzed.

ABO blood-group genotyping

- 2 µl Rsa I (=20 U) (GT!AC)
- + 2 µl NEB1 buffer
- + 3 µl Ampuwa water
- + 5–7 µl amplification product
- 37°C for 3 h

- 2 µl HpyCH4IV (=20 U) (A!CGT)
- + 2 µl NEB1 buffer
- + 3 µl Ampuwa water
- + 5–7 µl amplification product
- 37°C for 3 h

[92] This is commonly supplied with the restriction endonuclease.

- 2 µl Nla III (=20 U) (CATG!)
- + 2 µl NEB4 buffer
- + 3 µl Ampuwa water
- + 1.5 µl BSA (10 mg/ml)
- + 5–7 µl amplification product
- 37°C for 3 h

- 2 µl Mnl I (=10 U) (!NNNNNNNGGAG)
- + 2 µl NEB2 buffer
- + 3 µl Ampuwa water
- + 1.5 µl BSA (10 mg/ml)
- + 5–7 µl amplification product
- 37°C for 3 h
- 65°C for 1 h (breaks down unspecific bands that occur during digestion)

Species identification I (sheep, goats, cattle, red deer, deer)

- 1.2 µl Tsp 509 (= 12 units) (!AATT)
- + 2 µl NEB1 buffer
- + 7 µl amplification product
- 65°C for 2 h

Species identification II (cattle, pig, sheep, goats, chicken, turkey)

- 1.5 µl Bfa I (12 units) (C!TAG)
- + 2 µl NEB4 buffer
- + 2 µl Ampuwa water
- + 7 µl amplification product
- 37°C for 2 h

8.10 Basic chemistry and reagents

Many of the instructions to set up buffers, reagents or gel preparation can be found in laboratory manuals, and they can also be found in the instructions supplied by the manufacturers. However, they were included in the protocol section of this book for convenience.

In general, all chemicals and reagents that are in use prior to the PCR amplification should be made up into aliquots in convenient amounts in order to enable the necessary checks for contamination.

Some of the reagents need to have a certain molarity and to be adjusted to a particluar pH value by titration using either HCl_{conc} or $NaOH_{(l)\,conc}$. The mixture of HCl_{conc} or $NaOH_{(l)conc}$ slightly decreases molarity. The alternative is therefore titrating before reaching the final volume. However, this will lead to a deviation in the pH value. Since the reagents are not for use in quantitative chemical analysis, the recipes are based on precise pH values and take into account slightly lower molarities.

Materials: disposable gloves, face mask, glasses, head dress gown, lab coat, spatula, chemical spoon, beakers, pipettes and pipette tips, heatable magnetic stirrer, magnetic stirring bar, balance, weighing paper, pH meter, pH sticks, graduated cylinders, volumetric flasks, 2-ml and 0.5-ml Eppendorf tubes, tube rack, glass bottles, Pasteur pipettes, graduated pipettes, dispenser or pelaeus ball funnel

40% Acrylamide-Bisacrylamide (29:1) (Merck)

- is readily purchased
- store refrigerated

Agarose gel 2.5% (NEEO Agarose, Roth)

- 2 g agarose
- + 80 ml 1×TBE

Agarose gel 3.75% (NEEO Agarose, Roth)

- 3 g agarose
- + 80 ml 1×TBE

Agarose gel 4.4% (NEEO Agarose, Roth)

- 3.5 g agarose
- + 80 ml 1×TBE

Ammonium persulfate (APS >98%, Aldrich)

- prepare 10 mg aliquots
- store at -20°C
- before use, add 100 µl H_2O (HPLC grade)

AmpliTaqGold DNA polymerase (Applied Biosystems)

AmpliTaqGold DNA polymerase is inhibited at room temperature. This allows the enzyme to be added directly to the master mix without beginning its activation. The activation of the enzyme is carried out through an initial heating step at 94°C. The duration of this step influences the amount of enzyme that is released to the reaction in its active form. Further enzymes will successively be activated throughout the amplification process at each denaturation phase. Applied Biosystems recommends storage at -18°C, although RT should be suitable, at least theoretically.

Ampuwa (Fresenius)

Ampuwa is certified to be sterile H_2O (p.i.) that, from our experience, has proven to be DNA-free, even if subjected to an extremely high number (>60) of cycles. It should be stored at RT.

Bfa I (New England Biolabs)

This is readily purchased.

The manufacturer recommends storage at -70°C. However, experience in our laboratory has shown that the enzyme may also be stored at -20°C for at least up to 6 months without noticeable loss of enzyme activity.

BigDye Terminator Cycle Sequencing Reaction Kit (Applied Biosystems)

This consists of all reagents that are necessary for a well-balanced cycle sequencing reaction. It should be stored refrigerated.

Bromphenol blue/sucrose mix (Merck)

- 0.025 g bromphenol blue
- + 4 g sucrose
- fill up to 10 ml with H_2O_{bidest}
- store refrigerated

BSA (100μg/ml) (Bovine serum albumin, New England Biolabs)

- is readily purchased
- store at –20°C

Buffer II (amplification reaction buffer, Applied Biosystems)

- an amplification reation buffer delivered with AmpliTaqGold
- consists of 500 mM KCl and 100 mM Tris-HCl

To prepare the same buffer for the reaction in the laboratory:

- 1.211 g Tris
- + 3.728 g KCl
- make up to 100 ml using Ampuwa water
- adjust (titrate) pH to 8.3 using HCL (conc.)
- store refrigerated

Chelex 100 (5%) (BioRad)

- 5 g Chelex100
- + 100 ml Ampuwa
- store at RT or refrigerated

Chloroform (Applied Biosystems)

- is readily purchased
- store refrigerated

DNA marker standard (1KB ladder, Life Technologies Gibco BRL)

- 100 µl 1KB ladder (stock solution)
- + 100 µl bromphenol blue/sucrose mix
- + 800 µl H_2O_{bidest}
- store at -18°C

DNA marker standard (LS V, Roche)

- 100 µl LS V (stock solution)
- + 100 µl bromphenol blue/sucrose mix
- + 400 µl ddH_2O
- store at -18°C

0.5 M EDTA (pH 8.3) (Merck)

- 46.53 g EDTA
- make up to 250 ml using Ampuwa water
- adjust (titrate) pH to 8.3 using $NaOH_{(l)conc}$

(EDTA will not solve until pH 8 is reached)

- store at RT

Ethanol (99.5%) (Merck)

- is readily purchased
- store at RT

Ethanol (~80 %) (Merck)

- 800 ml ethanol (99.5%)
- make up to 1,000 ml using Ampuwa water
- store at RT

Ethanol (~70 %) (Merck)

- 700 ml ethanol (99.5%)
- make up to 1,000 ml using Ampuwa water
- store at RT

Ethidium bromide (1%) (Merck)

- is purchased as a 1% solution
- caution: EtBr is suspected of being a strong mutagen and carcinogen; wear double disposable gloves during handling
- store refrigerated

Formamide (99.5%) (Amresco)

- is readily purchased
- store refrigerated

GS 500Rox (Applied Biosystems)

- suitable for analysis of amplification products up to ≅ 450 bp
- store refrigerated

GS 350Rox (Applied Biosystems)

- suitable for analysis of amplification products up to ≅ 300 bp
- store refrigerated

H_2O (HPLC grade) (Merck)

- is readily purchased
- store at RT

HpyCH4IV (New England Biolabs)

- is readily purchased

The manufacturer recommends storage at -70°C. However, experience in our laboratory has shown that the enzyme may also be stored at -20°C for up to at least 6 months without noticeable loss of enzyme activity.

Liquid soap (Alconox) (Sigma-Aldrich)

- 20 g Alconox
- + 500 ml ddH_2O
- store at RT

Loading buffer (Applied Biosystmes)

- is readily purchased with GS 500Rox or GS 350Rox
- store refrigerated

2xLysis buffer (Applied Biosystems)

- is readily purchased
- store refrigerated

The buffer consists of NaCl, n-lauryl sarcosine, CDTA, Tris-HCl (pH 7.9). The concentrations of the components are not declared.

Master Mix

This is any type of solution in which different components are prepared for a number of samples within one assay before the aliquots are distributed to the single samples. This strategy ensures uniform concentrations for all samples. Master mixes are recommendable for any type of preparation where small volumes must be handled (e.g., PCR reaction mix, cycle sequencing reaction mix, primer mixes for multiplex assays, etc.).

$MgCl_2$ (stock 25 mM, Applied Biosystmes)

This is an amplification reaction component delivered with AmpliTaqGold.

If this component is self prepared:

- 0.238 g $MgCl_2$
- fill up to 100 ml by Ampuwa

or

- 0.508 g $MgCl_2 \times 6H_2O$
- fill up to 100 ml by Ampuwa
- store refrigerated

Mineral oil (Nujol, Applied Biosystems)

- is readily purchased
- store at RT

Mnl I (New England Biolabs)

- is readily purchased

The manufacturer recommends storage at -70°C. However, experience in our laboratory shows the enzyme may also be stored at -20°C for up to at least 6 months without noticeable loss of enzyme activity.

2 M NaAc (pH 4.5) (Applied Biosystems)

- is readily purchased

If this component is self prepared:

- 8.203 g CH_3COONa
- make up to 50 ml with Ampuwa water
- adjust (titrate) pH to 4.5 using HCl_{conc}
- store at RT

3 M NaAc (pH 4.6) (Merck)

- 12.30 g CH_3COONa
- make up to 50 ml using Ampuwa water
- adjust (titrate) pH to 4.6 using HCl_{conc}
- store at RT

NEB1 buffer (New England Biolabs)

- is readily purchased with certain restriction endonucleases from New England Biolabs
- store at -20°C

NEB2 buffer (New England Biolabs)

- is readily purchased with certain restriction endonucleases from New England Biolabs
- store at -20°C

NEB4 buffer (New England Biolabs)

- is readily purchased with certain restriction endonucleases from New England Biolabs
- store at -20°C

Nla III (New England Biolabs)

- is readily purchased

The manufacturer recommends storage at -70°C. However, experience in our laboratory shows the enzyme may also be stored at -20°C for up to at least 6 months without noticeable loss of enzyme activity.

dNTPs (solutions of dATP, dCTP, dGTP, dTTP, each 100 mM, Sigma-Aldrich)

- are readily purchased
- preparation of a working solution (e.g., 1.25 mM each dNTP)
- 25 µl dATP
- + 25 µl dCTP
- + 25 µl dGTP
- + 25 µl dTTP
- 1.9 ml Ampuwa
- mix thoroughly and store in small aliquots
- store at -20°C

Phenol (phenol:chloroform:isoamylalcohol =25:24:1, Roth)

- is readily purchased
- store refrigerated

POP 6 (Performance-optimized polymer, Applied Biosystems)

- store refrigerated

This is readily purchased for use with capillary electrophoresis. It is particularly suitable for fragment length and sequence analysis.

Primers (lyophilised)

All primers should be HPLC purified. This improves amplification reactions with respect to specificity of results. It is essential for legible sequence reads in direct sequencing applications.

If primers are intended for use in a multiplex PCR, it is advisable to order them lyophilised, since this enables preparation of stock solutions with high concentrations (≥20 pmol/µl).

Example:

- preparation of a solution of 20 pmol/µl from 15nmol lyophilised primer
- 15 nmol=15,000 pmol
- 15,000 pmol/X µl=20 pmol/µl
- X=750 µl must be added to the lyophilised primer
- store at -20°C

Alternatively, the total amount of primer in the tube is shown on the side of the tube in nanomoles. For example, the tube might contain 35.5 nmol. To make an appropriate concentration of primer, add 1,000 µl of Ampuwa water to the tube; the ultimate concentration of the primer will now be 35.5 pmol/µl (or 35.5 µM).

Primers (in solution)

All primers should be HPLC purified. This improves amplification reactions with respect to the specificity of results. It is essential for the readability of sequences in direct sequencing.

Example:

if the primers have arrived in solution, to make a 10 pmol/µl from 500 µl consisting of 12 pmol/µl

- 12 pmol/µl×500 µl=10 pmol/µl×X µl
- X=600 µl, and therefore, 100 µl must be added to the original solution
- store at -20°C

2-Propanol (100%) (Merck)

- is readily purchased
- store at RT

Proteinase K (lyophilized, Applied Biosystems)

Proteinase K is sold by weight, usually lyophilized. Since the enzyme activity (units) varies by batch, it is necessary to vary the amount of buffer that is used to prepare a particular concentration.

Example:

10 mg proteinase K of a certain batch reveals 2,180 U

- + 21.8 ml of TE-buffer
- store refrigerated

QIAquick PCR purification kit (Qiagen)

- store at RT

This is readily purchased. Contains all components (except Ampuwa) that are necessary for sample preparation and subsequent Taq cycle sequencing.

Rsa I (New England Biolabs)

- is readily purchased

The manufacturer recommends storage at -70°C. From the experience of our lab, the enzyme may also be stored at -20°C for up to at least 6 months without noticeable loss of enzyme activity.

Sequencing loading mix (Merck, Sigma-Aldrich, Applied Biosystems)

- 0.0465 g EDTA
- + 5 ml formamide
- + 625 μl loading buffer
- store refrigerated

Silica particle solution (Glassmilk, QBiogene)

- is readily purchased
- aliquot to ensure uniform silica concentrations are used
- store at RT

10xTBE (stock for electrophoresis buffer) (Merck)

- 109.03 g Tris
- + 55.65 g $B(OH)_3$
- + 9.31 g EDTA
- make up to 1,000 ml with ddH_2O
- store at RT

1xTBE (electrophoresis buffer)

- 100 ml 10×TBE
- + 900 ml ddH_2O
- store at RT

TE buffer (100 mM Tris, 20 mM EDTA)

- 1.211 g Tris
- + 0.744 g EDTA
- make up to 100 ml using Ampuwa
- store solution at RT

TEMED (>99%) (N,N,N',N'-Tetra-methyl-ethylendiamine, Merck)

- is readily purchased
- store refrigerated

Tsp 509 (New England Biolabs)

- is readily purchased.

  The manufacturer recommends storage at -70°C. From the experience of our laboratory, the enzyme may also be stored at -20°C for up to at least 6 months without noticeable loss of enzyme activity.

TSR buffer (Applied Biosystems)

- is readily purchased
- store refrigerated

References

Applied Biosystems Amp*FlSTRProfilerPlus*™. User Manual

Bartels A (2002) Verwandtschaftsrekonstruktion am Rotfuchs (Vulpes vulpes) durch molekulargenetische Analysen an degradierten Gewebeproben. Diplomarbeit, Georg August-Universität, Göttingen

Bramanti B, Sineo L, Vianello M, Caramelli D, Hummel S, Chiarelli B, Herrmann B (2000) The selective advantage of cystic fibrosis heterocygotes tested by aDNA analysis. Int J Anthropol 15:255–262

Bollongino R (2000) Bestimmung humanökologisch relevanter Tierarten aus historischen, musealen und forensischen Materialien durch Sequenzierung mitochondrialer Genorte. Diplomarbeit, Georg August-Universität, Göttingen

Burger J (2000) Sequenzierung, RFLP-Analyse und STR-Genotypisierungen alter DNA aus archäologischen Funden und historischen Werkstoffen. Dissertation Georg August-Universität, Göttingen

Gerstenberger J, Hummel S, Herrmann B (1998) Assignment of an isolated skeletal element to the skeleton of Duke Christian II. Ancient Biomol 2: 63–68

Gerstenberger J, Hummel S, Schultes T, Häck B, Herrmann B (1999) Reconstruction of a historical genealogy by means of STR analysis and Y-haplotyping of ancient DNA. Europ J Hum Genet 7:469–477

Heyen DW, Beever JE, DA Y, Evert RE, Green C, Bates SR, Ziegele JS, Lewin HA (1997) Exclusion probabilities of 22 bovine microsatellite markers in fluorescent multiplexes for semiautomated parentage testing. Anim Genet 28:21–27

Hummel S, Schmidt D, Kahle M, Herrmann B (2002) ABO blood group genotyping of ancient DNA by PCR-RFLP. Int J Legal Med (in press)

Kayser M, Cagliá A, Corach D, Fretwell N, Gehrig C, Graziosi G, Heidorn F, Herrmann S, Herzog B, Hidding M, Honda K, Jobling M, Krawczak M, Leim K, Meuser S, Meyer E, Oesterreich W, Pandya A, Parson W, Penacino G, Perez-Lezaun A, Piccinini A, Prinz M, Schmitt C, Roewer L (1997) Evaluation of Y-chromosomal STRs: a multicenter study. Int J Legal Med 110:125–129

Kimpton C, Walton A, Gill P (1992) A further tetranucleotide repeat polymorphism in the vWF gene. Hum Mol Genet 1:287

Kimpton C, Fisher D, Watson S, Adams M, Urquhart A, Lygo J, Gill P (1994) Evaluation of an automated DNA profiling system employing multiplex amplification of four tetrameric STR loci. Int J Legal Med 106:302–311

Kogan SC, Doherty M, Gitschier J (1987) An improved method for prenatal diagnosis of genetic diseases by analysis of amplified DNA sequences - Application to Hemophilia A. New Engl J Med 317:985–990

Lareu MV, Barral S, Salas A, Pestoni C, Carracedo A (1998) Sequence variation of a hypervariable short tandem repeat at the D1S1656 locus. Int J Legal Med 111:244–247

Mannucci A, Sullivan KM, Ivanov PL, Gill P (1994) Forensic application of a rapid and quantitative DNA sex test by amplification of the X-Y homologous gene amelogenin. Int J Legal Med 106:190–193

Müller B (2002) Design einer Multiplex-PCR zur Typisierung von Y-STR Haplotypen aus degradierter DNA. Diplomarbeit, Georg August-Universität, Göttingen

Panzer SW, Hammond HA, Stephens L, Chai A, Caskey CT (1993) Trinucleotide repeat polymorphism at D6S366. Hum Mol Genet 2:1511

Polymeropoulos MH, Rath DS, Xiao H, Merril CR (1991a) Tetranucleotide repeat polymorphism at the human c-fes/fps proto-oncogene (FES). Nucleic Acids Res 19:4018

Polymeropoulos MH, Xiao H, Rath DS, Merril CR (1991b) Tetranucleotide repeat polymorphism at the human tyrosine hydroxylase gene (TH) Dinucleotide repeat polymorphism at the human non-histone chromosomal protein HMG14 gene. Nucleic Acids Res 13:3753

Roewer L, Arnemann J, Spurr NK, Grzeschik KH, Epplen JT (1992) Simple repeat sequences on the human Y chromosome are equally polymorphic as their autosomal counterparts. Hum Genet 89:389–394

Sambrook J, Fritsch EF, Maniatis T (1989) Molecular cloning. A laboratory manual (2nd edn). Cold Spring Harbour Laboratory Press, New York

Schmidt D (2001) DNA-Extraktion aus konservierten Lebensmitteln. Staatsexamensarbeit, Georg August-Universität, Göttingen

Schultes T (2000) Typisierung alter DNA zur Rekonstruktion von Verwandtschaft in einem bronzezeitlichen Skelettkollektiv. Cuvellier, Göttingen

Schultes T, Hummel S, Herrmann B (1999) Amplification of Y-chromosomal STRs from ancient skeletal remains. Hum Genet 104:164–166.

Urquhart A, Oldroyd NJ, Kimpton CP, Gill P (1995) Highly discriminating heptaplex short tandem repeat PCR system for forensic identification. BioTechniques 18:116–121

Walsh PS, Metzger DA, Higuchi R (1991) Chelex100 as a medium for simple extraction of DNA for PCR-based typing from forensic material. BioTechniques 10:506–513

Wrischnik LA, Higuchi RG, Stoneking M, Erlich HA, Arnheim N, Wilson AC (1987) Length mutations in human mitochondrial DNA: direct sequencing of enzymatically amplified DNA. Nucleic Acids Res 15:529–541

Appendix

I Laboratory equipment

Working on ancient DNA basically means running three independent laboratories: a pre-PCR laboratory, a PCR laboratory and a post-PCR laboratory. If the documentation of agarose gels is performed using Polaroid photography, an additional dark room may be necessary.

The pre-PCR laboratory is used for sample preparation, DNA extraction and PCR set up. The PCR laboratory is dedicated to the thermocycling machines only. The post-PCR laboratory is used for all types of electrophoresis, Taq cycle sequencing and RFLP digestions. All rooms must be fully equipped with assigned instruments, chemicals, reagents and disposable materials necessary for the respective analysis processes. None of these should be used in another laboratory.

It is often mentioned that all modern sample processing (from sample preparation to electrophoretic analysis) should be carried out separately. In some cases it has been suggested that this be done in separate buildings. This is neither a necessity nor does it make much sense. Although it may decrease the risk of cross-contamonations in the pre-PCR analysis steps, there is still the possibility of contamination with highly intact DNA through laboratory personal (cf. chapter 6). However, such measures are totally ineffective in post-PCR analysis. A carry-over contamination originating from a successful aDNA amplification is precisely as dangerous as one generated from modern DNA. Therefore, it seems rather important to strictly maintain all precautions that minimize all possible contamination risks.

These precautions should obviously include never sharing instruments, machines or any disposable material. Any item that has once been used for a post-PCR analysis step can never be allowed to return to a pre-PCR environment. This also means that once people have entered the post-PCR lab, they cannot go back to the pre-PCR lab on the same day. What seems to be often overlooked is the fact that people often go into these areas for cleaning and maintenance. The personal may also cause contamination by working in the wrong direction, from post-PCR to pre-PCR. From experiments carried out by Szibor (1998), it is known that cleaning cloths and cleaning water carry so many amplification products that after cleaning a post-PCR laboratory it was easily possible to amplify all kinds of sequences from μl-aliquots of the claening water in regular cycle number PCRs. Therefore, moving in the wrong direction by cleaning and maintenance staff also presents a serious risk of transmitting contamination.

Due to the fact that it is possible to avoid or minimize risks but never to fully exclude contamination events, a basic prerequisite for any reliable aDNA analysis is the generation of individual-specific control information for any ancient sample whenever possible (cf. section 6.2). This does not necessarily mean that additional amplifications need to be carried out by multiplexing independent loci, as shown by the many examples given throughout this book.

I.1 Pre-PCR laboratory equipment

Instruments and tools for sample preparation, DNA extraction and PCR set up: Refrigerator and freezer, hood, electric hand saw, forceps, mortar, mill, over-head rotator, pipettes, thermostatic mixer, incubator, vortex machine, bench top centrifuge (14,000 rpm and standard rotor), water bath, UV hand lamp (λ 254 nm), spatulas, tube racks, organic waste bottle

Instruments and tools for preparing chemicals and reagents: magnetic heater/stirrer, magnetic stirring bars, balance, weighing paper, pH meter, pH sticks, chemical spoons and spatulas, pipettes and pipette tips, graduated glass pipettes, dispenser or pelaeus ball, Pasteur pipettes, funnel, beakers, graduated cylinders, volumetric flasks, glass bottles

Disposables: gloves, face masks, glasses, head-dress gowns, cotton tips, disposable scalpels, 15-ml falcon tubes, 2-ml Eppendorf tubes, 0.5-ml Eppendorf tubes, pipette tips, Kleenex

I.2 PCR laboratory

Instrument: thermocycling machine

Disposables: none

I.3 Post-PCR laboratory

Instruments and tools: Refrigerator and freezer, capillary electrophoresis unit (e.g., Type 310 or 3100, Applied Biosystems),[93] horizontal electrophoresis unit (ca. 11×16 cm), power supply (150 V/300 mA), bench top centrifuge (14,000 rpm and standard rotor), magnetic heater/stirrer, magnetic stirrer bar, pipettor, balance, tube racks, ice box, watch glass dishes, 100-ml beakers, Erlenmeyer flasks (200 ml), graduated cylinders (100 ml and 1,000 ml), titration plates, ethidium bromide waste bottle, waste bin for acrylamide

Disposables: gloves, 2-ml Eppendorf tubes, 0.5-ml Eppendorf tubes, regular and gel loading pipette tips, Kleenex

I.4 Photographic documentation

Instruments: UV transilluminator (λ 254 nm), UV-protection face shield, photographic documentation unit (either a Polaroid or a digital camera system)

Disposables: none

[93] Or a slab gel electrophoresis unit (e.g., Types 373 or 377, Applied Biosystems) with 12-cm and 24-cm glass plate sets.

II PCR trouble shooting

This chapter aims to address the most common analysis failures and artifacts of aDNA analysis. It is not intended for those failures that are "homemade," such as mistakes made by pipetting incorrectly, in the software programing or from miscalculating the concentrations of reagents and chemicals. This section also will not address false results that are due to contamination through the handling of the samples. The possible reasons for contamination that may occur in aDNA analysis and the ways in which this can be avoided are described in detail in chapter 6.

The following table deals with:

- total amplification failure
- partial amplification failure
- unspecific PCR products (discrete bands)
- unspecific PCR products (smears)
- stutter bands (STR analysis)
- artifacts in automated fragment length determination (STR analysis)
- noisy/unreadable sequences (direct sequencing)

Table II-1. Trouble shooting for common aDNA analysis artifacts

Artifact	Reason	Recognize/check	Help
Total amplification failure	DNA degradation	• Shorter products are amplified from the respective extract • Electrophoresis of aDNA extract reveals small fragments only	• Add more aDNA extract • Design primers for shorter product • Section 4.4
	DNA degradation of extracts that used to work		• Improve extract storage by adding silica • Section 3.3.4
	Inhibiting substances	• Total lack of primer dimer • Turquoise fluorescence on unstained agarose under 254-nm UV light • Still no product if modern DNA is added	• Add less extract to amplification • Clean up extract • Use more Taq polymerase • Use more cycles • Section 4.6.2
	Denaturation temperature	• Check actual temperature in the cycler	• Repair • Section 4.3.2
	Annealing temperature		• Decrease annealing temperature
	Mismatch at 3'-ends of primers	• Check for possible mismatch in primer	• Check GenBank for more sequences and carry out new alignment and primer design • Sections 4.4.4, 7.2.7
	Cycle number		• Increase cycle number • Section 4.2
Partial amplification failure	Amplification of nuclear DNA fails, mtDNA is amplified		• Check DNA extraction method • Section 3.1
	Allelic dropout		• Increase DNA amount • Increase cycle number • Section 2.2.2
	Null alleles	• Check for possible mismatch in primer	• Redesign primers • Sections 2.2.2, 4.4

Artifact	Reason	Recognize/check	Help
Unspecific PCR products (discrete bands)	Annealing temperature		• Increase anneling temperature • Section 4.3.1
	Multiple primer match	• Check if primers match elsewhere (only possible for some thousand base pairs as in mtDNA)	• Redesign primers • Section 4.4
	Energy profile of primer	• Check for bad profile, i.e., weak 5'-end and strong 3'-end	• Redesign primers • Section 4.4
Unspecific PCR products (smear)	DNA overload	• No bands visible in smear that reaches high molecular weight • Determine DNA yield	• Decrease DNA amount • Increase primer amount • Sections 3.2, 4.6.3
	Overcycling	• Bands are visible in smear	• Reduce cycle number • Sections 3.2, 4.6.3
	Diverging annealing temperatures (≥5°C) of primers	• Check annealing temperatures	• Redesign primers • Section 4.4
	Excessive primer dimer in multiplex	• Smear does not reach high molecular weight • Check for all possible primer dimer formation	• Separate critical primer pair • Redesign primers • Sections 4.4, 4.5
Stutter bands (STRs)	Excessive slippage		• Use tetranucleotide repeat STRs • Increase DNA amount • Decrease elonagation temperature • Sections 2.2.2, 4.3.3
Artifact in fragment length determination	Pull-up peaks	• Additional peaks of exactly same size occur in neighboring dye colors (may only disturb in multiplex PCR allele determination)	• Decrease product load • Redo calibration by matrix standards • Change to technology that does not use filter wheel (capillary electrophoresis) • Sections 2.2.2, 5.1.2

Artifact	Reason	Recognize/check	Help
Artifact in fragment length determination	Product overload	• Flattened and inverted peaks • Large peak slopes • Imprecise size determination	• Decrease product amount • Sections 2.2.2, 5.1.2
	Extremely deviating peak sizes of short and long amplification products	• Flattened and inverted peaks for short products (cf. overload and pull-ups), too small peaks for long products	• Load different amounts of the sample in neighboring lanes • Sections 3.3.3, 5.1.2
Unreadable direct sequencing	Sequencing primer reveals a "population" of oligonucleotides of different lengths	• If the entire sequence is noisy or unreadable, ask manufacturer for oligo-synthesis standards	• Use only HPLC purified primers • Section 5.2
	Nuclear insertions	• If parts of the sequence are unclear or unreadable	• Check by amplification of sperm DNA of same species • Redesign primers • Section 2.1
	Contamination of human and domestic animal origin in tubes	• In case of noisy/unreadable or otherwise ambiguous results of conservative mtDNA analysis, e.g., cytochrome b	• Check different brands by amplification and sequence analysis of numerous no-template controls • Try out cleaning measures • Mismatches against contaminating species to primers at 3'-end if possible • Decrease cycle number • Change to multiplex PCR if possible, e.g., run with a HVR amplification • Chapter 6

III Internet sites of interest

Ancient DNA research

Australia/New Zealand

http://biochem.otago.ac.nz/staff/hagelberg/ehagelberg.htm (Dr. *Erika Hagelberg*, Biochemistry Department, University of Otago, Dunedin, New Zealand)

http://www.perthzoo.wa.gov.au (Dr. *Peter Spencer*, Perth Zoological Gardens, Australia)

http://www.austmus.gov.au/thylacine/index2.htm (Dr. *Mike Archer*, Australian Museum, Sydney, Australia)

Belgium

http://www.kuleuven.ac.be/upers/abl188.htm (Dr. *Els Jehaes*, University of Leuven, Belgium)

Canada

http://www.lakeheadu.ca/~lucas/pdnamain.htm (Dr. *El Molto*, Paleo-DNA Laboratory, Lakehead University, Thunderbay, Canada)

http://socserv.socsci.mcmaster.ca/~adna/ (*Palaeogenetics Institute*, McMaster University, Hamilton, Canada; links to GenBank, PubMed and a simple primer design program *Whitehead/MIT 3*)

France

http://www.inrp.fr/Acces/biotic/genetic/adn/html/emg.htm (Dr. *Eva-Maria Geigl*, UFR des Sciences de la vie, Université Pierre et Marie Curie, Paris, France)

http://www.cgmc.univ-lyon1.fr/ (Dr. *Catherine Hänni*, CGMC, University of Lyon, France)

Germany

http://www.uni-mainz.de/FB/Biologie/Anthropologie/ (Dr. *Joachim Burger*, Prof. *Kurt Alt*, Institute of Anthropology, Johannes Gutenberg-University, Mainz, Germany)

http://www.anthro.uni-goettingen.de (Dr. *Susanne Hummel*, Prof. *Bernd Herrmann*, Department of Anthropology, Georg August-University, Göttingen, Germany)

http://www.eva.mpg.de/ (Prof. *Svante Pääbo*, Max Planck-Institute for Evolutionary Anthropology, Leipzig, Germany)

http://www.uni-tuebingen.de/uni/afa/people/scholz.html (Dr. *Michael Scholz*, Osteologische Sammlung, University of Tübingen, Germany)

Great Britain

http://www.bi.umist.ac.uk/staff/user.asp?item=Research&id=55796 (Dr. *Terry Brown*, UMIST, Manchester, UK)

http://nrg.ncl.ac.uk/research/resareas/ancient-biomols.html (Dr. *Mathew Collins*, The Ancient Biomolecules Research group, University of Newcastle, UK)

http://evolve.zps.ox.ac.uk/ABC/ (Dr. *Allan Cooper*, Ancient Biomolecules Centre University of Oxford, UK)

http://www.ucl.ac.uk/biology/goldstein/Marti1.html (Dr. Martin Richards, Galton Laboratory, University College, London, UK)

http://immwww.jr2.ox.ac.uk/groups/cellgen.html (Dr. *Bryan Sykes*, Oxford, UK)

http://www.nhm.ac.uk/science/intro/palaeo/project1/ (The *Natural History Museum*, London, UK)

Israel

http://kuvin.huji.ac.il/sci_ant/index.html (Prof. *Charles Greenblatt*, Prof. *Mark Spigelman*, Kuvin Centre for the Study of Infectious and Tropical Diseases, Hebrew University of Jerusalem, Israel) http://dental.huji.ac.il/newEsite/institute/instituteindex.html (Dr. *Marina Faerman*, Prof. *Patricia Smith*, Dental School of Medicine, Hebrew University, Jerusalem, Israel)

Italy

http://www.uniroma2.it/biologia/lab/anthromol/home.htm (Prof. *Olga Rickards*, Prof. *Gianfranco DeStefano*, Molecular Anthropology Lab, Tor Vergata, University of Rome, Italy)

http://www.unicam.it/~fscienze/Piano_di_Sviluppo/Cv/curruculum_Rollo.PDF (Dr. *Franco Rollo*, Dr. *Isolina Marota*, Dr. *Massimo Ubaldi*) and http://web.unicam.it/ (University of Camerino, Italy)

http://www.unifi.it/unifi/antrop/adna.htm
http://www2.unife.it/genetica/Cristiano/research.htm (Dr. *Cristiano Vernesi*, Institute of Anthropology, Univerity of Florence, Italy)

http://www.urp.cnr.it/concorsi2000/310290RM138-1.htm (Dr. *Marilena Cipollaro*, Institute of Molecular Biology, 2nd University of Naples, Italy)

Spain

http://www.bio.ub.es/bioani/lalueza/main2.htm (Dr. *Carles Lalueza-Fox*, Dr. *Jaume Betranpetit*, Dr. *Daniel Turbon*, Department of Animal Biology, University of Barcelona, Spain)

http://www.ucm.es/info/antropo/trancho/eduardo/eduardo.htm (Dr. *Eduardo Arroyo*, Laboratory of Forensic Biology, University de Complutense, Madrid, Spain)

Sweden

http://hem.spray.se/gother/sida1.html (Dr. Anders Gotherstrom, Insitute of Archaelogy, University of Stockholm, Sweden)

http://www.hum.gu.se/ark/anl/per_persson.htm (Dr. *Per Persson*, Insitute of Archaelogy, University of Göteborg, Sweden)

Switzerland

http://www.unibas.ch/archbot/adna.htm (Dr. *Angela Schlumbaum*, Department of Archaeobotany, University of Basel, Switzerland)

United States

http://www.lms.si.edu/research.html (Dr. *Michal Braun*, Dr. *Noreen Tuross*, Laboratory of Analytical Biology, The National Museum of Natural History, Smithsonian Institution, Washington DC, USA)

http://www.indiana.edu/~anthro/faculty/kaestle.html (Dr. *Frederika Kaestle*, Department of Anthropology, Indiana University, Bloomington, USA)

http://www.unm.edu/~acstone/lab/molecular_anthropology_lab.htm (Dr. *Anne Stone*, Department of Anthropology, University of New Mexico, Albuquerque, USA)

http://gsbs.gs.uth.tmc.edu/tutorial/lawlor.html (Dr. *David Lawlor*, Health Science Center, University of Texas, San Antonio, USA)

http://www-personal.umich.edu/~andym/ (Dr. *Andrew Merriwether*, Department of Anthropology, University of Michigan, Ann Arbor, USA)

http://microbiology.byu.edu/Faculty/woodward.html (Dr. *Scott Woodward*, Cancer Research Center, University of Utah, USA)

Biostatistics

http://www.dna-view.com/ (*Biomathmatics*, Dr. Charles Brenner)

Ring exercises (Europe)

http://medweb.uni-muenster.de/institute/remed/ (*GEDNAP – STR ring exercise*; Prof. Dr. Bernd Brinkmann)

http://www.med.uni-magdeburg.de/image/e32.htm (*Mitochondrial HVRs ring exercise*; Prof. Dr. Dieter Krause)

http://www.d-loop-base.de/index1.htm (C*ontact for HVRs ring trials*, Dr. M. Wittig, Otto von Guericke-University, Magdeburg, Germany)

Ancient DNA services

http://www.ancientdna.com/ (*Services of the Paleo-DNA Laboratory*, Dr. El Molto, Lakehead University, Thunderbay, Canada)

http://www.paleoscience.com/ (*Paleoscience Inc.*, Miami, USA)

http://www.oxfordancestors.com/ (M*atrilineage and patrilineage* analysis on modern DNA, Oxford Ancestors, Oxford, UK)

Databases

General

http://www.ncbi.nlm.nih.gov/ (Home page of the *National Center of Biotechnology Information*; access to PubMed, GenBank, OMIM, Taxonomy Browser etc.; NIH)

http://www.ncbi.nlm.nih.gov/Genbank/GenbankSearch.html (Home page of *GenBank*)

http://www.embl.de/ (Home page of the *European Molecular Biology Laboratory* (EMBL), Heidelberg, Germany)

http://www.umd.necker.fr/core.html (*Mutation database* and links to other major databases ; Hôpital Necker-Enfants Malades, Paris, France)

Short tandem repeats (STR)

http://www.cstl.nist.gov/biotech/strbase (*Short tandem repeat* (STR) database, NCBI, NIH, USA)

http://ystr.charite.de (*Y-chromosomal STR* database; Charité Berlin)

http://ixdb.molgen.mpg.de/ (*X-chromosomal marker* database, Max Planck Insitute for Molecular Genetics, Berlin; Germany) http://www.uni-duesseldorf.de/WWW/MedFak/Serology/dna.html (*World wide STR allele frequencies*, Dr. Walter Huckenbeck, Dr. Hans-Georg Scheil)

Mitochondrial DNA

http://www.gen.emory.edu/mitomap.html (*Mitochondrial genome* database, Center for Molecular Medicine, Emory University, Atlanta, USA)

http://www.hvrbase.de/ (*Mitochondrial DNA database for human and great apes* HVR I and HVR II, Max Planck-Institute for Evolutionary Anthropology, Leipzig, Germany)

http://www.d-loop-base.de/index1.htm (*Mitochondrial DNA* database and contact for HVRs ring trials, Dr. M. Wittig, Otto von Guericke-University, Magdeburg, Germany)

Single nucleotide polymorphisms (SNP)

http://www.ncbi.nlm.nih.gov/SNP/ (*Single Nucleotide Polymorphism* database; NCBI, NIH, USA)

http://snp.cshl.org/ (*Single nucleotide polymorphism* database)

aDNA References

General

http://www.comic.sbg.ac.at/staff/jan/ancient/titel.htm (*Ancient DNA references*; Dr. Jan Kieslich, Salzburg, Austria)

http://www.ncbi.nlm.nih.gov/entrez/query.fcgi?CMD=search&DB=PubMed (*PubMed*, NCBI, NIH, USA)

up-to-date *reference lists* are also found on almost all home pages of ancient DNA research laboratories (cf. above)

aDNA Journal

http://www.tandf.co.uk/journals/titles/13586122.html (*Ancient Biomolecules*)

Journals regularly publishing aDNA research

http://www.journals.uchicago.edu/AJHG/home.html (*American Journal of Human Genetics*)

http://www.interscience.wiley.com/jpages/0002-9483/ (*American Journal Physical Anthropology*)

http://www.schweizerbart.de/j/anthropologischer-anzeiger/ (*Anthropologischer Anzeiger*; most publications in English)

http://www.wiley-vch.de/publish/en/journals/alphabeticIndex/2027/ (*Electrophoresis*)

http://www.naturesj.com/ejhg/contacts.html (*European Journal of Human Genetics*)

http://www.elsevier.com/locate/forsciint (*Forensic Science International*)

http://link.springer.de/link/service/journals/00414/index.htm (*International Journal Legal Medicine*)

http://www.academicpress.com/jas (*Journal of Archaeological Science*)

http://www.blackwell-science.com/~cgilib/jnlpage.asp?Journal=MECOL&File=MECOL&Page=aims (*Molecular Ecology*)

http://www.nature.com/nature/ (*Nature*)

http://link.springer.de/link/service/journals/00114/index.htm (*Naturwissenschaften,* most publications in English)

http://nar.oupjournals.org/ (*Nucleic Acids Research*)

http://www.sciencemag.org/ (*Science*)

Miscellaneous

http://linkage.rockefeller.edu/wli/glossary/genetics.html (*Glossary of genetics*)

http://www.kruglaw.com/f_dna.htm (A *collection of links* ranging from ABO blood group genetics to STR databases)

http://www..medfac.leidenuniv.nl/~fldo/hptekst.html (On Y-chromosomal STRs, Dr. Peter de Knijff, University of Leiden, Netherlands)

http://gsu.med.ohio-state.edu/ppt/primer_design/index.htm (On *primer design*, Department of Human Cancer Genetics, Ohio State University, Columbus, USA)

http://www.ebi.ac.uk/index.html (offers tools as alignment programs, software downloads and many links to useful databases, *European Bioinformatics Institute*, EMBL)

http://www.ornl.gov/hgmis/ (*Human Genome Project* Information, many useful links)

http://www.uky.edu/ArtsSciences/Geology/webdogs/amber/people/george.htm (homepage of *George and Roberta Poinar*)

http://www.neanderthal-modern.com/index.html (On *Neanderthals...*, Dr. S. Brown)

http://www.cr.nps.gov/aad/kennewick/tuross_kolman.htm (About the Kennewick Man and ancient DNA; Park Net, National Park Service, USA)

http://www.comp-archaeology.org/ (links to informations about the *Ötztaler Iceman* and the *Kennewick Man*, Dr. MO Baldia)

http://forensic.shef.ac.uk/anthro.html (On *forensic anthropology and ancient DNA*, University of Sheffield, UK)

http://www.apsnet.org/education/feature/ancientdna/top.htm (On *palaeogenetics...*, Dr. K. McCluskey, University of Kansas, USA)

http://mailbox.univie.ac.at/elisabeth.trinkl/forum/forum1298/09dna.htm (On *ancient DNA*, Dr. Jan Kiesslich) http://spinner.lab.nig.ac.jp/aDNA/index.html (*Ancient DNA research links*, Japan)

http://www.corpus-delicti.com/DNA.html (*Forensic science bookstore*)

Subject index